ANALYSIS OF ENGINEERING CYCLES

Power, Refrigerating and Gas Liquefaction Plant

FOURTH EDITION

**THE PERGAMON TEXTBOOK
INSPECTION COPY SERVICE**

An inspection copy of any Pergamon textbook will gladly be
sent to academic staff without obligation for their consideration
for course adoption or recommendation.
Copies may be retained for a period of 60 days from receipt and
returned if not suitable. When a particular title is adopted or
recommended for adoption for class use and the recommendation results in a sale of 12 or more copies the inspection copy
may be retained with our compliments.
The Publishers will be pleased to receive suggestions for revised
editions and new titles.

THERMODYNAMICS AND FLUID MECHANICS SERIES
General Editor: W. A. WOODS

Pergamon Titles of Related Interest

DIXON
Fluid Mechanics, Thermodynamics of Turbomachinery, 3rd Edition

DUNN & REAY
Heat Pipes, 2nd Edition

HAYWOOD
Analysis of Engineering Cycles – Worked Problems

JAPAN SOCIETY OF MECHANICAL ENGINEERS
Visualized Flow

LAI et al
Introduction to Continuum Mechanics, Revised Edition

REAY & MACMICHAEL
Heat Pumps: Design and Applications, 2nd Edition

XIN MINGDAO
Advances in Phase Change Heat Transfer

Pergamon Related Journals

(*free specimen copy gladly sent on request*)

Chinese Journal of Mechanical Engineering

Computer Fluids

Computer and Industrial Engineering

International Communications in Heat and Mass Transfer

International Journal of Engineering Sciences

International Journal of Heat and Mass Transfer

International Journal of Multiphase Flow

Heat Recovery Systems and CHP

Previews of Heat and Mass Transfer

ANALYSIS OF ENGINEERING CYCLES

Power, Refrigerating and Gas Liquefaction Plant

FOURTH EDITION

R. W. HAYWOOD
University of Cambridge, UK

PERGAMON PRESS

OXFORD · NEW YORK · SEOUL · TOKYO

UK	Pergamon Press plc, Headington Hill Hall, Oxford OX3 0BW, England
USA	Pergamon Press, Inc., 395 Saw Mill River Road, Elmsford, New York 10523, U.S.A.
KOREA	Pergamon Press Korea, KPO Box 315, Seoul 110-603, Korea
JAPAN	Pergamon Press, 8th Floor, Matsuoka Central Building, 1-7-1 Nishi-Shinjuku, Shinjuku-ku, Tokyo 160, Japan

First edition 1967

Second edition 1975

Third edition 1980, reprinted (with corrections) 1985

Fourth edition 1991

Library of Congress Cataloging in Publication Data
Haywood, R. W. (Richard Wilson)
Analysis of engineering cycles: power, refrigerating and gas liquefaction plant/R. W. Haywood.—4th ed.
p.   cm.
Includes bibliographical references and index.
1. Power-plants.   2. Refrigeration and refrigerating machinery. 3. Gases—Liquefaction.   I. Title.
TJ164.H384 1991   621.4—dc20   90-26251

British Library Cataloguing in Publication Data
Haywood, R. W. (Richard Wilson)
Analysis of engineering cycles.—4th ed.
1. Prime movers
I. Title
621.4
ISBN 0-08-040739-0 (Hardcover)
ISBN 0-08-040738-2 (Flexicover)

Printed in Great Britain by BPCC Wheatons Ltd., Exeter.

Contents

Part I. Simple Power and Refrigerating Plants

Part II. Advanced Power and Refrigerating Plants

6 Advanced gas-turbine plant   81

7 Advanced steam-turbine plant   99

10   Advanced refrigerating and gas-liquefaction plant   229

Preface to the Fourth Edition

Though the author of a textbook which goes into its Fourth Edition may be entitled to feel gratification at its continuing popularity, he might also be excused for occasionally wishing that his subject, rather than technology, had been one such as theology, with a much less rapid rate of change. As for the Third Edition published in 1980, the major changes are additions, rather than alterations, to Chapters 7, 8 and 9, with the further addition of two new appendices.

In *Chapter 7 (Advanced Steam-Turbine Plant)*, the increasing importance of energy "saving" has led to revived attention being given to the topic of *Combined Heat and Power (CHP)*, which is increasingly given the title of *Cogeneration*. This fact has led to the addition of a new section on **COGENERATION (CHP)**. This has six sub-sections dealing in greater detail than hitherto with both theory and practice. An addition has also been made to Section 7.18 dealing with reheating in conventional steam plant.

In *Chapter 8 (Nuclear Power Plant)*, the existing material has been retained, with some updating. However, an extensive new section has been added under the title **PWR PLANT—FURTHER STUDIES**, containing five sub-sections. With the increasing dominance of the pressurised-water reactor (PWR) in the industrial nuclear field, the design of the Sizewell B steam generators is discussed in some detail. That is followed by a discussion of two recent proposals for "safe" reactors, the British *SIR (Safe Integral Reactor)* and the Swedish *PIUS (Process Inherent Ultimate Safety)*. The concluding section discusses briefly the operator errors and design faults that led to four major reactor disasters and which stimulated the search for safer designs.

In *Chapter 9 (Combined and Binary Power Plant)*, there is an addition of ten new sections. These deal exhaustively with modern developments in combined gas–steam plant, which are discussed in considerably greater detail than in the earlier sections of the chapter. Topics include both simple recuperative plant and high-efficiency combined plant with supplementary firing. Studies of both dual-pressure and triple-pressure cycles in the steam plant lead on to a brief study of *STIG (Steam-Injection Gas-Turbine)* plant, in which steam is injected into the combustion chamber of the gas turbine in order to reduce the emission of noxious oxides of nitrogen, NO_x, from the plant. Finally, there is a detailed study of a conceptual *FBC (Fluidised-bed Combustion)* plant which serves the similar purpose of reducing the emission of NO_x and SO_2 from a coal-burning combined plant. That study includes more detailed consideration of an entire boiler plant than has hitherto been given in the book.

Appendix D (Boiler Circulation Theory) gives a detailed treatment of a topic not previously covered in the book. There were two stimuli for including this. Firstly, in the more extensive studies of both nuclear and combined plant in Chapters 8 and 9, it was striking that the physical process of *natural circulation* in the steam generators (boilers) played a prominent part in both types of plant. Secondly, in an early research paper, I had succeeded in unifying some extant conflicting theories of circulation after eliminating errors in them. It therefore seemed to be particularly appropriate to treat the matter in some detail in Appendix D.

Eight new problems have been added to the book. Fully worked solutions to all of these are given in *Appendix E (Solutions to Additional Problems)*. Fully worked solutions to all the original problems set in this and the Third Edition are given in my companion book, *Analysis of Engineering Cycles—Worked Problems*,[A] published in 1986. The reader may also like to know of two volumes of *Worked Solutions*[B,C] recently published by Krieger in Florida as companion volumes to the reprinting by Krieger in two volumes[D,E] of my book on *Equilibrium Thermodynamics*[5] originally published by Wiley in 1980. Indeed, the two Pergamon books and the four new Krieger books may be said to constitute a uniquely comprehensive treatment of both theoretical and applied thermodynamics in an up-to-date form, useful both to students and to practising engineers and scientists. My *Thermodynamic Tables*,[1] published by Cambridge University Press, have been used throughout as the source of data required in the calculation of all these worked problems, 82 in number in Ref. A and 140 in total number in the two volumes of Refs. B and C. The student will probably find those Tables, and the accompanying diagrams, quite the most convenient source of the required numerical data. There are a further 8 worked problems in Appendix E herein, making a total of 230 fully worked solutions available to students and teachers.

For the convenience of readers, all references have been retained and an appreciable list of *Additional References* (labelled alphabetically) has been added, relating chiefly to Chapters 7, 8 and 9.

I have finally to acknowledge, with heartfelt thanks, the great encouragement and support given by Dr J. H. Horlock, my good friend and former colleague at Cambridge. It was he, when a General Editor of this series, who originally invited me to write the book, nearly 30 years ago. Whilst he was still Vice-Chancellor of the Open University, he nevertheless found time to give me great assistance with the provision of new source material and in invaluable discussions, both written and in person. Without that, it is doubtful whether I could have brought the task of updating to a successful conclusion.

At Dr Horlock's suggestion, I have made one rather important alteration throughout the book, in that I have changed the term *performance* **parameter** to *measure of performance* or, for short, *performance* **measure**. He pointed out to me, quite rightly, that my use of the word *parameter* conflicted with its commonly accepted usage in relation to plant parameters such as pressure, temperature, etc.

I must not end without an expression of immense gratitude to my wife, Sylvia, for the unstinting and long-suffering support which she has given me during the writing of all my books.

October 1990   RICHARD WILSON HAYWOOD

Preface to the Third Edition

The principal modifications in this Third Edition arise from the updating and expansion of material on nuclear plant in Chapter 8 and on combined and binary plant in Chapter 9. In addition, in view of increased importance and topicality, new material has been added in Chapter 6 on gas-turbine plant for Compressed Air Energy Storage systems and in Chapter 7 on steam-turbine plant for the combined supply of power and process steam, including plant for district heating. The use of gas-turbine plant in association with district-heating schemes is also discussed in Chapter 8, in which the treatment of high-temperature and fast-breeder gas-cooled nuclear reactors has been extended. The material on combined gas-turbine/steam-turbine plant in Chapter 9 has also been expanded and updated, together with that on combined steam plant with magnetohydrodynamic and thermionic topping respectively.

Additions to the material on the simple steam cycle in Chapter 2, and the Problem in Chapter 9 relating to the ideal super-regenerative steam cycle, give an opportunity to apply the important principles and concepts of thermodynamic availability, which feature prominently in a new book by the author.[5] New problems have been added in Chapters 2, 6, 7, 8 and 9, and there has been an appreciable addition to the list of quoted references.

In order to emphasise the fact that, in practice, economic assessments are just as important as thermodynamic analyses, a new Section dealing with Discounted Cash Flow has been added in Appendix C. For this I am indebted to my colleague, Dr. M. D. Wood, of Gonville and Caius College.

Cambridge   R. W. HAYWOOD
July 1979

Preface to the Second Edition (SI units)

Although the material in much of the text of the First Edition was not dependent on any particular set of units, the problems at the end of each chapter were set in British units. In this Second Edition, the problems have been reset (and a few new ones added) in the metric units of the *Système International d'Unités* (SI). In general, the answers given are based on data taken from the author's small volume of *Thermodynamic Tables*[1] in these units, but it has also been found convenient occasionally to use more extensive compilations of the properties of steam[2] and carbon dioxide.[3] In such cases, this is indicated in the text.

In preparing this Second Edition, the opportunity has been taken to bring the material up to date, where that was needed. To this end, a considerable part of Chapter 8 (Nuclear Power Plant) has been rewritten and some additions have been made to Chapters 9 and 10. Appendix A has also been completely rewritten and considerably extended, so that it now gives a more detailed treatment of thermodynamic availability and irreversibility in non-cyclic processes.

R. W. HAYWOOD

The Second Edition is available in Russian translation—

Р. В. Хейвуд, Анализ циклов в технической термодинамике. Перевод с английского Е. Я. Гадаса. Москва, « Энергия », 1979.

Preface to the First Edition

This book deals principally with an analysis of the overall performance, under design conditions, of work-producing power plants and work-absorbing refrigerating and gas-liquefaction plants, most of which are either cyclic or closely related thereto. The consideration of off-design performance is beyond the scope of the series in which this volume appears. Likewise, no attempt is made to describe the mechanical construction of the different kinds of plant considered, for it is assumed that the reader already has a knowledge of the associated "hardware".

The division of the work into two parts, dealing first with simple and then with more complex plants, has several advantages over possible alternative ways of handling the subject matter. It would have been possible to deal with cyclic steam plant, both simple and complex, and to follow this by a treatment in turn of cyclic gas-turbine plant, internal combustion plant and finally refrigerating plant. The scheme adopted, however, enables attention to be drawn to the close similarities, and the differences, between steam and gas plant, before the reader is immersed in the complexities of advanced plant of either kind. A further advantage is the ease with which individual treatment, in Part II, of complex gas-turbine and steam plant can be followed by chapters on combined plant and nuclear power plant. Study of both power and refrigerating plants in the one volume makes it possible to apply the same techniques to an analysis of the performance of both without repetitive elaboration. Finally, division of the subject matter in this way makes the two parts suitable for study in consecutive academic years.

In a work which claims to deal with the analysis of engineering cycles, some justification is required for the inclusion of a chapter on non-cyclic internal-combustion plant, namely open-circuit gas-turbines and reciprocating internal-combustion engines. The inclusion of this chapter arises directly from the long-established practice of linking the performance of this type of **non-cyclic** plant with that of ideal **cyclic** plant operating on hypothetical cycles such as those of Joule, Otto and Diesel. The reader is not allowed, as in so many books, to skate over the highly arbitrary nature of this kind of exercise. As a result, he may well find Chapter 4 rather more difficult at first reading than the other chapters of Part I. However, the more rigorous treatment will bring to the persevering reader his own reward. The material in §4.16 has little to do with cycle analysis, but it has been included to maintain the student's interest by enabling him to apply the preceding work to the problems of a more practical nature set at the end of the chapter. Whilst the gas-liquefaction plants discussed in Chapter 10 are also non-cyclic, they bear such a close affinity to cyclic refrigerating plant that their inclusion calls for no justification.

Throughout the work, emphasis is placed on the distinction between performance **parameters**, which merely provide a **measure** of plant performance, and performance **criteria**, which provide a **yardstick** against which the actual performance can be judged. Having designed, built and tested his plant, any engineer worth his salt will be curious to know how much better its performance could have been. Performance criteria, not performance parameters, provide the answers, and it is important to realise that these criteria do not result from practical experiment, since all real-life processes are in some degree imperfect. Only in Thermotopia, that idyllic land of the thermodynamicist in which all processes are reversible, are there no lost opportunities for producing work. Thus, to set up performance criteria against which to judge the excellence of performance of the plant that he has built, the engineer can only call upon the resources of the human intellect, unaided by experiment; this he does through the science and laws of thermodynamics. As an example of the power of abstract thought, such an exercise is of particularly high educational value. This exercise will have added value if it is accompanied by a realisation that a striving towards Thermotopian perfection of man-made devices and machines, without regard to their social context, can be as much the mark of a well-informed barbarian as that of an educated engineer. The latter needs more than devilish ingenuity, for he must be as aware of the impact of his devices on society around him as he is to the fact, brought out clearly in this book, that their imperfections irretrievably leave their mark upon the environment. Said the judge, after a peroration by the great advocate F. E. Smith, first Lord Birkenhead: "I am afraid, Mr. Smith, that I am none the wiser." "True, my Lord," snapped back F. E. Smith, "but you are better informed." It is the hope of the author that this book will not merely leave the reader in the same state as the learned judge.

Throughout this text, the aim has been to replace "custom and wont" by sound scientific argument and rigorous analytical treatment. If in this it succeeds, the book may be of value not only to students of engineering thermodynamics but also to other teachers of the subject and to practising engineers. A good textbook should not leave the reader with the impression, when he has reached the end, that all has been said; it should, instead, act as a stimulant to further reading, since the most exciting recent developments will always be found in periodicals and papers of the time rather than in standard texts. The reader is therefore well furnished, particularly in Part II, with a wide range of references into which he can dip as they take his fancy. They are in no way comprehensive.

This work is the result both of extensive practical experience in the power-plant industry and of experience gained in many years of lecturing to Cambridge undergraduates at all levels. Many of the problems originated in the same way, and the answers given are based on data taken from the author's small volume of *Thermodynamic Tables.*[†] Throughout the book, unless otherwise stated, the word "pressure" denotes absolute pressure when a numerical value is given.

In accumulating, over the years, the store of knowledge now put into this book, he owes much to help received from many people, in ways both direct and indirect, and not least from those whom he has taught. In not naming any, for fear of omitting the names of some, he thanks all who have in any way contributed towards its production.

R. W. HAYWOOD

[†]*Thermodynamic Tables and Other Data,* edited by R. W. Haywood, Cambridge University Press, 2nd edition, 1960.
  Second Edition footnote: These Tables were in British units. A more extensive set of Tables in SI units prepared by the Author is quoted under reference 1 on page 293 (page 317 of this Third Edition).

Editorial Introduction

The books in the Pergamon Thermodynamics and Fluid Mechanics Series were originally planned as a series for undergraduates, to cover those subjects taught in a three-year course for Mechanical Engineers. Subsequently, the aims of the series were broadened and several volumes were introduced which catered not only for undergraduates, but also for postgraduate students and engineers in practice. These included new editions of books published earlier in the series.

The present volume is the fourth edition of the most popular book in the series.

With the enormous interest now being shown in new and combined power cycles, the appearance of the new edition is particularly appropriate and timely.

June 1991   W. A. WOODS

PART I

Simple Power
and Refrigerating Plants

CHAPTER 1

Power plant performance measures and criteria

In this chapter we consider the general mode of operation of simple steam and gas-turbine power plant, study what they have in common and in what respects they differ, and finally discuss the quantities used to measure the performance of cyclic plant. Performance criteria are studied in greater detail in succeeding chapters.

1.1. Operation of the simple steam plant

Figure 1.1 gives a diagrammatic arrangement of a simple steam plant. The diagram shows a turbine as the work-producing component because turbines are universally used in the large-scale generation of electrical power.

The fluid enters the boiler as high-pressure water at low temperature and leaves as high-pressure steam at high temperature, heat being transferred to the fluid while it remains at approximately **constant pressure**. By virtue of the low pressure existing in the condenser, from which at starting the air is withdrawn by an air pump or ejector, the steam flows from the boiler through the turbine, falling in temperature and performing work on the shaft as the fluid expands from boiler pressure to condenser pressure. The low-temperature exhaust steam, which is invariably a wet-steam mixture, is condensed at very nearly **constant pressure** on the outer surface of the condenser tubes while heat is transferred to the circulating water passing through the tubes. The resulting condensate is withdrawn from the condenser by a pump which acts both as condensate extraction pump and boiler feed pump, since it delivers the condensate as feed water at high pressure to the boiler.

The boiler and condenser pressures are kept approximately constant at all loads, so that opening of the throttle valve at turbine inlet increases the flow rate of steam to the turbine. The rate at which steam is produced in the boiler is simultaneously increased by increasing the rates of fuel and air supply to the boiler furnace, the pressure at which steam is produced being

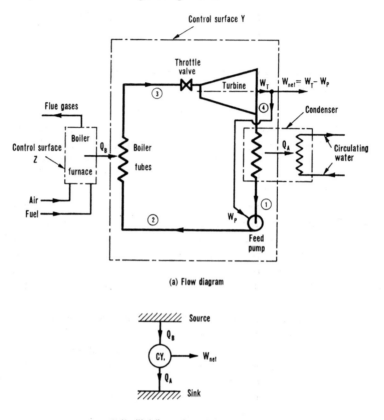

(a) Flow diagram

(b) Simplified diagram for control surface Y (CHPP)

FIG. 1.1. Simple steam plant.

thereby kept constant. As throughout this volume, the plant performance will be studied only at full load. The throttle valve is then wide open and the pressure drop across it very small.

It is seen that the fluid passes continuously round a closed circuit, or *thermodynamic cycle*. Since heat reception and rejection both occur when the fluid is at essentially constant pressure, the plant is said to operate on a *constant-pressure cycle*, although the pressure is far from being constant in either the turbine or the pump.

The complete assembly within control surface *Y* of Fig. 1.1(a) comprises a *cyclic heat power plant* (CHPP), or, as it is frequently called, a cyclic *heat engine*. Its object is to produce positive network output when heat is transferred from a source at high temperature (the boiler furnace) and rejected to a sink at low temperature (the condenser circulating water). A study of the energy quantities crossing this control surface shows that the plant may be represented more simply by the diagrammatic sketch of a CHPP given in Fig. 1.1(b).

1.2. Internal-combustion and external-combustion gas-turbine plant

Most practical gas-turbine plant are of the *internal-combustion* (or *open-circuit*) type, with the products of combustion passing directly through the turbine. Figure 1.2 shows the simplest arrangement of such a plant, and a study of the quantities crossing the control surface drawn round the plant shows that this is not a cyclic heat power plant (CHPP), for there is no heat source and no heat sink. It is, instead, a *non-cyclic open-circuit steady-flow work-producing device* in which *reactants* (fuel and air) cross the control surface at inlet and *products* of combustion (exhaust gases) leave it at exit, while only work, but no heat (other than stray heat loss to the environment), crosses the control surface. (Note that the exhaust gases, though hot, convey energy but not heat. Heat does not reside in a body; it is energy in transit **as the result of a temperature difference**.)

As the internal-combustion gas-turbine plant does not operate on a cycle, it might be said to have no place in a volume on engineering cycles. Nevertheless, because it is of such practical importance it will receive further mention in Chapter 4. In the literature it is frequently described as an *open-cycle* plant. Since the plant is not cyclic, and a cycle is always closed, this term is misleading and confusing. It is avoided throughout this volume by using the term *open circuit*.

In the *external-combustion* (or *closed-circuit*) gas-turbine plant, the products of combustion do not pass directly through the turbine. Instead, after leaving the combustion chamber, they pass through a heat exchanger in which they drop in temperature while transferring heat to a gaseous fluid (usually air) supplied by the compressor. This fluid circulates continuously round the closed circuit of the gas-turbine plant, as shown in Fig. 1.3(a).

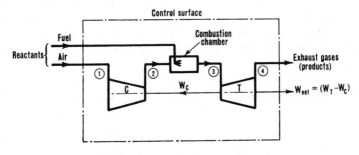

Fig. 1.2. Simple open-circuit gas-turbine plant.

Comparison of Figs. 1.1 and 1.3 shows the close identity of the simple steam and gas-turbine plants; the combustion chamber and heater in the latter are the counterpart of the boiler in the steam plant, while the cooler takes the place of the condenser and the compressor the place of the feed pump. The only essential difference is that the gas-turbine uses a fluid which remains gaseous throughout the cycle, with the result that, unlike the steam plant, heat rejection does not occur at constant temperature. The plant contained within control surface Y of Fig. 1.3(a) comprises a CHPP, of which Fig. 1.3(b) is a simple representation.

The closed-circuit gas-turbine plant is used when burning a type of fuel from which the products of combustion, if passed through the turbine, would rapidly foul the turbine passages. Because of the fouling problem,

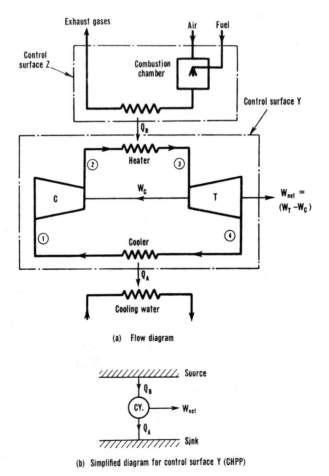

(a) Flow diagram

(b) Simplified diagram for control surface Y (CHPP)

FIG. 1.3. Simple closed-circuit gas-turbine plant.

the use of coal in open-circuit gas-turbine plant has not reached a successful stage of development, although oil and gas find ready use. The present chapter, however, is concerned solely with closed-circuit plant, although some simple problems involving open-circuit plant are given at the end of the chapter.

1.3. Operation of the simple gas-turbine plant

At starting, the turbo-compressor set is run up to a fraction of normal running speed by means of an electric motor or other suitable starting device engaged through a clutch. At this speed, fuel and air supplied to the combustion chamber are ignited by an electrical starter plug and there is sufficient pressure difference across the compressor and turbine to enable the plant speed to be self-sustaining. The machine can then be run up to full speed without the further aid of the starting motor. As load is applied to the generator, the speed is maintained by automatically increasing the fuel and air supplied. Only at one particular set of running conditions does the plant operate at its point of maximum efficiency; this is the full-load condition for which the plant has been designed at the normal running speed. The plant performance will be studied only at this design load.

1.4. Performance measures for cyclic steam and gas-turbine plant

The steam and gas-turbine power plants of Figs. 1.1 and 1.3 are both external-combustion plant in which the working fluid flowing through the turbine passes round a closed thermodynamic cycle. The methods of assessing their performance are consequently similar. The internal-combustion gas-turbine plant of Fig. 1.2 is of a different kind, and discussion of its performance will be deferred to Chapter 4.

Those quantities which give information on the actual performance of a plant, as determined from experimental measurements, are here termed *performance measures*. In defining such quantities it is necessary carefully to define that part of the plant to which a particular performance measure relates. For this purpose it is convenient to consider three separate cases.

(1) Control surface Y—thermal or cycle efficiency and heat rate

As already noted, the plant within control surface Y of Figs. 1.1 and 1.3 constitutes a CHPP. Its performance is expressed in terms of the *thermal or cycle efficiency*, η_{CY}, defined as the ratio of the net work output to the heat input. Thus:

$$\eta_{CY} \equiv \frac{W_{net}}{Q_B}. \tag{1.1}$$

An alternative and widely used performance measure for the plant within this control surface is the so-called *heat rate*, or rate of heat supply per unit rate of net work output. With Q_B and W_{net} expressed in the same units, the heat rate is simply the reciprocal of η_{CY}. In the British system of units, however, the heat rate was usually expressed as so many Btu of heat input per kW h of net work output. In these units:

$$\text{Heat rate} = \frac{3412}{\eta_{CY}} \text{ Btu/kW h.} \tag{1.2}$$

If η_{CY} were to equal 100%, W_{net} would equal Q_B and there would be no heat rejected to the sink. Unless the sink were at absolute zero of temperature this would contravene the Second Law of Thermodynamics, since the plant would then constitute a perpetual motion machine of the second kind. Thus η_{CY} is necessarily less than 100%. It will be seen shortly that, for an ideal plant, its value depends only on the mean temperatures of heat reception and rejection, and that even for an ideal simple steam plant, η_{CY} is only of the order of 20–30%.

(2) Control surface Z—heating device efficiency

The plant within this control surface constitutes a *steady-flow heating device* which is supplied with *reactants* (fuel and air) and discharges *products* of combustion (flue gases) while supplying heat to the working fluid passing round the cycle. If the products were to leave the heating device at the same temperature as the incoming reactants, the quantity of heat transferred per unit mass of fuel supplied would equal the calorific value (CV) of the fuel at that temperature. [*Note:* $CV \equiv (H_R - H_P)_T$, where H_R and H_P are the enthalpies of the reactants and products respectively at the specified temperature T.] In the heating device, the products in fact always leave at a temperature in excess of that of the incoming reactants because it is uneconomic to provide sufficient tube surface in the boiler or heat exchanger to effect complete cooling. Hence Q_B is always less than the calorific value and the performance is expressed in terms of the *heating-device efficiency*, η_B, defined as the ratio of the heat supplied by the device, per unit mass of fuel burned, to the calorific value of the fuel. Thus

$$\eta_B \equiv \frac{Q_B}{CV}. \tag{1.3}$$

It must be noted that, although the word "efficiency" is used to describe both η_{CY} and η_B, these are entirely different in kind. η_B can theoretically reach a value of 100%, and would do so if the products resulted from complete combustion and were cooled to the temperature of the incoming reactants. On the other hand, as has already been noted, the greatest possible value of η_{CY} is very much less than 100%.

(3) The entire plant—overall efficiency

The entire plant contained within both control surfaces Y and Z constitutes a simple power plant designed to produce work from the chemical energy released in the combustion of the fuel. Because W_{net} is related to Q_B through η_{CY}, and Q_B is itself related to CV through η_B, it is a rational procedure to define the *overall efficiency*, η_o, of the plant as the ratio of the net work produced, per unit mass of fuel supplied, to the calorific value of the fuel. Thus

$$\eta_o \equiv \frac{W_{net}}{CV}, \tag{1.4}$$

so that

$$\eta_o = \eta_{CY}\eta_B. \tag{1.5}$$

Because η_{CY} is always very much less than 100%, so also is η_o.

This volume is concerned with the analysis of engineering cycles, so that η_{CY} will be studied in detail while η_B and η_o will receive less attention. It must be pointed out, however, that exclusive attention to the improvement of η_{CY} by the power-station designer must be avoided, for steps taken to improve η_{CY} may cause η_B to fall if remedial measures are not adopted. An improvement in η_{CY} would not then be reflected in an equal improvement in η_o, and it is this figure which is of importance to the power station engineer. Such a situation is shown to arise in the more advanced cycles considered in Part II.

1.5. Performance criteria

The performance measures discussed in the preceding section merely provide a means of expressing the **measured** performance of the plant. Having expressed the measured performance in this way, the engineer will be curious to know how much better the performance could have been had he been more expert in designing and building the plant. He will thus need not just a performance measure, which is solely a measure of performance, but a *performance criterion* against which the measured values of η_{CY}, η_B and η_o can be compared.

Since it has already been noted that η_B would have a value of 100% if, before leaving the plant, the exhaust products were brought to the temperature of the entering reactants, this figure of 100% provides the criterion against which to judge the measured value of η_B. In practice, η_B is less than 100% only because to reach this figure the tubular surface area of the boiler or heat exchanger would need to be of infinite extent. Consequently, economic considerations limit the value of η_B for even the largest steam boilers to a figure of 85–90%.

To provide a criterion against which to judge the measured cycle efficiency, η_{CY}, requires a detailed study of ideal cycles operating under comparable conditions, and this study will be made in succeeding chapters.

Problems

1.1. In a test of a cyclic steam power plant, the measured rate of steam supply was 7.1 kg/s when the net rate of work output was 5000 kW. The feed water was supplied to the boiler at a temperature of 38 °C, and the superheated steam leaving the boiler was at 1.4 MN/m² and 300 °C.

Calculate the thermal efficiency of the cycle and the corresponding heat rate. What would be the heat rate when expressed in Btu of heat input per kW h of work output?

Answer: 24.4%, 4.09; 13 960 Btu/kW h.

1.2. In the test of the power plant of Problem 1.1, the coal supply rate was 3250 kg/h, and the calorific value of the coal was 26 700 kJ/kg. Calculate the boiler efficiency, and the overall efficiency of the complete plant.

Answer: 84.9%; 20.7%.

1.3. A cyclic steam power plant is designed to supply steam from the boiler at 10 MN/m² and 550 °C when the boiler is supplied with feed water at a temperature of 209.8 °C. It is estimated that the thermal efficiency of the cycle will be 38.4% when the net power output is 100 MW. Calculate the steam consumption rate. The enthalpy of the feed water may be taken as being equal to the enthalpy of saturated water at the same temperature.

The boiler has an estimated efficiency of 87%, and the calorific value of the coal supplied is 25 500 kJ/kg. Calculate the rate of coal consumption in (a) kg/s, (b) ton/min.

Answer: 100 kg/s; 11.74 kg/s, 0.693 ton/min.

1.4. Air flows round a closed-circuit gas turbine plant, entering the compressor at 18 °C and leaving it at 190 °C. The air temperature is 730 °C at turbine inlet and 450 °C at turbine outlet. The values of the mean specific heat capacity of air over the temperature ranges occurring in the compressor, heater and turbine are respectively 1.01, 1.08 and 1.11 kJ/kg K. Calculate the cycle efficiency, neglecting mechanical losses.

Answer: 23.5%.

1.5. Calculate the required rate of air circulation per second round the circuit of Problem 1.4 if the net power output is to be 5000 kW. Calculate also the required rate of fuel supply when oil of calorific value 44 500 kJ/kg is supplied to the combustion chamber, the efficiency of which as a heating device is 75%.

Answer: 36.5 kg/s; 0.637 kg/s.

1.6. In the plant shown diagrammatically in Fig. 1.4, fuel and air at the pressure and temperature of the environment, namely 1 atm and 25 °C, are fed to an imperfectly lagged combustion chamber. The products of combustion then pass through a heat exchanger in which they transfer heat to the fluid of a cyclic heat power plant and finally exhaust to the atmosphere. Stray heat loss from the heat exchanger is negligible.

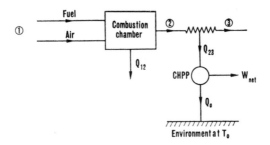

FIG. 1.4.

In a test on the plant, in which W_{net} was 11 MJ per kg of fuel burnt, the values of enthalpy and entropy at the points indicated were found to be as listed in the following table, all quantities being expressed per kg of fuel supplied:

State point	1	2	3	Products at 1 atm and 25 °C
Enthalpy, MJ	61.00	58.50	22.63	10.99
Entropy, MJ/K	0.2398	0.3070	0.2698	0.2429

Calculate the following quantities:

(a) The stray heat loss, Q_{12}, to the environment from the combustion chamber, expressed as a percentage of the calorific value of the fuel.
(b) The *heating-device efficiency* of the combined combustion chamber and heat exchanger.
(c) The *thermal efficiency* of the cyclic heat power plant.
(d) The *thermal efficiency* of an ideal, fully reversible, cyclic heat power plant which takes in heat by cooling the products of combustion between the same states 2 and 3 and in which heat rejection takes place at the temperature of the environment.
(e) The *overall efficiency* of the complete plant.
(f) The *rational efficiency* of the complete plant.
(g) The *lost work due to irreversibility* (1) in the combustion-chamber process and (2) in consequence of the discharge of the gases from the plant at a temperature in excess of the environment temperature.
Note: Before attempting (f) and (g), study Appendix A.

Answer: (a) 5.0%; (b) 76.7%; (c) 30.7%; (d) 69.1%; (e) 22.0%; (f) 21.6%; (g) 22.53 MJ, 3.62 MJ.

CHAPTER 2

Simple steam plant

A steam power station exists to produce electrical power at the least possible cost. The resulting drive for efficiency has led to the construction of very complicated plant. In this chapter, however, the performance of only the simplest kind of cyclic steam plant is studied. This is still of interest for small-scale plant, though not for the plants of very large output that are found in central power stations. These are discussed in Chapter 7. At this point, the reader should familiarise himself with the simple steam plant described in Chapter 1 and illustrated in Fig. 1.1.

2.1. Performance measures

In Chapter 1 it was seen that the measured performance of a simple steam plant is expressed in terms of the following three *performance measures*:

Thermal or cycle efficiency: $\qquad \eta_{CY} \equiv \dfrac{W_{net}}{Q_B}.$

Heating device efficiency: $\qquad \eta_B \equiv \dfrac{Q_B}{CV}.$

Overall efficiency: $\qquad \eta_o \equiv \dfrac{W_{net}}{CV} = \eta_{CY}\eta_B.$

Having noted in §1.5 that a figure of 100% provides the criterion against which to judge the measured value of η_B, we now proceed to set up a criterion against which to judge the measured value of η_{CY}.

2.2. Performance criterion for the efficiency of the simple steam cycle—Rankine cycle efficiency

The appropriate *performance criterion* against which to judge the measured value of η_{CY} will be the cycle efficiency of an **ideal** steam plant supplied with steam at the same temperature and pressure, and exhausting to the same condenser pressure. Such an ideal plant must produce the greatest possible net work output for a given heat input when operating under the specified steam conditions. From a knowledge of the Second Law and its corollaries, it is known that *all irreversible processes result in lost opportunities for producing work, so that all processes in the ideal plant must be reversible.* The resulting reversible cycle is called the *ideal Rankine cycle*, and its efficiency the *Rankine cycle efficiency* for the specified steam conditions. (It should be noted that when a cycle is described as being reversible, it is meant that all the *processes* that go to make up the cycle are reversible.)

In the present context, we shall be considering only the *internal reversibility* of the cycle. We shall not concern ourselves with the temperature of the source from which heat is transferred to the working fluid, nor the temperature of the sink to which heat is transferred from the working fluid; that is, in the present context we shall not concern ourselves with any *external irreversibility* due to temperature differences between the working fluid and the source and sink.

2.3. The ideal Rankine cycle

Since the ideal cycle is to be internally reversible:

(1) there must be no frictional pressure drops in the boiler, condenser and piping;
(2) the flow through the turbine and feed pump must be frictionless.

In an ideal plant there must also be no loss of energy through stray heat loss from any item of the plant to the environment. In that case the expansion of the fluid in the turbine and its compression in the feed pump will be adiabatic as well as frictionless, so that these processes will be *isentropic*. The state of the fluid as it passes round the ideal Rankine cycle, when superheated steam leaves the boiler, will therefore be as shown in the diagrams of Fig. 2.1. In these diagrams the liquid specific volumes, and the temperature and enthalpy rises in the feed pump, have been greatly exaggerated in order to indicate clearly the nature of the process occurring in the pump; for example, in reality the temperature rise in the feed pump would be so small that point 2 in Fig. 2.1(b) would be indistinguishable from point 1.

The diagrams should be studied while bearing in mind the following relations, which may readily be written down from the Steady-flow Energy Equation and from the $T\,ds$ equation ($T\,ds = dh - v\,dp$) for unit mass of fluid when it undergoes an infinitesimal change of state.

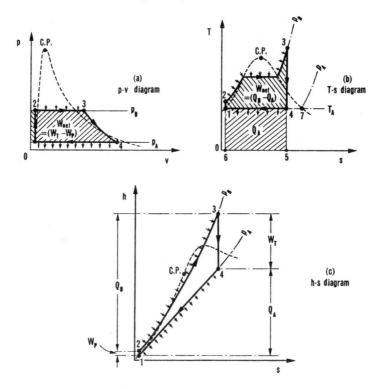

Fig. 2.1. Rankine cycle state diagrams.

(1) In the ideal turbine and feed pump

For reversible, adiabatic (namely, isentropic) expansion or compression in steady flow,

$$\int \delta W_x = -\int dh = -\int v \, dp, \tag{2.1}$$

since $ds = 0$ in an isentropic process.

(2) In the ideal boiler and condenser

For a reversible, steady-flow process at constant pressure with heat transfer to or from the fluid,

$$\int \delta Q = \int dh = \int T \, ds, \tag{2.2}$$

since $dp = 0$ in an isobaric process.

The temperature–entropy diagram is particularly instructive as it gives a direct visual representation of the magnitude of the cycle efficiency, since

$$\eta_{CY} \equiv \frac{W_{net}}{Q_B} = \frac{Q_B - Q_A}{Q_B} = \frac{\text{Area 23412}}{\text{Area 23562}}.$$

If drawn to scale, the pressure–volume and enthalpy–entropy diagrams serve to show the relative magnitudes of the turbine work output and feed pump work input.

2.4. Expressions for the Rankine cycle efficiency

In symbols, the thermal efficiency of the Rankine cycle may be expressed as follows:

(a) *Exact expression*

$$\text{Exact } \eta_{RANK} = \frac{W_T - W_P}{Q_B} = \frac{(h_3 - h_4) - (h_2 - h_1)}{(h_3 - h_2)}. \tag{2.3}$$

(b) *Approximate expression*

Equation (2.3) may be rewritten as

$$\text{Exact } \eta_{RANK} = \frac{(h_3 - h_4) - (h_2 - h_1)}{(h_3 - h_1) - (h_2 - h_1)}.$$

The quantity $(h_2 - h_1)$ is the work input to the feed pump, and is equal to $\int_1^2 v \, dp$. Because the specific volume of the water passing through the pump is so much smaller than that of the steam passing through the turbine, this term is so small that, at the operating pressures encountered in simple steam plant, it may be neglected in relation to the other terms in the expression.[†] Hence, neglecting this feed pump term:

$$\text{Approx. } \eta_{RANK} = \frac{h_3 - h_4}{h_3 - h_1}. \tag{2.4}$$

This avoids the need to evaluate h_2, the enthalpy of compressed liquid water after isentropic compression in the feed pump from saturation state 1 at the condenser pressure.

EXAMPLE 2.1. Calculate the turbine work output and the feed pump work input per kg of fluid for a Rankine cycle in which steam leaves the

---

[†]In large, high-pressure plant, however, the work input to the feed pump is by no means negligible. For example,[(4)] the steam turbine driving the feed pump of a 1300 MW plant is itself of no less than 70 MW output. As will be seen from Table B.1 in Appendix B, this is as great as the output of the largest *power-station* turbine of only thirty years ago.

boiler at $2\,\mathrm{MN/m^2}$ and $350\,^\circ\mathrm{C}$, and in which the steam exhausts to the condenser at $7\,\mathrm{kN/m^2}$. Calculate the exact and approximate cycle efficiencies.

Using the notation of Fig. 2.1(b): $h_3 = 3138.6\,\mathrm{kJ/kg}$, $s_3 = 6.960\,\mathrm{kJ/kg\,K}$, $s_4 = s_3$.

If x_4 is the dryness fraction at point 4,

$$(1 - x_4) = \frac{s_7 - s_4}{s_7 - s_1} = \frac{8.277 - 6.960}{8.277 - 0.559} = 0.1707$$

and

$$h_4 = h_7 - (1 - x_4)(h_7 - h_1)$$
$$= 2572.6 - 0.1707 \times 2409.2$$
$$= 2161.4\,\mathrm{kJ/kg}.$$

[Note that greater slide-rule accuracy is obtained by calculating $(1 - x_4)$ instead of x_4.]

$\therefore$ turbine work output: $W_T = (h_3 - h_4) = \mathbf{977.2\,kJ/kg}$,

feed pump work input: $W_P = \int_1^2 v\,dp \doteq v_1(p_B - p_A)$
$$= 0.001\,007 \times 1993$$
$$= \mathbf{2.0\,kJ/kg}.$$

Net work output: $W_{\mathrm{net}} = 975.2\,\mathrm{kJ/kg}.$
$$h_1 = 163.4\,\mathrm{kJ/kg},$$
$$h_2 = 163.4 + 2.0 = 165.4\,\mathrm{kJ/kg}.$$

$\therefore$ Heat input: $Q_B = (h_3 - h_2) = 2973.2\,\mathrm{kJ/kg}.$

$$\therefore \text{Exact } \eta_{\mathrm{RANK}} \equiv \frac{W_{\mathrm{net}}}{Q_B} = \frac{975.2}{2973.2} \times 100$$
$$= \mathbf{32.80\%}.$$

Alternatively, $Q_A = (h_4 - h_1) = 1998.0\,\mathrm{kJ/kg}$ and

$$\text{Exact } \eta_{\mathrm{RANK}} = \left(1 - \frac{Q_A}{Q_B}\right) = \left(1 - \frac{1998.0}{2973.2}\right) \times 100 = \mathbf{32.80\%}.$$

$$\text{Approx. } \eta_{\mathrm{RANK}} = \frac{h_3 - h_4}{h_3 - h_1} = \frac{977.2}{2975.2} \times 100 = \mathbf{32.84\%}.$$

2.5. Comparison of actual and ideal performance — the efficiency ratio

Since the Rankine cycle efficiency is the rational criterion of excellence against which to compare the measured cycle efficiency of an actual steam plant, the ratio of the latter to the former will give a measure of the excellence of performance of the actual plant. This ratio is called the *efficiency ratio*. It is a more informative measure of the plant performance than is the cycle efficiency, since a statement of the latter conveys no information as to how much better, in theory, the performance could have been. Since the net work output is nearly equal to the turbine work output when the feed pump work input is small, and since the effects of pressure drops in the boiler, condenser and piping are relatively small, the efficiency ratio is approximately equal to the isentropic efficiency of the turbine.

EXAMPLE 2.2. Tests on a simple steam plant taking steam from the boiler at the same condition as in Example 2.1, and exhausting to the same condenser pressure, showed that the steam consumption rate, expressed as kg of steam per MJ of turbine work output, was 1.35 kg/MJ. Determine the efficiency ratio for the plant.

Neglecting the feed pump work input, and using the notation of Fig. 2.1(b):

$$\text{Per kg of steam: } Q_B \doteq (h_3 - h_1) = 2975.2 \text{ kJ/kg}.$$

∴ Per MJ of turbine work output, heat supplied to fluid

$$= 2975.2 \times 1.35 \times 10^{-3}$$

$$= 4.017 \text{ MJ}.$$

$$\therefore \eta_{CY} = \frac{1}{4.017} \times 100 = 24.90\%.$$

Comparing this with the approximate Rankine cycle efficiency:

$$\text{Efficiency ratio} = \frac{24.90}{32.84} = \mathbf{0.758}.$$

2.6. Imperfections in the actual steam plant — the effect of irreversibilities

All irreversible processes result in lost opportunities for producing work, so it is these for which search must be made if the poor performance of a plant is to be explained. All frictional processes are irreversible, and it is the purpose here to study the reduction in work output due to some of these. At this point the frictional dissipation of mechanical energy in bearings and the like will be ignored since these effects are external to the fluid passing round the thermodynamic cycle, and due account can later be taken of them through the mechanical efficiency of the turbine and pump. Attention will here be confined to imperfections within the cycle; that is, to internal irreversible processes. For purposes of illustration, only two will be considered and the processes in the rest of the plant will be imagined to be ideal. The two considered are illustrated in Fig. 2.2 and are:

(1) Frictional pressure drop in the steam pipe and across the governing throttle valve between boiler and turbine

If stray heat losses are neglected, this is an *adiabatic throttling process*. Applied to such a process, the Steady-flow Energy Equation shows that the

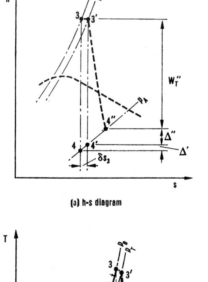

(a) h-s diagram

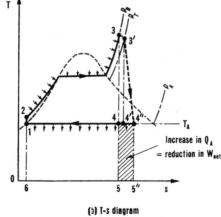

(b) T-s diagram

FIG. 2.2. Effect of irreversibilities in the simple steam cycle.

enthalpy of the steam after the throttle valve at point 3′ will be the same as the enthalpy of the steam leaving the boiler at point 3, for the difference between the kinetic energies of the steam at these two points is negligibly small. The **ideal** work output obtainable from the turbine is thereby reduced from $(h_3 - h_4)$ to $(h_{3'} - h_{4'})$, the amount of this reduction being shown as Δ' in Fig. 2.2(a).

(2) Frictional effects in flow through the turbine

Because of frictional effects in the turbine nozzle and blade passages, the exit enthalpy is greater than it would be in the ideal case and the turbine work output is consequently less. The exit state of the steam is thus at 4″ instead of 4′. [Note that the entropy at exit is also increased, in accordance with the Second Law corollary that $(\Delta S_{adiab})_{irrev} > 0$.] The resulting reduction in turbine work output is indicated as Δ'' in Fig. 2.2(a). Its magnitude may be specified by defining a *turbine isentropic efficiency*, η_T, as the ratio of the actual enthalpy drop across the turbine to the isentropic enthalpy drop when expanding from the same initial conditions to the same exhaust pressure. Thus

$$\text{Turbine isentropic efficiency, } \eta_T = \frac{h_{3'} - h_{4''}}{h_{3'} - h_{4'}}.$$

For an ideal turbine this would be 100%.

2.7. Lost work due to irreversibility

Since the heat input to the cycle is unaffected by these irreversibilities, being still equal to $(h_3 - h_2)$, the reduction in W_{net} is equal to the increase in heat rejected. If the exhaust steam is wet, as it always is in practice, then the *lost work due to irreversibility* may be simply expressed as

$$\text{Lost work = Increase in } Q_A = T_A(s_{4''} - s_4).$$

This quantity is equal to the area 4″455″4″ in Fig. 2.2(b).

The lost work due to irreversibility is thus equal to the product

$$T_A \times [\text{entropy increase of steam due to irreversibilities}]. \qquad (2.5)$$

If the condenser temperature T_A is taken to be the same as the temperature T_0 of the environment, this is a special case of a more general and important theorem in the study of availability which is discussed in Appendix A.

EXAMPLE 2.3. A simple steam plant takes steam from the boiler at the same condition as in Example 2.1 and exhausts to the same condenser pressure. There is a pressure drop of 0.1 MN/m^2 between boiler and turbine. Estimate the entropy increase between boiler and turbine, and the resulting reduction in the ideal turbine work output.

If the turbine has an isentropic efficiency of 80%, determine the thermal efficiency of the cycle and the efficiency ratio for the plant.

Referring to Fig. 2.2, since $h_{3'} = h_3$, it would be possible to find $s_{3'}$ from the steam tables and so to calculate δs_3. This would be a little tedious, and since δs_3 is only small a better method is to use the $T ds$ equation, $T ds = dh - v\, dp$.

From 3 to 3', $\delta h = 0$, so that $T\, \delta s = -v\, \delta p$, and

$$\delta s_3 \doteq \frac{v_3(p_B - p_T)}{T_3} = \frac{0.1386 \times 10^5}{623.2} = 22.2 \text{ J/kg K}.$$

$\therefore$ Lost work $= T_A\, \delta s_3 = 312.2 \times 22.24 \times 10^{-3} = 6.9 \text{ kJ/kg}.$

$\therefore$ Ideal turbine work for expansion from point 3'

$$= 977.2 - 6.9$$

$$= 970.3 \text{ kJ/kg}.$$

$\therefore$ Actual turbine work: $W_T'' = 970.3 \times 0.8 = 776.2 \text{ kJ/kg}.$

Neglecting feed-pump work input:

$$\text{Thermal efficiency} = \frac{776.2}{2975.2} \times 100 = 26.09\%.$$

$$\text{Efficiency ratio} = \frac{26.09}{32.84} = 0.794.$$

It is seen that the lost work due to the pressure drop between boiler and turbine is small compared with that due to inefficiency of the turbine.

2.8. Alternative expressions for Rankine cycle efficiency and efficiency ratio in terms of available energy

It has already been noted that, in practice, the exhaust steam entering the condenser is always wet. The steam is then condensed at constant temperature in the ideal condenser. In these circumstances an alternative expression for the exact Rankine cycle efficiency may be written down from a study of availability. If, for purposes of analysis, the environment temperature T_0 is taken as being the same (or, more strictly, an infinitesimal amount less than) the condenser temperature T_A, then the turbine, condenser and feed pump of the ideal Rankine cycle together constitute an *ideal non-cyclic open-circuit steady-flow work-producing*

device exchanging heat reversibly with the environment at temperature T_A.
In this device, all internal processes are reversible (giving *internal reversibility*), and the heat exchange with the environment is reversible (giving *external reversibility* here) since there is zero (or, more strictly, negligible) temperature difference between the fluid and the environment in the heat exchange process.

It is shown in Table A.1 of Appendix A that, under these conditions of internal and external reversibility, the work produced by an ideal device such as this, when the fluid is in state 3 at inlet and in state 2 at outlet and the absolute temperature of the environment is T_A, is given by the expression

$$\text{Ideal } W_{net} = (b_3 - b_2), \text{ the } \textit{steady-flow available energy}, \qquad (2.6)$$

where $\qquad b \equiv (h - T_A s)$, the *steady-flow availability function*. $\qquad (2.7)$

Hence an alternative expression for the exact Rankine cycle efficiency is:

$$\text{Exact } \eta_{RANK} = \frac{b_3 - b_2}{h_3 - h_2}. \qquad (2.8)$$

It is left to the reader to show, with the aid of eqn. (2.7), that eqn. (2.8) can be transformed readily into eqn. (2.3).

It is seen from Theorem 1 in Appendix A that no other device which worked between the given states 3 and 2, while exchanging heat with an environment at T_A, could produce an amount of work greater than $(b_3 - b_2)$, thus confirming that the Rankine cycle efficiency is a rational criterion for the excellence of performance of a simple steam plant. Because $(b_3 - b_2)$ is the greatest net work output in the given situation, it is called *the steady-flow available energy for the given change of state in the presence of the given environment*. Expressing the Rankine cycle efficiency in the form of eqn. (2.8) makes it clearly evident that it relates to an ideal plant which produces a net work output equal to the available energy. A further advantage of expressing the efficiency in this form is that, as will be seen in Chapter 7, the thermal efficiency of ideal regenerative steam cycles of much greater complexity can be expressed in precisely the same form. A similar type of analysis will also be needed when dealing with dual-pressure steam cycles for nuclear power plant in Chapter 8.

We may use the above analysis to note a connection between the *efficiency ratio* defined in §2.5 and the *rational efficiency* (as defined in §A.12 of Appendix A) of what we may term the *work-producing steam circuit*. By the latter, we mean that part of the complete steam circuit of Fig. 1.1(a) which contains the turbine, condenser and feed pump; this is depicted afresh in Fig. 2.3.

From eqn. (2.6) and eqn. (A.14) of §A.12, the rational efficiency η_R of this work-producing steam circuit is seen to be given by

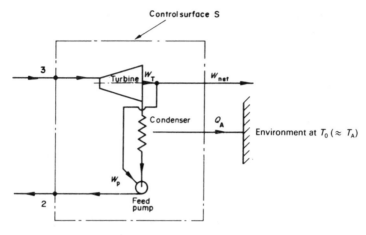

FIG. 2.3. Work-producing steam circuit.

$$(\eta_R)_{\text{steam circuit}} \equiv \frac{\text{Actual } W_{\text{net}}}{\text{Ideal } W_{\text{net}}} = \frac{\text{Actual } W_{\text{net}}}{b_3 - b_2}. \qquad (2.9)$$

Now, for the actual steam cycle, we have

$$\text{Actual } \eta_{\text{CY}} = \frac{\text{Actual } W_{\text{net}}}{h_3 - h_2}, \qquad (2.10)$$

while, for the ideal Rankine cycle under the idealised conditions defined in the first paragraph of this section, η_{RANK} is given by equation (2.8). Thus, from eqns. (2.10) and (2.8), the efficiency ratio will be given by

$$\text{Efficiency ratio} \equiv \frac{\text{Actual } \eta_{\text{CY}}}{\text{Ideal } \eta_{\text{RANK}}} = \frac{\text{Actual } W_{\text{net}}}{b_3 - b_2}. \qquad (2.11)$$

Hence, from eqns. (2.9) and (2.11)

$$\text{Efficiency ratio} = (\eta_R) \text{ steam circuit.} \qquad (2.12)$$

Thus, if we want to know the value of the efficiency ratio of the actual plant, it is evidently simpler to obtain this by calculating directly the rational efficiency of the open-circuit work-producing plant within control surface S of Fig. 2.3. The engineer should thus be encouraged, wherever possible, to apply the concepts of thermodynamic availability to a study of cyclic as well as non-cyclic plant and devices. These concepts are treated fully in a second book by the author[5][†] and also in the author's critical review paper on the subject.[6]

†See also **Additional References** B to E.

2.9. Variation in cycle efficiency with change in the design steam conditions

Because the working fluid in a steam cycle does not have a simple equation of state, the effect on the cycle efficiency of a change in the conditions for which the plant is designed cannot be given simple mathematical expression. Such a study must consequently be made by graphical presentation after a series of detailed calculations. This question is therefore left over to Chapter 7.

Problems

2.1. For the conditions given in Example 2.1, express the Rankine cycle efficiency as a percentage of the Carnot cycle efficiency for the same extreme limits of temperature.

Answer: 65.7%.

2.2. Calculate the thermal efficiency of a Rankine cycle in which the steam is initially dry saturated at 2 MN/m^2 and the condenser pressure is 7 kN/m^2. Express this efficiency as a percentage of the Carnot cycle efficiency for the same limits of temperature. Explain why the Rankine cycle efficiency is less than in Example 2.1, and why the ratio of the efficiencies of the respective Rankine and Carnot cycles is greater.

Answer: 31.5%; 88.3%.

2.3. In a steam power plant the turbine has an isentropic efficiency of 80% when the conditions are otherwise as in Problem 2.2. Neglecting the work input to the feed pump, determine the thermal efficiency of the plant if the condensate is returned to the boiler at the saturation temperature corresponding to the condenser pressure. Calculate the steam consumption rate, in kg per MJ of turbine work output.

If the steam is supplied from a boiler of 84% efficiency, and the calorific value of the coal is $28\,000 \text{ kJ/kg}$, calculate the overall efficiency of the plant and the coal consumption rate in kg/MJ of turbine work output.

Answer: 25.2%; 1.506 kg/MJ; 21.2%; 0.169 kg/MJ.

2.4. For the plant of Problem 1.1, determine the efficiency ratio, as defined in §2.5, and the rational efficiency of the work-producing steam circuit, as calculated from eqn. (2.9) of §2.8. Take the environment temperature as being equal to the saturation temperature of the steam in the condenser, namely 38 °C.

Answer: 0.803; 80.3%.

2.5. For the plant of Problem 2.3, write down the values of the efficiency ratio and the rational efficiency of the work-producing steam circuit.

Answer: 0.8; 80%.

2.6. Steam is expanded isentropically in the high-pressure cylinder of a steam turbine from a pressure of 2 MN/m^2 at 350 °C to 0.1 MN/m^2. Isentropic expansion is then continued down to 7 kN/m^2 in the low-pressure cylinder. Calculate the percentage of the total work output that is performed by the HP cylinder. (See Example 2.1 for the total work output of both cylinders.)

Answer: 62.7%.

2.7. The expansion in a turbine is adiabatic and irreversible. The entropy of the steam at inlet is 6.939 kJ/kg K, and the turbine exhausts at a pressure of 7 kN/m². If the dryness fraction of the steam at exhaust is 0.91, calculate the lost work due to irreversibility per kilogram of steam flowing through the turbine. If the inlet pressure is 4 MN/m², what is the isentropic efficiency of the turbine?

Answer: 200.7 kJ/kg; 82.9%.

CHAPTER 3

Simple closed-circuit gas-turbine plant

As in the case of simple steam plant, the efficiency of a simple gas-turbine plant is often too low to be commercially attractive. This results in the construction of plants of some complexity, but the performance of only the simplest kind of cyclic gas-turbine plant is studied in this chapter. At this point the reader should refer back to Fig. 1.3 and the description of that plant given in Chapter 1.

3.1. Performance measures

In Chapter 1 it was seen that the measured performance of a simple closed-circuit gas-turbine plant is expressed in terms of the same three *performance measures* as were applicable to the simple steam plant, namely:

Thermal or cycle efficiency: $\quad \eta_{CY} \equiv \dfrac{W_{net}}{Q_B}.$

Heating device efficiency: $\quad \eta_B \equiv \dfrac{Q_B}{CV}.$

Overall efficiency: $\quad \eta_o \equiv \dfrac{W_{net}}{CV} = \eta_{CY}\eta_B.$

Having again noted that a figure of 100% provides the criterion against which to judge the measured value of η_B, we now proceed to set up a criterion against which to judge the measured value of η_{CY}.

3.2. Performance criterion for the efficiency of the simple gas-turbine cycle—Joule cycle efficiency

For reasons similar to those given in §2.2 for the simple steam plant, the appropriate *performance criterion* against which to judge the measured value of η_{CY} will be the thermal efficiency of an ideal reversible cycle operating under comparable conditions, in this case the ideal *Joule cycle* (sometimes called the *Brayton cycle*). Its efficiency will be shown to be a function only of the pressure ratio. The ratio of the actual cycle efficiency to the corresponding Joule cycle efficiency will then again be described as the *efficiency ratio*, and this will give an indication of the closeness to perfection of the actual plant (only an indication, because in the Joule cycle the working substance is assumed to behave as a perfect gas).

3.3. The ideal Joule cycle

As in the ideal Rankine cycle, there must be no frictional pressure drops in the heat exchangers and ducting, and the expansion and compression in the turbine and compressor respectively must be isentropic (adiabatic and reversible). The state of the gas as it passes round the ideal Joule cycle will therefore be as shown in the diagrams of Fig. 3.1. The Joule cycle is defined to be one in which the working substance is assumed to behave as a perfect gas. Since the change in enthalpy of a perfect gas is directly proportional to its change in temperature, there is no need in this case to draw a separate temperature-entropy diagram, for it will be similar to the enthalpy–entropy diagram. The diagrams should again be studied with eqns. (2.1) and (2.2) in mind.

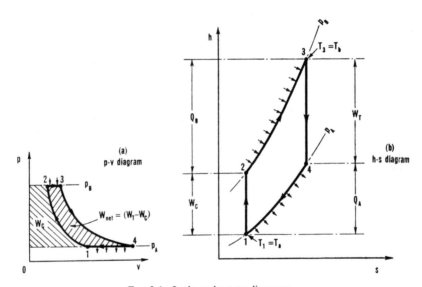

Fig. 3.1. Joule cycle state diagrams.

3.4. Expression for the Joule cycle efficiency

The best way of evaluating the thermal efficiency of any gas-turbine cycle is to work in terms of the temperatures at the various points round the cycle.

For the isentropic processes 1–2 and 3–4 when the working substance is a perfect gas:

$$\frac{T}{p^{(\gamma-1)/\gamma}} = \text{const.} \tag{3.1}$$

Hence

$$\frac{T_2}{T_1} = \frac{T_3}{T_4} = \left(\frac{p_B}{p_A}\right)^{(\gamma-1)/\gamma} = r_p^{(\gamma-1)/\gamma} \equiv \rho_p, \tag{3.2}$$

where $r_p \equiv p_B/p_A$, the pressure ratio, and $\rho_p \equiv r_p^{(\gamma-1)/\gamma}$, the isentropic temperature ratio for the cycle.

The thermal efficiency of the Joule cycle is then given by

$$\eta_{\text{JOULE}} = \left(1 - \frac{Q_A}{Q_B}\right) = 1 - \frac{c_p(T_4 - T_1)}{c_p(T_3 - T_2)}. \tag{3.3}$$

Whence, from eqns. (3.2) and 3.3),

$$\eta_{\text{JOULE}} = \left(1 - \frac{1}{\rho_p}\right) = \left[1 - \frac{1}{r_p^{(\gamma-1)/\gamma}}\right]. \tag{3.4}$$

Thus the thermal efficiency of the ideal Joule cycle with a perfect gas as the working fluid is a function only of the pressure ratio, being independent of both the compressor and turbine inlet temperatures. If the working fluid is air, the cycle is described as an *air-standard Joule cycle*.

EXAMPLE 3.1. Tests on a simple closed-circuit, gas-turbine plant using air as the working fluid showed that its thermal efficiency was 20% when operating at the design pressure ratio of 5.4. Calculate the thermal efficiency of the corresponding air-standard Joule cycle, and the efficiency ratio for the plant.

$$\rho_p \equiv r_p^{(\gamma-1)/\gamma} = 5.4^{(0.4)/(1.4)} = 1.62.$$

$$\therefore \ \eta_{\text{JOULE}} = \left(1 - \frac{1}{1.62}\right) \times 100 = \mathbf{38.2\%}.$$

$$\therefore \ \text{Efficiency ratio} = \frac{20.0}{38.2} = \mathbf{0.523}.$$

3.5. Variation of η_{JOULE} with pressure ratio

It is of interest to study the variation of η_{JOULE} with variation in the **design** pressure ratio; that is, how the cycle efficiencies of different ideal plants at their design load would compare when they were designed for different pressure ratios. In practice, this would not be the same as the variation in efficiency of a given plant as the pressure ratio changed with varying load.

For the present purpose, it will be assumed that the plants with different pressure ratios are all designed for the same compressor inlet temperature $T_1 = T_a$, and the same turbine inlet temperature $T_3 = T_b$. The former would be determined by the cooling water temperature and the economic size of the cooler-heat-exchanger, and the latter by the metallurgical temperature limit of the first row of turbine blades. Although η_{JOULE} is independent of these temperatures and dependent only on the pressure ratio, there is a limit to this pressure ratio when T_a and T_b are both fixed. That limit is illustrated in Fig. 3.2, and it is seen to occur when the temperature after isentropic compression from T_a is equal to the turbine inlet temperature T_b.

Equation (3.4) shows that η_{JOULE} increases continuously with r_p right up to the limiting value, its variation with ρ_p being shown by the curve labelled "reversible" in Fig. 3.3. This graph relates to values of t_a and t_b equal respectively to 15 °C and 800 °C, and to air treated as a perfect gas. It is more convenient to plot against ρ_p than against r_p, but key values of r_p are indicated along the top of the graph.

The limiting pressure ratio of 100, corresponding to $\rho_p = T_b/T_a$, is

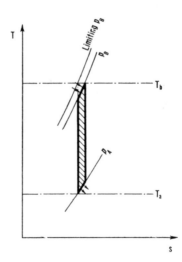

F$_{\text{IG}}$. 3.2. Limiting pressure ratio for given T_a and T_b in the Joule cycle.

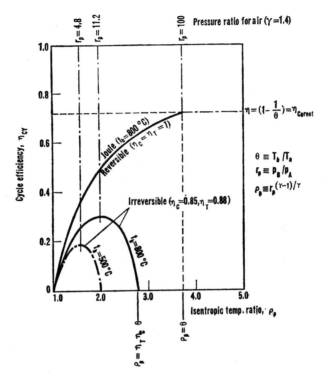

FIG. 3.3. Variation of cycle efficiency with isentropic temperature ratio ($t_a = 15\,°C$).

impracticably large. Study of Fig. 3.2 also shows that as r_p approaches this figure the area enclosed by the cycle on the temperature–entropy diagram tends towards zero. This means that the work output per unit mass of air circulated tends to zero, so that the physical size of a plant designed for such a condition would also become impracticably large. However, neither of these considerations is of any real practical significance since, as will be seen next, a pressure ratio as high as this is never used because the performance is altered beyond recognition when the effects of irreversibilities are taken into account.

3.6. Imperfections in the actual plant—the effect of irreversibilities

All irreversible processes result in lost opportunities for producing work, so these will cause the performance of an actual plant to fall below that of the ideal. Hence, for the best possible performance, frictional pressure drops in the heat exchangers and ducting must be minimised to an extent which is economically profitable. Larger ducting will give smaller gas velocities and hence smaller parasitic pressure drops, but it will also cost more and will be an embarrassment if too large. Frictional effects in the turbine and compressor must also be minimised; this means that the isentropic efficiencies of both must be as high as possible.

For the purpose of illustrating the effects of irreversibilities on the plant performance, only inefficiencies in the turbine and compressor will be considered. This is a matter of convenience only, and it must not be taken to imply that frictional effects in the heat exchangers and ducting are, in practice, unimportant; such is far from being the case. The cycle is now as shown in Fig. 3.4, the turbine work output being less and the compressor work input being greater than under ideal conditions, when both processes were isentropic; they are still assumed to be adiabatic, but in both the entropy now increases. Thus, whereas the *turbine isentropic efficiency* η_T is defined as the ratio of the actual to the isentropic enthalpy drop, the *compressor isentropic efficiency* η_C is defined as the ratio of the isentropic to the actual enthalpy rise. Thus

$$\eta_T = \frac{h_3 - h_{4'}}{h_3 - h_4} \quad \text{and} \quad \eta_C = \frac{h_2 - h_1}{h_{2'} - h_1}.$$

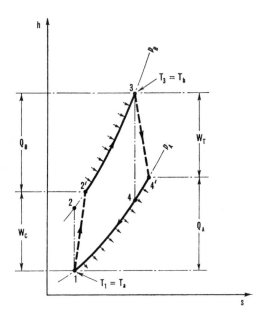

FIG. 3.4. Enthalpy–entropy diagram for irreversible cycle, taking account of inefficiency of turbine and compressor.

For given values of $T_1 = T_a$ and $T_3 = T_b$, it is a simple matter to calculate $T_{2'}$ and $T_{4'}$ for various values of ρ_p when η_T and η_C are specified. Thus, to calculate $T_{2'}$ when the working fluid is treated as a perfect gas,

$$(T_2 - T_a) = T_a(\rho_p - 1)$$

and

$$(T_{2'} - T_a) = \frac{T_2 - T_a}{\eta_C} = \frac{T_a(\rho_p - 1)}{\eta_C}, \qquad (3.5)$$

whence $T_{2'}$ can be calculated.

Similarly, to calculate $T_{4'}$,

$$(T_b - T_4) = T_b\left(1 - \frac{1}{\rho_p}\right)$$

and

$$(T_b - T_{4'}) = \eta_T(T_b - T_4) = \eta_T T_b\left(1 - \frac{1}{\rho_p}\right), \qquad (3.6)$$

whence $T_{4'}$ can be calculated.

Having determined the temperatures at all points round the cycle, the thermal efficiency can be calculated from the expression

$$\eta_{CY} = \left(1 - \frac{Q_A}{Q_B}\right) = 1 - \frac{T_{4'} - T_a}{T_b - T_{2'}}. \qquad (3.7)$$

It is possible therefrom to write down an analytical expression for η_{CY}, but this is left as an exercise for the reader in Problem 3.3. The curves labelled "irreversible" in Fig. 3.3 show the resulting variation of η_{CY} with ρ_p when the turbine and compressor isentropic efficiencies are respectively 88% and 85%, $t_a = 15\,°C$ and t_b has alternative values of 800 °C and 500 °C. The reader is advised to check two or three points on these curves with the aid of the above expressions.

Comparison of the curves labelled respectively "reversible" and "irreversible" in Fig. 3.3 reveals the very great influence on the plant thermal efficiency of inefficiencies in the turbine and compressor, and incidentally shows of what little practical interest is the curve giving the ideal plant performance. For the given values of $t_a = 15\,°C$ and $t_b = 800\,°C$, the optimum pressure ratio is reduced from 100 for the ideal plant to only 11.2 when inefficiencies in the turbine and compressor are taken into account. Furthermore, whereas η_{JOULE} was a function only of ρ_p, being bounded by but otherwise independent of the ratio $\theta \ (\equiv T_b/T_a)$ of the limiting upper and lower absolute temperatures in the cycle, the efficiency of the irreversible cycle is seen to be greatly dependent on θ, falling drastically with fall in turbine inlet temperature. Not only is the optimum cycle efficiency much smaller at 500 °C than at 800 °C, but the optimum pressure ratio is reduced still further to only 4.8 at the lower temperature. It is a compensating advantage that, for a single compressor, this is of a more practicable magnitude.

Further study of the variation of η_{CY} with ρ_p follows in §3.8 after the variation of W_{net} has first been examined.

3.7. Variation of W_{net} with ρ_p in the irreversible cycle

From eqn. (3.5) the compressor work input per unit mass of gas circulated is given by

$$W_C = c_p(T_{2'} - T_a) = \frac{c_p T_a}{\eta_C}(\rho_p - 1). \tag{3.8}$$

Similarly, the turbine work output is given by

$$W_T = c_p(T_b - T_{4'}) = c_p \eta_T T_b \left(1 - \frac{1}{\rho_p}\right). \tag{3.9}$$

Whence, the net work output is

$$W_{net} = (W_T - W_C) = \frac{c_p T_a}{\eta_C}\left(1 - \frac{1}{\rho_p}\right)(\alpha - \rho_p), \tag{3.10}$$

where

$$\alpha \equiv \eta_C \eta_T \theta \tag{3.11}$$

and

$$\theta \equiv T_b/T_a. \tag{3.12}$$

Thus

$$W_{net} \text{ is zero when } \rho_p = 1 \text{ and } \rho_p = \alpha \tag{3.13}$$

and, by differentiation of eqn. (3.10) with respect to ρ_p,

$$W_{net} \text{ is a maximum when } \rho_p = \sqrt{\alpha}. \tag{3.14}$$

The variation of W_{net} with ρ_p is thus as shown in Fig. 3.5, in which one of the graphs shows W_{net} plotted non-dimensionally as

$$\frac{W_{net}}{c_p(T_b - T_a)}$$

against ρ_p.

3.8. Variation of η_{CY} with ρ_p in the irreversible cycle

It would be possible, by differentiating with respect to ρ_p the expression for η_{CY} given in Problem 3.3, to obtain an analytical expression for the value of ρ_p at the point of maximum cycle efficiency, and thence deduce the value of that efficiency. However, the resulting expressions are involved, and a more instructive procedure is to follow a graphical method due to Hawthorne and Davis,[7] since this also gives a picture of the variation of Q_B, W_T, W_C and W_{net} with variation in ρ_p for fixed values of T_a and T_b. These quantities are plotted in dimensionless form against ρ_p in Fig. 3.5. Since W_{net} falls to zero at $\rho_p = \alpha$, values of ρ_p greater than α are of no practical interest, and the graph is extended to higher values than this for constructional purposes only.

The straight line showing the variation in W_C with ρ_p is obtained by noting from eqn. (3.8) that W_C varies linearly with ρ_p and would have a value equal to $c_p(T_b - T_a)$ when the compressor outlet temperature $T_{2'}$ equalled T_b; from eqn. (3.5) it is seen that this would occur at a value of ρ_p equal to β, where

$$(\beta - 1) = \eta_C(\theta - 1). \tag{3.15}$$

Thus

$$\frac{W_C}{c_p(T_b - T_a)}$$

is equal to unity when $\rho_p = \beta$.

The straight line showing the variation of the heat input Q_B with ρ_p is obtained by noting that Q_B also varies linearly with ρ_p, since it is given by

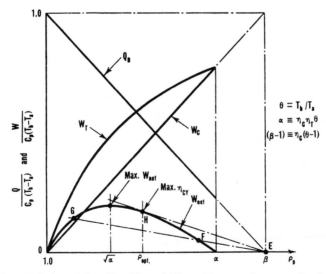

FIG. 3.5. Variation of Q_B, W_T, W_C and W_{net} with ρ_p, and construction for the point of maximum η_{CY}. After Hawthorne and Davis, *Engineering* **181**, 361 (1956).

$$Q_B = c_p(T_3 - T_{2'}) = c_p(T_b - T_a) - W_C. \tag{3.16}$$

Thus
$$\frac{Q_B}{c_p(T_b - T_a)}$$

falls linearly from a value of unity at $\rho_p = 1$ (when $W_C = 0$) to zero at $\rho_p = \beta$ [when $W_C = c_p(T_b - T_a)$].

The curve for W_T may be plotted from eqn. (3.9), while that for W_{net} may be obtained either by subtracting W_C from W_T, or direct from eqn. (3.10).

The value of ρ_p at which the maximum value of η_{CY} occurs may be obtained by a simple construction. If any straight line is drawn from the point E in Fig. 3.5 to cut the curve for W_{net} at the two points F and G, then from the similar triangles resulting from this line and the line representing Q_B:

$$\frac{W_{net}}{Q_B} \text{ at } F = \frac{W_{net}}{Q_B} \text{ at } G,$$

so that

$$\eta_{CY} \text{ at } F = \eta_{CY} \text{ at } G.$$

Hence the maximum efficiency will be obtained at that value of ρ_p corresponding to the point H at which a straight line from E is a tangent to the curve for W_{net}, namely $\rho_p = \rho_{opt.}$ in Fig. 3.5. It is at once seen that the points of maximum η_{CY} and maximum W_{net} do not coincide, the value of ρ_p being greater for the former than for the latter. It may be shown that, if ρ_w and $\rho_{opt.}$ are the values of ρ_p for maximum W_{net} and maximum η_{CY} respectively, then $(\rho_w/\rho_{opt.}) = \sqrt{(1 - \eta_m)}$, where η_m is the maximum value of η_{CY} (Problem 3.4). Since the physical size of the plant for a given output is dependent on the value of W_{net}, the net work output per unit mass of gas circulated, there may be occasions when a value of ρ_p nearer to that for maximum W_{net} than to that for maximum η_{CY} may be preferred.

3.9. Comparison of gas and steam constant-pressure cycles

Both the simple gas-turbine cycle and the simple steam cycle are described as *constant-pressure cycles* because in both of them heat supply and rejection occur while the fluid is at constant pressure. The close similarity of the two cycles is evident from a comparison of Figs. 2.1(c) and 3.1(b). The essential differences arise simply from the fact that the steam cycle uses a condensible fluid which is liquid during the compression process. The compression work input in ideal reversible steady-flow is $\int v \, dp$, so that, because the specific volume of a gas is so much greater than that of a liquid, the ratio of negative (compression) work to positive (expansion) work is much greater in the ideal gas-turbine cycle than in the ideal steam cycle. The situation is further aggravated when turbine and compressor efficiencies are taken into account, the net work decreasing much more rapidly with decrease in these efficiencies for the gas-turbine plant than for the steam plant. The drastic effect on the performance of the former has already been seen in Fig. 3.3. The far less serious effect on the steam plant is well illustrated by comparing the following tables, in which the reversible turbine work output is taken as being 100 units in each plant:

GAS TURBINE

$t_a = 15\,°C$; $t_b = 800\,°C$; $\rho_p = 2.0$

	Rev.	Irrev.	
	$\eta_T = \eta_C = 1$	$\eta_T = \eta_C = 0.8$	$\eta_T = \eta_C = 0.6$
W_T	100	80	60
W_C	53.7	67.1	89.5
W_{net}	46.3	12.9	(-29.5)

STEAM TURBINE

$p_B = 2\,MN/m^2$; $t_b = 350\,°C$; $p_A = 7\,kN/m^2$

	Rev.	Irrev.	
	$\eta_T = \eta_P = 1$	$\eta_T = \eta_P = 0.8$	$\eta_T = \eta_P = 0.6$
W_T	100	80	60
W_C	0.20	0.25	0.33
W_{net}	99.8	79.8	59.7

The difference between the two types of plant in their respective sensitivities to inefficiency in the compression process is seen to be most marked. It is this difference that accounted, in its early development, for the much greater difficulty in producing a successful gas-turbine plant, and which for long militated against the gas-turbine plant as an efficient prime-mover in the large-scale production of electrical power. A further contributory factor at the late development of gas-turbine plant was the early limitation on gas-turbine inlet temperature imposed by metallurgical considerations; the marked drop in cycle efficiency with fall in turbine inlet temperature has already been noted in Fig. 3.3. The problem of developing blade materials suitable for continuous operation at elevated temperatures came under active study, and both ceramic and liquid-cooled blades have been used.

Problems

In these Problems, air is to be treated as a perfect gas with $\gamma = 1.4$ and $c_p = 1.01$ kJ/kg K.

Closed-circuit Plant

3.1. In an air-standard Joule cycle the temperatures at compressor inlet and outlet are respectively 60 °C and 170 °C, and the temperature at turbine inlet is 600 °C.

Calculate (a) the temperature at turbine exhaust; (b) the turbine work and compressor work per kg of air; (c) the thermal efficiency of the cycle; (d) the pressure ratio.

Answer: (a) 383.3 °C; (b) 218.9, 111.1 kJ/kg; (c) 24.8%; (d) 2.71.

3.2. In a closed-circuit gas-turbine plant using a perfect gas as the working fluid, the thermodynamic temperatures at compressor and turbine inlets are respectively T_a and T_b. The plant is operating with an isentropic temperature ratio of compression of ρ_p and the isentropic efficiencies of the compressor and turbine are respectively η_C and η_T. Show that the ratio of the compressor work input to the turbine work output is given by

$$\frac{W_C}{W_T} = \frac{\rho_p}{\alpha}, \text{ where } \alpha \equiv \eta_C \, \eta_T \, \theta \text{ and } \theta \equiv T_b/T_a.$$

Evaluate this ratio when $t_a = 20$ °C, $t_b = 700$ °C, $\eta_C = \eta_T = 85\%$, the pressure ratio = 4.13 and the working fluid is air.

Answer: 0.625.

3.3. For the plant described in Problem 3.2, show that the thermal efficiency of the cycle is given by

$$\eta_{CY} = \frac{(1 - 1/\rho_p)(\alpha - \rho_p)}{(\beta - \rho_p)},$$

where

$$\alpha \equiv \eta_C \eta_T \theta,$$
$$\beta \equiv [1 + \eta_C(\theta - 1)],$$
$$\theta \equiv T_b/T_a.$$

For the values given in Problem 3.2, calculate the thermal efficiency (a) using this expression, (b) by first calculating the temperatures at outlet from the compressor and turbine.

Answer: 20.3%.

3.4. In a design study for the plant described in Problem 3.2, T_a, T_b, η_C and η_T are kept constant while the pressure ratio of compression for which the plant is to be designed is varied.

Show that, when the design pressure ratio is changed by a small amount, the changes in heat rejected and heat supplied are related by the expression $(\delta Q_{out}/\delta Q_{in}) = \alpha/\rho_p^2$. Also show that W_{net} has its maximum value when $\delta Q_{out} = \delta Q_{in}$, while η_{CY} has its maximum value η_m when $(\delta Q_{out}/Q_{out}) = (\delta Q_{in}/Q_{in})$.

Hence show that, if ρ_w and $\rho_{opt.}$ are the values of ρ_p for maximum W_{net} and maximum η_{CY} respectively, then $(\rho_w/\rho_{opt.}) = \sqrt{(1 - \eta_m)}$.

Open-circuit Plant

3.5. In a method of providing air for the cabin of a piston-engined aircraft flying at high altitude, the air supply is taken from the engine supercharger at a pressure of 1.17 bar and a temperature of 55 °C, and is then further compressed in a centrifugal compressor. The compressed air is then passed through an air cooler, in which it is cooled by atmospheric air before entering a turbine, which drives the centrifugal compressor. The air from the turbine exhaust is supplied to the cabin at a pressure of 0.86 bar. The isentropic efficiencies of the turbine and compressor are each 75%. Mechanical losses, and pressure drops in the ducting and air cooler, may be neglected.

Sketch a temperature–entropy diagram for the air in its passage through the plant. If the air is to be supplied to the cabin at a temperature of 15 °C, show that the compressor must be designed for a pressure ratio of about 1.35. What is then the required temperature at turbine inlet, and how much heat must be transferred in the air cooler per kg of air flowing through it?

Answer: 54 °C; 40.4 kJ/kg.

3.6. A petrol engine is fitted with a turbo-supercharger, which comprises a centrifugal compressor driven by an exhaust gas-turbine. The gravimetric ratio of air to fuel supplied to the engine is 12, the fuel being mixed with the air between the compressor and the engine. The air is drawn into the compressor at a pressure of 1 bar and a temperature of 15 °C, and is delivered to the engine at a pressure of 1.5 bar. The exhaust gases from the engine enter the turbine at a pressure of 1.3 bar and a temperature of 510 °C and leave the turbine at a pressure of 1 bar. The isentropic efficiencies of the turbine and compressor are each 80%.

Assuming that the thermal properties of the exhaust gases are the same as for air, calculate:

(a) the temperature of the air leaving the compressor;
(b) the temperature of the gases leaving the turbine;
(c) the power loss in the turbo-supercharger, due to external friction, expressed as a percentage of the power generated in the turbine.

Answer: 59.2 °C; 464.8 °C; 9.7%.

CHAPTER 4

Internal-combustion power plant

4.1. Introduction

This chapter is principally concerned with performance measures and criteria for reciprocating internal-combustion engines. However, it has already been noted in §1.2 that an open-circuit gas-turbine plant is an internal-combustion (IC) device, and some reference to it will consequently be made here. Figure 1.2 showed that the latter is a *non-cyclic open-circuit steady-flow work-producing device* which does not operate on a thermodynamic cycle, so that it is not a cyclic heat power plant such as is depicted in Fig. 1.3(b). The same is true of a reciprocating internal-combustion engine, which is depicted diagrammatically in Fig. 4.1, although there is less justification for calling it a steady-flow device. In so classifying it, we are thinking in terms of a multi-cylinder engine, with the effect of the pulsations reduced by allowing the fluids to pass through smoothing boxes at inlet and outlet, any property measurements being made at those locations. The control surface depicted in Fig. 4.1 is imagined to enclose the flow passages and cylinders through which the working substances pass, and not to include within it any part of the engine or piston walls.

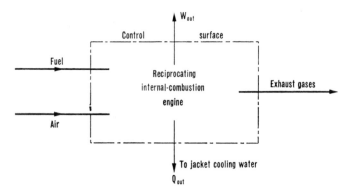

FIG. 4.1. Flow diagram for reciprocating internal-combustion engine.

So far as the quantities crossing the control surfaces of Figs. 1.2 and 4.1 are concerned, the only difference between the turbine and reciprocating plants is that in the latter there is a large transfer of heat across the control surface from the engine cylinder to the jacket cooling water. In a well-lagged turbine plant there is only a small stray heat loss, and this has been assumed to be negligible in Fig. 1.2. In principle, both plants are *non-cyclic open-circuit steady-flow work-producing devices* which can exchange heat with only one reservoir—*the environment.* Since, in the thermodynamic sense, they are not cyclic work-producing devices (which, according to the Second Law, must exchange heat with at least two reservoirs), some explanation of their appearance in a volume on the analysis of engineering cycles is necessary. This arises from the fact that the rational criterion of performance for non-cyclic devices such as these is not of a form which has appealed to the engineer, for reasons which will become apparent. He has therefore chosen in the past to compare the performance of these non-cylic devices with that of cyclic devices which are comparable in certain limited respects. It is the latter which bring the consideration of internal-combustion plant into the present volume, but it is first essential to study their true character. It is also essential to note the differences between a **thermodynamic** cycle and the **mechanical** cycles of operation which cause reciprocating internal-combustion engines to be classed as operating on either a *two-stroke* or a *four-stroke* cycle.

In addition to open-circuit turbine and reciprocating plant, both the *electrical storage battery* and the *fuel cell* come within the category of work-producing devices which exchange heat with only a single reservoir, the environment. The fuel cell operates in steady-flow to produce electrical work directly when supplied with reactants (e.g. H_2 and O_2) and delivering products (e.g. H_2O), while being able to exchange heat with the environment, so that Fig. 4.1 could equally well apply to it. The next section consequently applies equally to fuel cells.

4.2. A rational performance criterion for IC plant—W_{REV}

For the purpose of setting up a criterion of performance, it is assumed that the reactants (e.g. fuel and air) enter the plant at the pressure p_0 and the temperature T_0 of the environment (e.g. the atmospheric or ambient temperature). If discharged into the environment, the exhaust products will finally also be at p_0, but experience shows that the products from an actual internal-combustion device are frequently well in excess of the environment temperature. There is here a lost opportunity for producing work, for further work could have been obtained by using the exhaust products as a source of heat for a cyclic heat power plant which rejected heat to the lower temperature environment. It would not be possible to cool the products continuously in this way to a temperature less than T_0

without the aid of a work-absorbing cyclic device (a refrigerator or heat pump), so that for maximum work production the products must leave the plant at T_0 as well as at p_0. In addition, for maximum work production, all processes, including heat exchange with the environment, must be reversible, since all irreversible processes result in lost opportunities for producing work.

It is shown in Table A.1 of Appendix A that the work output from such an ideal, reversible, non-cyclic steady-flow work-producing device in which the reactants and products are both at the temperature and pressure of the environment is given by

$$W_{REV} = (G_{R_0} - G_{P_0}) \equiv -\Delta G_0, \qquad (4.1)$$

where $G_{R_0} \equiv G$ of the reactants at p_0 and T_0,

$\qquad G_{P_0} \equiv G$ of the products at p_0 and T_0,

$\qquad \Delta G_0 \equiv (G_{P_0} - G_{R_0})$,

and the *Gibbs function G* is defined by the expression

$$G \equiv H - TS. \qquad (4.2)$$

For convenience, **all quantities relate to unit quantity of fuel entering the reaction**, although they could equally well be expressed per unit time.

Since the maximum quantity of work that is theoretically obtainable from the given chemical reaction in such a device is that given by eqn. (4.1), W_{REV} will be a rational criterion against which to judge the work output of the actual internal-combustion plant.

4.3. A rational performance measure for IC plant—the rational (exergetic) efficiency

From the foregoing a rational measure of the excellence of performance of the type of non-cyclic device under consideration will clearly be the ratio of the actual work output W to W_{REV}. This is here called the *rational* (or *exergetic*) *efficiency,* though elsewhere it is frequently and misleadingly called the thermal efficiency, a term exclusively reserved in this volume for the cycle efficiency of a cyclic device. The reason for calling it the exergetic efficiency will be apparent from a study of §A.13 in Appendix A. Thus

$$\text{Rational efficiency: } \eta_R \equiv \frac{W}{W_{REV}} = \frac{W}{-\Delta G_0}. \qquad (4.3)$$

Unlike the thermal efficiency of a cyclic device, the upper limit to this rational efficiency is seen to be 100%. In this respect, it is therefore more akin to the efficiency **ratio** of a cyclic device (§§2.5 and 3.2) than to the thermal efficiency.

In spite of the fact that the upper limit to η_R is 100%, the actual value for an internal-combustion plant, whether turbine or reciprocating, is always well below 100% and, for reasons which will be seen later, is usually of much the same order as the thermal efficiency of a cyclic plant. Fuel cells, however, have been made with a rational efficiency as high as 90% at light current loading and about 50% at useful current loadings.

4.4. An arbitrary performance measure for IC plant—the overall efficiency

The quantity $-\Delta G_0$ for a chemical reaction cannot readily be determined by simple experiment, as can the calorific value. The latter is defined as

$$\text{CV} \equiv (H_{R_0} - H_{P_0}) \equiv -\Delta H_0, \tag{4.4}$$

where $\Delta H_0 \equiv (H_{P_0} - H_{R_0})$ is the difference between the enthalpies of products and reactants when both are at the temperature T_0 at which the calorific value is specified. Thus $-\Delta H_0$ is equal to the heat transferred to the cooling water in a simple, steady-flow calorimeter.

Because $-\Delta G_0$ is not readily determinable, the engineer has found little use for the rational efficiency η_R. Instead, he frequently chooses to use a purely arbitrary performance measure by comparing the work output (per unit quantity of fuel used) with the calorific value of the fuel. This we shall call the *arbitrary overall efficiency*, η_o, defined by

$$\eta_o \equiv \frac{W}{\text{CV}} \equiv \frac{W}{-\Delta H_0}. \tag{4.5}$$

Whilst, in §1.4, η_{CY} and η_B were seen to give a direct connection between the work output of a cyclic plant and the calorific value of the fuel, so that η_o was a rational performance measure in that context, there is no such rational connection between the work output and the calorific value in non-cyclic IC plant. The engineer's use of η_o in this context is consequently an arbitrary, if convenient, procedure. Unfortunately, he frequently refers to it misleadingly as the "thermal efficiency" of the plant.

The work delivered by the gases to the piston is described as the *indicated work output,* and that delivered at the engine coupling the *brake work output,* their ratio being the *mechanical efficiency* of the engine, η_M, so that:

$$\text{Mechanical efficiency: } \eta_M \equiv \frac{\text{brake work output}}{\text{indicated work output}}. \tag{4.6}$$

The overall efficiency may consequently be quoted either as an *indicated overall efficiency* or as a *brake overall efficiency*, according to the work output to which it relates.

4.5. Comparison of the rational and overall efficiencies

That the engineer does not run into trouble with his use of η_0 instead of η_R is due to two factors—the relatively poor performance of practical IC plant compared with that of ideal plant, and the fact that the difference between $-\Delta H_0$ and $-\Delta G_0$ is frequently not large. Using eqn. (4.2), this difference is given by

$$-\Delta G_0 = -\Delta H_0 - T_0(S_{R_0} - S_{P_0}).\qquad(4.7)$$

$T_0(S_{R_0} - S_{P_0})$ is the heat transferred to the environment in the ideal, reversible work-producing process. This is positive for some reactions and negative for others, and is usually fairly small compared with the calorific value. The magnitude of the difference for some simple fuels at 25 °C and 1 atm is given in Table 4.1.

TABLE 4.1. *Comparison of $-\Delta G_0$ and $-\Delta H_0$ for some simple fuels at 1 atm/25 °C*

Fuel	Reaction	$-\Delta G_0$	$-\Delta H_0$ (Cal. value)	Maximum theoretical value of η_0
		MJ/kg of fuel		$= (-\Delta G_0/-\Delta H_0)$
C	$C + O_2 \rightarrow CO_2$	32.84	32.77	100.2%
CO	$CO + \frac{1}{2}O_2 \rightarrow CO_2$	9.19	10.11	90.9%
H_2 $\Big\{$	$H_2 + \frac{1}{2}O_2 \rightarrow H_2O$ liq.	117.6	142.0	82.8%
	$H_2 + \frac{1}{2}O_2 \rightarrow H_2O$ vap.	113.4	120.0	94.5%

Since the ideal work output is $-\Delta G_0$, the greatest possible value of the arbitrary overall efficiency η_0 is as given in the last column of the table. The resulting anomalous situation of the ideal efficiency being in one case slightly in excess of 100% does not worry the engineer unduly, since, for reasons which will be seen shortly, turbine and reciprocating plant have actual values of η_0 always well below 100%. For the fuel cell, however, which has a much closer approach to complete reversibility and consequently a very high efficiency at light current loadings, the engineer is better advised to use the rational efficiency η_R in preference to the arbitrary η_0.

4.6. A practical performance measure—the specific fuel consumption

To the practising engineer the foregoing discussion would appear academic. A client who purchases an internal-combustion plant, whether turbine or reciprocating, is not so much concerned with the quoted efficiency, however defined, as with the *specific fuel consumption* that is guaranteed by the manufacturer at the design or full load. This is the rate of fuel consumption per unit power output, and is consequently expressed in kg/MJ, accompanied by a specification of the calorific value of the fuel on which the guarantee is based (although, rationally, $-\Delta G_0$ should be specified rather than $-\Delta H_0$).

The specific fuel consumption is inversely proportional to the overall efficiency. Since the purchaser of an engine is interested in the brake output rather than in the indicated output, the specific fuel consumption is generally stated in terms of the former, and is related to the overall efficiency by the expression:

$$\text{Specific fuel consumption} = \frac{1}{-\Delta H_0 \, (\eta_o)_b} \text{ kg/MJ}, \qquad (4.8)$$

where $(\eta_o)_b$ is the brake overall efficiency and $-\Delta H_0$ is equal to the calorific value, CV, of the fuel in MJ/kg.

4.7. The performance of turbine and reciprocating IC plant

This section relates to the performance of simple IC plant at their design load. The relation between the temperature of the exhaust products and the overall efficiency in such plant can be studied by reference to Fig. 4.2, which relates to the combustion of a typical hydrocarbon fuel of chemical formula C_8H_{18} with varying amounts of excess air. The latter acts as a dilutant and so, other things being equal, leads to a lower products temperature.

In the equation to the right of Fig. 4.2, the quantity $(H_{P_2} - H_{P_1})$ is the

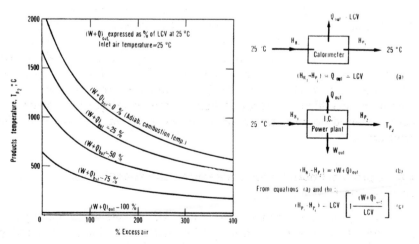

FIG. 4.2. Products temperature for complete combustion of octane (C_8H_{18}).

excess enthalpy of the products of combustion relative to their enthalpy in the calorimeter process. In an IC power plant this quantity may be loosely described as the waste energy in the exhaust products, and is often erroneously described as the "waste heat". In a plant of low overall efficiency the excess enthalpy of the exhaust products is high, and so therefore is the exhaust temperature. In a reciprocating IC engine, the large amount of energy transferred as heat to the jacket cooling water represents a considerable proportion of the energy released in the chemical reaction. Typical values of the various quantities for the different types of plant at their design load are given below.

(a) IC gas-turbine plant

The simple gas-turbine plant operates very nearly adiabatically. Hence the temperature at exit from the combustion chamber and at inlet to the turbine is obtainable from Fig. 4.2 by reading along the line $(W + Q)_{out}$ = 0%. A metallurgically permissible turbine inlet temperature may be about 700 °C, and it is seen that in this case as much as 300% excess air would be needed. The overall efficiency of a simple gas-turbine plant operating with this turbine inlet temperature might be about 25% and this would also be the value of $(W + Q)_{out}$ since Q_{out} is zero. Figure 4.2 shows that, with 300% excess air, the exhaust temperature would then be about 550 °C.

(b) Reciprocating IC engine

For most economical operation a spark-ignition (petrol) engine requires to be supplied with an approximately stoichiometric air–fuel mixture; that is, with zero excess air. A compression-ignition (diesel) engine requires between 25% and 50% excess air for greatest economy. It is seen from Fig. 4.2 that if combustion were adiabatic, as in a gas-turbine plant, temperatures would then be excessive, so that the engines have to be cooled. Even then the peak temperature in the mechanical cycle of operations of a reciprocating engine is much higher than the steady temperature at inlet to a gas turbine. This is possible because the peak is only of momentary duration, and to this high peak temperature is attributable the higher efficiency of the reciprocating plant in spite of the energy transfer to the jacket cooling water. A typical spark-ignition engine might have an overall efficiency of 30% and the heat transferred to the jacket cooling water might amount to 45% of the calorific value of the fuel, so that $(W + Q)_{out}$ would be 75%. Figure 4.2 shows that, with zero excess air, the exhaust temperature would then be about 630 °C.

4.8. An arbitrary performance criterion for IC plant—the thermal efficiency of a corresponding ideal air-standard cycle

In spite of the fact that IC plants are not cyclic work-producing devices, the values quoted in §4.7 for the arbitrarily defined overall efficiency of turbines and reciprocating IC plant are of the same order as the rationally defined overall efficiency η_o of cyclic, closed-circuit gas-turbine plant. For the latter $\eta_o = \eta_{CY}\eta_B$, and the low value of η_o is due principally to the low value of the cycle efficiency, η_{CY}. For IC plants, which are not cyclic devices, the low value of the overall efficiency is not explainable in terms of an η_{CY}. A qualitative explanation of the fact that the values of η_o are roughly comparable in the two types of plant can be given in the following terms.

Even when an IC plant uses little or no excess air, as in a spark-ignition engine, 77% of the air used is nitrogen, and it would make no difference to the operation of the plant if this nitrogen were regarded as passing round a thermodynamic cycle in which heat was first transferred to it in the plant as a result of the energy released in combustion of the fuel with oxygen, and in which heat was later transferred from it to the atmosphere before it re-entered the plant. If, as in a gas-turbine plant, 300% excess air were used, no less than 94% of the air would remain unchanged in chemical composition in passing through the plant, and all of this could be regarded as passing round a thermodynamic cycle. The relation between the work output per unit mass of fuel burnt and the calorific value in an IC plant might consequently be expected to be not greatly different from that in a cyclic plant working under conditions which were as nearly as possible comparable. This gives the engineer further reassurance in his arbitrary use of η_o as a performance measure for IC plant. The first requirement for conditions to be comparable is clearly that the compression ratio should be the same.

By virtue of this argument, as an elementary and academic study the engineer evaluates the cycle efficiency η_{CY} of a comparable ideal cyclic plant using air as the working fluid and having the same compression ratio as the IC plant. Such a plant is said to work on the comparable *air-standard cycle*. Using this efficiency as a criterion against which to judge the measured overall efficiency of the IC plant, the *efficiency ratio* ($\equiv \eta_o/\eta_{CY}$) is then taken as a measure of the excellence of performance of the plant, the indicated overall efficiency clearly being preferred to the brake overall efficiency in this evaluation of the efficiency ratio. This is a subterfuge which avoids the computation of the less readily determinable rational efficiency defined in §4.3. It should be noted, however, that it is a somewhat arbitrary procedure based on distinctly shaky foundations. Nevertheless, it enables something to be learnt of the effect of change in compression ratio on IC plant performance; though there, also, care is necessary in applying the results deduced therefrom, since in §3.6 the effect of irreversibilities on plant performance was seen to be very great.

4.9. Air-standard cycle for gas-turbine plant—the Joule cycle

For an IC gas-turbine plant, in which combustion takes place at approximately constant pressure and in which air intake and gas exhaust are normally at the same (usually atmospheric) pressure, the corresponding ideal air-standard cycle is clearly the constant-pressure Joule cycle operating with the same **pressure** ratio of compression r_p as the actual plant. This has already been treated fully in Chapter 3. Study of Figs. 1.2, 1.3 and 3.1 shows that external heat supply in the constant-pressure process 2–3 of Fig. 3.1 replaces the actual combustion process, and heat rejection in the constant-pressure process 4–1 replaces the gap which exists in the open-circuit IC plant between turbine exhaust at atmospheric pressure and compressor intake, also at atmospheric pressure. It is true that, in the actual combustion process in the IC plant, fuel addition increases the mass flow rate of the fluid stream, but with 300% excess air the gravimetric air–fuel ratio is about 18:1, so that neglect of this increase does not cause undue concern in the rather gross type of comparisons under consideration; furthermore, this increase in mass flow rate through the turbine is frequently offset in practice by the necessity to bleed air from the compressor outlet to cool the turbine discs.

Thus, within the limitations pointed out in §4.8, the studies of cyclic gas-turbine plant in Chapter 3 may be used as a guide to the behaviour of non-cyclic IC plant, and to the influence thereon of variation in the design pressure ratio of compression.

4.10. Air-standard cycles for reciprocating IC engines

There are two types of reciprocating IC engine, the *spark-ignition* and *compression-ignition,* and each type can work on either one of two modes of operation, *two-stroke* or *four-stroke.* In the present discussion the mode of operation is immaterial, but it is necessary to distinguish between the two types. Spark-ignition (petrol) engines use a fuel of high volatilty and relatively high ignition temperature, while the fuel for compression-ignition (diesel) engines is of lower volatility and ignition temperature and ignition of the fuel after injection is initiated solely as a result of the temperature rise during compression.

These engines are *non-cyclic open-circuit steady-flow work-producing devices* and do not operate on a thermodynamic cycle, yet, for academic purposes, the engineer has come to compare their performance with the thermal efficiency of what he calls a *corresponding air-standard cycle.* The correspondence arises from the similarity in appearance between the **indicator** diagram (pressure against **cylinder** volume) obtained on the actual engine and the **state** diagram (pressure against **specific** volume) of the hypothetical corresponding cycle. Two such cycles, the Otto and Diesel, will be studied; in both cases the compression ratio for the cycle is equal to the engine compression ratio, expressed in this case as a **volumetric** instead of a pressure ratio of compression.

4.11. The ideal air-standard Otto cycle

The shape of an indicator diagram from an engine varies greatly according to the load on the engine; this is particularly so of throttle-governed, spark-ignition engines. However, as heretofore, we are only considering the full-load design conditions for the engine, when the throttle will be wide open, and the indicator diagram obtained on a four-stroke engine will then be similar to that sketched in Fig. 4.3.

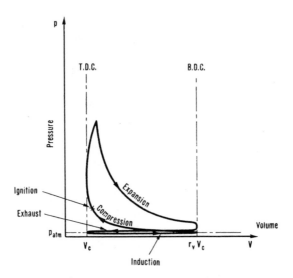

FIG. 4.3. Indicator diagram for four-stroke, spark-ignition engine at full load (throttle wide open).

In this diagram the four strokes (induction, compression, expansion and exhaust) are clearly indicated, the difference between the pressures on the exhaust and induction strokes having been exaggerated for the sake of clarity. This diagram may be represented approximately by the idealised indicator diagram of Fig. 4.4(a), and the path 12341 may be taken as a generally representative idealised indicator diagram for both spark-ignition and high-speed compression-ignition engines at full load.

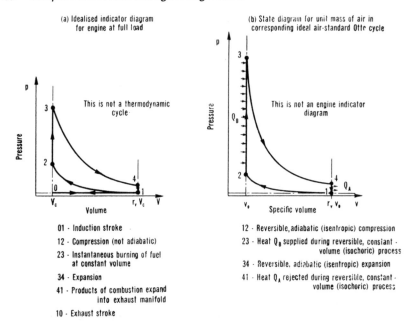

(a) Idealised indicator diagram
for engine at full load

(b) State diagram for unit mass of air in
corresponding ideal air-standard Otto cycle

This is not a thermodynamic cycle

This is not an engine indicator diagram

01 - Induction stroke

12 - Compression (not adiabatic)

23 - Instantaneous burning of fuel at constant volume

34 - Expansion

41 - Products of combustion expand into exhaust manifold

10 - Exhaust stroke

12 - Reversible, adiabatic (isentropic) compression

23 - Heat Q_B supplied during reversible, constant - volume (isochoric) process

34 - Reversible, adiabatic (isentropic) expansion

41 - Heat Q_A rejected during reversible, constant - volume (isochoric) process

FIG. 4.4. Idealised indicator diagram for engine, and state diagram for Otto cycle.

In view of the difficulty and length of realistic detailed calculations relating to the comparative performance of reciprocating IC engines designed for different compression ratios, the engineer takes refuge in the artifice of drawing attention to the similarity in shape between the idealised **indicator** diagram of Fig. 4.4(a) and the **state** diagram of Fig. 4.4(b) for what is known as the *corresponding ideal air-standard Otto cycle* having the same volumetric compression ratio r_v as the actual engine. In this, unit mass of air enclosed in a cylinder is made to execute a closed, reversible thermodynamic cycle. Unlike the corresponding processes in the engine, compression 1–2 and expansion 3–4 are adiabatic as well as reversible, and so isentropic. The actual engine processes of sudden pressure rise after ignition and sudden pressure drop after exhaust-valve opening are replaced respectively by pressure rise 2–3 at constant volume, with heat supply to the air, and pressure fall 4–1 at constant volume, with heat rejection. The artificiality of this procedure should be noted. It receives its principal justification from the simplicity of the resulting expression for the cycle efficiency, which is found to be a function only of the compression ratio.

As for the Joule cycle, the thermal efficiency of the Otto cycle can best be evaluated by first calculating the temperatures at the various points round the cycle. For a perfect gas, $Tv^{\gamma-1}$ is constant for the isentropic processes 1–2 and 3–4, so that

$$\frac{T_2}{T_1} = \frac{T_3}{T_4} = r_v^{\gamma-1} \equiv \rho_v, \tag{4.9}$$

where $r_v \equiv$ the volumetric compression ratio and $\rho_v \equiv r_v^{\gamma-1}$, the isentropic temperature ratio for the cycle.

The thermal efficiency of the Otto cycle is then given by

$$\eta_{\text{OTTO}} = \left(1 - \frac{Q_A}{Q_B}\right) = 1 - \frac{c_v(T_4 - T_1)}{c_v(T_3 - T_2)}. \tag{4.10}$$

Hence, from eqns. (4.9) and (4.10),

$$\eta_{\text{OTTO}} = \left(1 - \frac{1}{\rho_v}\right) = \left(1 - \frac{1}{r_v^{\gamma-1}}\right). \tag{4.11}$$

This expression may be compared with eqn. (3.4) for η_{JOULE}. It is particularly useful that η_{OTTO} is a function only of r_v because, for an actual engine, r_v is dependent only on the physical dimensions. Thus, if the engineer is to compare the overall efficiency, η_o, of the engine with the value of η_{OTTO} for the corresponding air-standard Otto cycle, he needs to specify nothing more than the compression ratio of the engine. He may also learn from a study of η_{OTTO} something about the effect of change in design compression ratio on the design performance of the actual engine. He will see from eqn. (4.11) that increase in r_v will lead to an increase in η_{OTTO}, and will correctly deduce from this that engines of higher compression ratio will also have a higher overall efficiency and lower specific fuel consumption.

For a spark-ignition petrol engine the compression ratio is limited to a value between about 6 and 9, since higher values of r_v lead to detonation of the burning charge and consequent rough running. For compression-ignition engines, on the other hand, r_v must be high since ignition of the injected fuel depends on the temperature rise achieved solely by compression. For these r_v will be of the order of 15. This higher compression ratio is seen from the foregoing to account for the better performance of the compression-ignition engine, which is revealed by a comparison of the specific fuel consumption figures in Problems 4.2 and 4.3.

An upper limit to the allowable compression ratio is set by the mechanical strength of the engine, which imposes a limit on the allowable peak pressure. This leads to consideration of an alternative ideal cycle, the air-standard Diesel cycle.

4.12. The ideal air-standard Diesel cycle

In compression-ignition engines, a combination of a high compression ratio and approximately constant-volume combustion leads to peak pressures which can be excessively high, particularly for large engines. The indicator diagram for these may be made less peaked by suitably delayed timing and suitable control of the fuel injection. In the largest engines, such as are used in marine applications, the combustion then occurs more nearly at constant pressure, as the piston moves down the cylinder, instead of at approximately constant volume. An Otto cycle consequently no longer provides a reasonable arbitrary criterion against which to judge the performance of such an engine. Hence for that purpose the engineer has devised what is known as the *corresponding ideal air-standard Diesel cycle* in which, as before, all processes are reversible, but in which heat is supplied in a constant-pressure process instead of in a constant-volume process. A state diagram for the cycle is shown in Fig. 4.5. This diagram should be compared with the Otto cycle diagram of Fig. 4.4(b), noting particularly the much greater pressure rise *during compression* in the Diesel cycle.

The thermal efficiency of the cycle is again best evaluated by first calculating the temperatures at the various points round the cycle, and for this purpose it is necessary to specify the ratio $\alpha \equiv v_3/v_2$. This ratio has variously been called the *cut-off ratio* and *load ratio*. The former name derives from steam-engine terminology as the result of the similarity in shape of Fig. 4.5 to that of a steam-engine indicator diagram, while the latter arises from the fact that, in actual indicator diagrams for the type of engine under discussion, the ratio of the volumes corresponding to v_2 and v_3 varies with the load on the engine.

For the isentropic process 1–2, as before:

$$T_2 = \rho_v T_1. \tag{4.12}$$

For the constant-pressure process 2–3:

$$\frac{T_3}{T_2} = \frac{v_3}{v_2} = \alpha, \quad \text{whence } T_3 = \alpha\rho_v T_1. \tag{4.13}$$

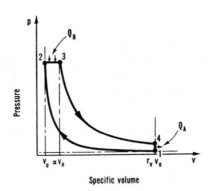

FIG. 4.5. State diagram for Diesel cycle.

For the isentropic process 3–4:

$$\frac{T_4}{T_3} = \left(\frac{v_3}{v_4}\right)^{\gamma-1} = \left(\frac{\alpha}{r_v}\right)^{\gamma-1}, \quad \text{whence } T_4 = \alpha^{\gamma} T_1, \quad (4.14)$$

since $r_v^{\gamma-1} \equiv \rho_v$.

The thermal efficiency of the Diesel cycle is then given by

$$\eta_{\text{DIESEL}} = \left(1 - \frac{Q_A}{Q_B}\right) = 1 - \frac{c_v(T_4 - T_1)}{c_p(T_3 - T_2)}. \quad (4.15)$$

Hence, from eqns. (4.12) to (4.15),

$$\eta_{\text{DIESEL}} = 1 - \frac{1}{\rho_v}\left[\frac{\alpha^{\gamma} - 1}{\gamma(\alpha - 1)}\right]. \quad (4.16)$$

4.13. Comparison of η_{OTTO} and η_{DIESEL}

Comparison of eqns. (4.11) and (4.16) shows that η_{DIESEL} is a less convenient performance criterion than η_{OTTO}, for it does not depend only on r_v (that is, on the physical dimensions of the engine), but also on α. The magnitude of α has to be specified in some arbitrary manner, such as that adopted in Problem 4.3(b).

Although eqns. (4.11) and (4.16) relate to ideal cycles and not to indicator diagrams of actual engines, a comparison of the equations enables something to be learnt, by inference, about the relative performance of engines of the same compression ratio whose indicator diagrams are similar to the state diagrams of the Otto and Diesel cycles respectively. The quantity α is always greater than unity and the expression within the square brackets in eqn. (4.16) is consequently always greater than unity, so that, **for the same compression ratio, η_{DIESEL} is always less that η_{OTTO}**; by inference, it may be expected that the overall efficiencies of the respective engines will bear a similar relation to each other. The Diesel cycle has been put forward as the academic yardstick, so to speak, for the large, slow-speed, marine compression-ignition engine. We thus conclude that the performance of such an engine can be expected to be poorer than that of a smaller engine of the same compression ratio since the latter, being smaller, will have less restriction on the allowable peak pressure, and its indicator diagram will consequently approximate more closely in appearance to the state diagram of the Otto cycle.

4.14. Comparison of the performance of petrol and diesel engines

Because compression-ignition engines are commonly called diesel engines, it is necessary to warn that it must not be inferred from the foregoing discussion that diesel engines are less efficient than petrol engines; the reverse is the case. The discussion in §4.13 relates to engines of the same compression ratio, but compression-ignition (so-called diesel) engines always have a much higher compression ratio than petrol engines and so a higher efficiency and lower specific fuel consumption (cf. Problems 4.2 and 4.3). It will be realised that further confusion can result from the rather free use of the word diesel, sometimes in reference to cycles and sometimes to engines, when it is pointed out that the indicator diagrams of both high-speed diesel engines and petrol engines resemble more closely the state diagram of the Otto cycle than that of the Diesel cycle. This leads to consideration of a third ideal cycle, the Dual cycle.

4.15. The ideal air-standard Dual cycle

It has been noted that, in operation at full load, combustion in petrol and high-speed diesel engines is followed by a rapid pressure rise at approximately constant volume, while in the large, slow-speed, marine diesel, it is followed by an initial expansion at approximately constant pressure. There is consequently an intermediate class of engine whose performance lies between these two extremes. On that account, a *corresponding ideal air-standard Dual cycle* appears in some textbooks. In this hypothetical cycle, some of the heat supply is at constant volume and some at constant pressure, while all heat rejection is at constant volume. This introduces further complication which it is difficult to justify in view of the artificiality which in any case exists in the arbitrary procedure of comparing the performance of actual non-cyclic engines with the thermal efficiency of hypothetical cycles. Such a procedure is largely an academic exercise and of little interest to the practical engineer, to whom the *specific fuel consumption* of an engine is of more immediate concern. It is also more profitable to study real engine processes, but this is beyond the scope of a book devoted to the analysis of thermodynamic cycles. For a study of such processes the reader may consult a specialist text.[8]

4.16. Other performance measures for IC engines

Although they may appear out of place in a volume on the analysis of thermodynamic cycles, three further measures of performance are widely used in the study of reciprocating engine performance. These performance measures are related to the effectiveness with which the engine draws in its fresh charge and to the work output of the engine.

(a) Volumetric efficiency

For a given air–fuel ratio, the smaller the amount of air or fresh charge drawn in by an engine on the induction stroke, the lower will be the power output of the engine. Although a carburetted spark-ignition engine draws a mixture of air and fuel into the cylinder, the effectiveness of the inhaling process of an engine is arbitrarily expressed in terms of the amount of *air* drawn in, by defining the *volumetric efficiency* as:

$$\eta_v \equiv \frac{\text{mass of air inhaled per suction stroke}}{\text{mass of air to occupy } \textbf{swept} \text{ volume at ambient } p \text{ and } T} \quad (4.17)$$

$$\equiv \frac{\text{volume of air of ambient density inhaled per unit time}}{\textbf{suction} \text{ volume swept through per unit time}}.$$

Thus, the volumetric efficiency will depend on the density of the gases in the cylinder at the end of the suction stroke, and this will depend on their temperature and pressure. Heat transfer from the induction manifold, and mixing of the fresh charge with hot residual gases left in the clearance volume at the end of the exhaust stroke, raise the temperature of the cylinder contents, and frictional pressure drop through the inlet passages lowers the pressure; both of these effects reduce the density, and so result in a volumetric efficiency less than unity. The value of η_v has a direct influence on the required engine capacity for a given output (Problem 4.5).

If the volumetric efficiency of a supercharged engine (Problem 3.6) were expressed in terms of the density at ambient p and T, and not, as is usual, in terms of the density at inlet to the engine from the supercharger, its value would exceed 100% if the pressure rise through the supercharger resulted in sufficiently high density at supercharger outlet. By increasing the amount of air and fuel inhaled by an engine, supercharging increases the power output obtainable from an engine of given physical dimensions, and this is its purpose.

(b) Indicated mean effective pressure (i.m.e.p.)

The i.m.e.p. being a measure of the work output **per unit swept volume**, is a convenient performance measure for expressing the indicated work output in a form which does not depend on the number and size of the cylinders. Thus it is a performance measure whose magnitude is representative of the **type**, rather than the size, of an engine. For example, the i.m.e.p., at the design rating, of unsupercharged engines of the same type is of the same order for engines of a wide range of sizes, and is appreciably less than that of similar supercharged engines.

The i.m.e.p. is defined by the relation

$$\text{i.m.e.p. } (\text{N/m}^2) \equiv$$

$$\frac{\text{indicated work output } (\text{J} \equiv \text{N m}) \text{ per cylinder per } \textbf{mechanical}}{\text{swept volume per cylinder } (\text{m}^3)} \cdot \quad (4.18)$$

Since the numerator of this expression is, to the appropriate scale, equal to the area enclosed on the indicator diagram, the i.m.e.p. is the mean ordinate enclosed on Fig. 4.6; that is, the mean Δp in Fig. 4.6. It would therefore be more accurately described as the mean indicated pressure **difference**. It bears no direct relation to the peak absolute pressure in the engine.

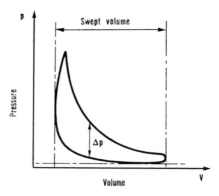

Fig. 4.6. Simplified indicator diagram.

(c) Brake mean effective pressure (b.m.e.p.)

Since the brake work output at the engine coupling is frequently of more concern than the indicated work, a similar measure of performance for the brake work is defined by the relation

$$\text{b.m.e.p. (N/m}^2) \equiv \frac{\text{brake work output (J} \equiv \text{N m) per cylinder per \textbf{mechanical} cycle of operations}}{\text{swept volume per cylinder (m}^3)}. \tag{4.19}$$

Thus the brake power of an engine whose b.m.e.p. is p_e is given by the relation:

$$\text{brake power} = p_e L A N', \tag{4.20}$$

where L = piston stroke,

A = piston area,

N' = mechanical cycles of operation per second,

$$= \left\{ \begin{array}{l} \text{rev/s for two-stroke engine} \\ \dfrac{\text{rev/s}}{2} \text{ for four-stroke engine} \end{array} \right\}.$$

Unlike the i.m.e.p., the b.m.e.p. cannot be thought of in terms of a mean Δp, and it derives its name from the fact that its units are those of pressure. It is important to realise that it is a measure of work output, and not of the actual pressures in the cylinder.

Problems

4.1. Calculate the temperature rise during reversible, adiabatic compression of air from a temperature of 18 °C when the volumetric compression ratio is (a) 6, (b) 15.

Answer: 305 K, 569 K.

4.2. The specific field consumption of a spark-ignition engine at full load is 0.093 kg per MJ of brake work output when the calorific value of the fuel is 44 MJ/kg. The mechanical efficiency is 80%. Calculate the indicated overall efficiency. Also express the specific fuel consumption in (1) kg/kW h, (2) lb/hp h.

Calculate the b.m.e.p. and i.m.e.p. of the engine at this load if the gravimetric air–fuel ratio is 18, the volumetric efficiency is 82% and ambient air conditions are 1 bar and 18 °C.

The engine has a volumetric compression ratio of 6. Determine the thermal efficiency of the comparable air-standard Otto cycle and thence the indicated efficiency ratio for the engine.

Calculate the i.m.e.p., maximum temperature and maximum pressure for the Otto cycle if the pressure and temperature at the beginning of compression are respectively 1 bar and 18 °C, and the heat supplied per unit mass of air is equal to the energy supplied in the engine by the fuel (in terms of its calorific value) per unit mass of air drawn in.

Answer: 30.5%; 0.335 kg/kW h, 0.550 lb/hp h; 0.587 MN/m^2, 0.733 MN/m^2; 51.2%, 0.596; 1.80 MN/m^2, 3720 °C, 8.23 MN/m^2.

4.3. The specific fuel consumption of a compression-ignition engine at full load is 0.068 kg per MJ of brake work output when the calorific value of the fuel is 44 MJ/kg. The mechanical efficiency is 80%. Calculate the indicated overall efficiency. Also express the specific fuel consumption in (1) kg/kW h, (2) lb/hp h.

Calculate the b.m.e.p. and i.m.e.p. of the engine at this load if the gravimetric air–fuel ratio is 28, the volumetric efficiency is 82% and ambient air conditions are 1 bar and 18 °C.

The engine has a volumetric compression ratio of 15. Determine the thermal efficiency of the following air-standard cycles having the same volumetric compression ratio as the engine, and thence calculate the indicated efficiency ratio for the engine with respect to each of these cycles:

(a) an Otto cycle;
(b) a Diesel cycle in which the temperature at the beginning of compression is 18 °C, and in which the heat supplied per unit mass of air is equal to the energy supplied in the engine by the fuel (in terms of its calorific value) per unit mass of air drawn in.

Answer: 41.8%; 0.245 kg/kW h, 0.402 lb/hp h; 0.516 MN/m^2, 0.645 MN/m^2; (a) 66.2%, 0.632; (b) 56.6%, 0.739.

4.4. Calculate the i.m.e.p., maximum temperature and maximum pressure for each of the cycles of Problem 4.3 when the temperature at the beginning of compression and the heat supplied per unit mass of air are the same for both cycles, and the pressure at the beginning of compression is 1 bar.

Tabulate the values of thermal efficiency, maximum temperature, maximum pressure and i.m.e.p. for the air-standard cycles of Problems 4.2 and 4.3, and comment on the figures.

Answer: (a) 1.33 MN/m^2;   2770 °C;   15.7 MN/m^2;   (b) 1.14 MN/m^2;   2143 °C; 4.43 MN/m^2.

4.5. A six-cylinder, four-stroke petrol engine is to develop 40 kW at 40 rev/s when designed for a volumetric compression ratio of 6.0. The ambient air conditions are 1 bar and 18 °C, and the calorific value of the fuel is 44 MJ/kg.

(a) Calculate the specific fuel consumption in kg per MJ of brake work output if the indicated overall efficiency is estimated to be 60% of the thermal efficiency of the corresponding air-standard Otto cycle and the estimated mechanical efficiency is 80%.

(b) The required gravimetric air-fuel ratio is 15.4 and the estimated volumetric efficiency is 82%. Determine the required total swept volume, and the cyclinder bore if the bore is to be equal to the stroke.

(c) Calculate the b.m.e.p.

Answer: (a) 0.0925 kg/MJ. (b) 2903 cm^3; 85 mm. (c) 0.689 MN/m^2.

4.6. The four-stroke petrol engine in an automobile has a total swept volume of V litres. The diameter of the road wheels is d m. At cruising speed the ratio of the engine speed to the speed of the road wheels is n, the volumetric efficiency of the engine is η_v, and the carburettor maintains a gravimetric air–fuel ratio of f. The specific volume of the ambient air is v m^3/kg and the specific gravity of the fuel is σ. Show that the "fuel consumption", in km/litre, is equal to

$$2\pi \frac{\sigma f d v}{n \eta_v V}.$$

Such an engine has a volumetric efficiency of 50% when the cruising speed of the automobile is 65 km/h. The brake overall efficiency of the engine (based on a calorific value of the fuel of 45 MJ/kg) is then estimated to be 28%. Calculate the "fuel consumption" and the power output of the engine under these conditions, given that $\sigma = 0.7$, $f = 17$, $d = 0.7$, $v = 0.8$, $n = 4.6$ and $V = 2.1$.

Answer: 8.67 km/litre; 18.4 kW.

4.7. The following particulars relate to a test on an open-circuit (internal combustion) gas-turbine plant, in which liquid n-octane (C_8H_{18}) of lower calorific value 44.43 MJ/kg was supplied to the adiabatic combustion chamber at a temperature of 25 °C:

Compressor:	Pressure ratio	4.13
	Air inlet temperature	290 K
	Air exit temperature	460 K
Turbine:	Gas inlet temperature	1000 K
	Gas exit temperature	750 K

Calculate:

(a) the isentropic efficiency of the compressor;
(b) the moles of air supplied to the combustion chamber by the compressor per mole of fuel burned, and thence the percentage of excess air;
(c) the turbine work output and compressor work input per kg of fuel burned;
(d) the overall efficiency of the plant, neglecting mechanical losses;
(e) the thermal efficiency of the corresponding air-standard Joule cycle, and thence the efficiency ratio for the plant.

The air passing through the compressor may be treated as a perfect gas for which $\gamma = 1.4$, $c_p = 1.01$ kJ/kg K and the molar mass is 29.0 kg/kmol. The enthalpies of the gases passing through the turbine, in MJ/kmol, are given in the following table:

Temperature K	O_2	N_2	CO_2	H_2O
1000	31.37	30.14	42.78	35.90
750	22.83	22.17	29.65	26.00
298	8.66	8.67	9.37	9.90

Answer: (a) 85.3%. (b) 285.3; 380%. (c) 21.03 MJ, 12.47 MJ. (d) 19.3%. (e) 33.3%; 0.580.

4.8. An isothermal, reversible fuel cell takes in hydrogen and oxygen, each at 1 atm and 25 °C, and delivers water at 1 atm and 25 °C. For this reaction, $-\Delta G_0 = 117.6$ and $-\Delta H_0 = 142.0$ MJ per kg of hydrogen consumed. Calculate the heat transfer from the cell to its environment at 25 °C, per kg of hydrogen consumed.

A hydrogen–oxygen fuel cell operating in an environment at 25 °C consumes hydrogen at a rate of 0.36 litre/min when delivering a current of 50 amperes at a p.d. of 0.8 V. The specific volume of the hydrogen is 12.14 m^3/kg. Calculate (a) the rational efficiency, (b) the arbitrary overall efficiency of the fuel cell, based on the higher calorific value of hydrogen.

Answer: 24.4 MJ; (a) 68.8%; (b) 57.0%.

4.9. In a steady-flow, reversible fuel cell operating isothermally at 25 °C and atmospheric pressure, hydrogen and oxygen gas streams enter and water leaves.

Calculate the e.m.f. of the cell at 25 °C, given that the electronic charge is 1.60×10^{-19} coulomb and that the molar number (Avogadro constant) is 6.02×10^{26} per kmol.

Determine the required rate of supply of hydrogen, in litres per minute at 1 atm pressure and 25 °C, for such a cell of 100 W output, and calculate the rate of heat transfer with the environment, in watts.

Answer: 1.22 V; 0.62 litre/min; 20.7 W.

CHAPTER 5

Simple refrigerating plant

5.1. Introduction

Previous chapters have dealt with **work-producing** devices, most of them cyclic. Figure 5.1(a) depicts the ideal Carnot cycle, in which the working fluid executes a reversible thermodynamic cycle while exchanging heat reversibly with two thermal-energy reservoirs, each at constant temperature. T_{23} is an infinitesimal amount **less** than T_B, so that Q_B is transferred **to** the working fluid between 1 and 2, while T_{41} is an

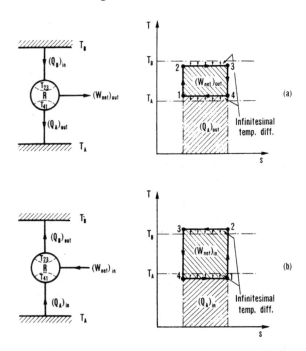

Fig. 5.1. Energy-flow and temperature–entropy diagrams for (a) Carnot cycle, (b) reversed Carnot cycle.

infinitesimal amount **greater** than T_A, so that Q_A is transferred **from** the working fluid between 4 and 1. As can be seen from the relevant areas on the temperature–entropy diagram, Q_B is greater than Q_A, and the net work delivered, $(W_{\text{net}})_{\text{out}}$, is equal to $(Q_B - Q_A)$.

Figure 5.1(b) depicts the ideal **reversed** Carnot cycle, in which the fluid circulates round the cycle in the opposite sense. T_{41} must not be **less** than T_A by an infinitesimal amount for the heat quantity Q_A to be transferred reversibly **to** the working fluid between 4 and 1, while T_{23} must be **greater** than T_B by an infinitesimal amount for the heat quantity Q_B to be transferred reversibly **from** the working fluid between 2 and 3. Since Q_B is greater than Q_A, the plant acts as a **work-absorbing** device, in which $(W_{\text{net}})_{\text{in}} = (Q_B - Q_A)$. It is seen to extract a certain quantity of thermal energy Q_A from a reservoir at low temperature, and to deliver a larger quantity of thermal energy Q_B to a reservoir at higher temperature, the balance being made up by the net work input $(W_{\text{net}})_{\text{in}}$. It is thus a *cyclic work-absorbing device*, functioning as a *refrigerating plant*.

5.2. Refrigerators and heat pumps

Refrigerating plant may be classified as either refrigerators or heat pumps. When interest centres on the extraction of thermal energy from the low-temperature reservoir A (e.g. when A is a refrigerating chamber and B is the atmosphere or a supply of cooling water), a refrigerating plant such as that described above is termed a *refrigerator*. On the other hand, when interest centres on the supply of heat to the high-temperature reservoir B (e.g. when B is a room to be heated and A is the atmosphere or a supply of river water at a temperature **lower** than that of the room), it is customary to describe the plant somewhat misleadingly as a *heat pump*.

5.3. Performance measures—coefficient of performance, and work input per tonne of refrigeration

The *performance measure* chosen in Chapter 1 for a cyclic **work-produ-cing** device (or CHPP) was the cycle or thermal efficiency η_{CY}, given by

$$\eta_{\text{CY}} \equiv \frac{(W_{\text{net}})_{\text{out}}}{Q_{\text{in}}}.$$

The corresponding performance measure for refrigerators and heat pumps is termed the *coefficient of performance* (CP), which relates the heat quantity of primary interest in each device to the net work input. Since interest centres on Q_A in a refrigerator and on Q_B in a heat pump, the definition of CP is not the same for both, the respective definitions being

$$\text{Refrigerator CP} \equiv \frac{(Q_A)_{\text{in}}}{(W_{\text{net}})_{\text{in}}}. \tag{5.1}$$

$$\text{Heat pump CP} \equiv \frac{(Q_B)_{\text{out}}}{(W_{\text{net}})_{\text{in}}}. \tag{5.2}$$

In order to avoid confusion, the coefficient of performance of a heat pump is sometimes alternatively described as the *performance energy ratio* (PER), but this usage is not universal. The PER of a heat pump is seen to be the reciprocal of the cycle efficiency of a cyclic work-producing device.

For refrigerators, an alternative performance measure encountered in refrigerating practice is the *power input per tonne of refrigeration*, where the latter is defined as the rate of thermal energy extraction corresponding to the production, in a period of 24 h, of 1 tonne (1000 kg) of ice at 0 °C from water at the same temperature; this corresponds to a rate of 3.86 kW. The power input per tonne of refrigeration is clearly inversely proportional to the coefficient of performance, and the reader is left to evaluate the constant of proportionality when the power input is expressed in kW (Problem 5.4).

5.4. The ideal reversed Carnot cycle

It is clearly desirable that, for a certain expenditure of work input, a refrigerator should extract as much thermal energy as possible from the low-temperature source, while a heat pump should deliver as much thermal energy as possible to the sink at higher temperature. Thus, in both cases, a high coefficient of performance is to be desired. It is readily shown (Problem 5.1), as a corollary of the Second Law, that the coefficient of performance of a cyclic refrigerating plant which operates between two thermal-energy reservoirs, each at uniform temperature, cannot be greater than that of an ideal reversible plant operating between the same two reservoirs. Since it has both internal and external reversibility, a plant operating on the reversed Carnot cycle of Fig. 5.1(b) is such an ideal plant, and it is evident that

$$\text{CP of reversed-Carnot refrigerator} = \frac{T_A}{T_B - T_A}. \tag{5.3}$$

$$\text{CP of reversed-Carnot heat pump} = \frac{T_B}{T_B - T_A}. \tag{5.4}$$

Thus, in both cases, the smaller the difference between the temperatures of the two reservoirs the higher will be the CP. This is the reverse of the requirement for high thermal efficiency of a cyclic work-producing device. It is evident that practical considerations would set a lower limit to this temperature difference, but in any case no practical plant works on such a cycle. Refrigerating plant using condensable vapours work, instead, on the vapour-compression cycle.

5.5. The ideal vapour-compression cycle

In principle, any working fluid that could be used in a cyclic heat power plant could also be used in a refrigerating plant **of a kind**, though not all such fluids would make suitable refrigerants. For example, a reversed ideal Joule cycle (§3.3) using air as the working fluid would operate as a refrigerating cycle (a practical example of an open-circuit plant is given in Problem 3.5 of Chapter 3), but since the temperatures of heat reception and rejection would be far from constant a gas would not generally make a very satisfactory working fluid for a refrigerating plant. The physical size of the plant would also tend to be large on account of the large specific volume of the fluid and the low heat transfer rates in the heat exchangers. For these reasons, condensable fluids are usually chosen as refrigerants, though gas refrigeration cycles have found application in low-temperature engineering at liquid-air temperatures and below. Further reference will be made to these in Chapter 10.

When the refrigerant is a condensable fluid, the plant operates on a cycle resembling a reversed Rankine cycle. Such a plant is said to operate on a *vapour-compression cycle*.

In Fig. 5.2, the path $4''3''2_s14''$ is that of an ideal reversible Rankine cycle (§2.3). If a refrigerating plant were to operate on the reversed Rankine cycle, the state of the fluid would trace out this path in the reverse

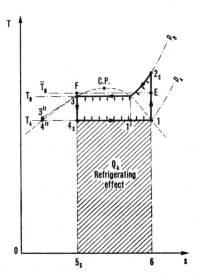

Fɪɢ. 5.2. The ideal vapour-compression cycle.

direction, the fluid now being compressed isentropically from 1 to 2_s in a compressor taking the place of the Rankine-cycle turbine, and being expanded isentropically from $3''$ to $4''$ in a liquid expander or turbine taking the place of the Rankine-cycle feed pump. Evaporation of the fluid would occur between $4''$ and 1, while cooling and condensation would take place in a condenser from 2_s to $3''$. A little thought will reveal that, although internally reversible, such a cycle would be ill-adapted to use in a refrigerating plant. This is because the external fluid to which heat would be transferred from the refrigerant passing through the condenser would have to be, at least at entry, at a temperature **lower** that that at $3''$, while the cold source from which heat would be transferred to the refrigerant passing through the evaporator would have to be at a temperature **higher** than that at $4''$. Consequently, any undercooling below saturation temperature of the refrigerant leaving the condenser must necessarily be very limited. For purposes of discussion hereafter we shall take it to be zero, so that saturated liquid will leave the condenser at state 3 and will be expanded isentropically to 4_s, the cycle then following the path 12_s34_s1.

Figure 5.2 shows the refrigerant leaving the evaporator as dry saturated vapour, and just as there could be some undercooling of the refrigerant leaving the condenser, there could also be a small degree of superheating of the refrigerant leaving the evaporator, or it might alternatively be wet. In practice, departures from the dry saturated condition would be small and for purposes of discussion hereafter we shall take the vapour leaving the evaporator to be dry saturated. We shall then describe the reversible cycle illustrated in Fig. 5.2, with zero undercooling at 3 and zero superheating at 1, as the *ideal vapour-compression cycle*. Whenever we use this term, the foregoing conditions will be implied.

5.6. CP of ideal vapour-compression cycle in terms of enthalpies

The coefficient of performance of the ideal vapour-compression cycle when it is operating as a refrigerator is given by

$$\text{CP} = \frac{Q_A}{Q_B - Q_A} = \frac{(h_1 - h_4)}{(h_{2_s} - h_3) - (h_1 - h_{4_s})}. \tag{5.5}$$

As a heat pump, its coefficient of performance would be given by

$$\text{CP} = \frac{Q_B}{Q_B - Q_A} = \frac{(h_{2_s} - h_3)}{(h_{2_s} - h_3) - (h_1 - h_{4_s})}. \tag{5.6}$$

By referring back to somewhat similar calculations in Chapter 2, the reader will have no difficulty in evaluating these expressions for given conditions. A typical calculation is given in Example 5.2 later. Ideal vapour-compression cycles are, however, of no practical interest, but for purposes which will become evident in §5.7 it is instructive to write down an alternative expression for the CP of a refrigerator in a form which is academically useful though of no practical utility. This is in terms of the mean temperatures of heat reception and rejection.

5.7. CP of the ideal vapour-compression refrigerator cycle in terms of mean temperatures

On the temperature–entropy diagram of Fig. 5.2 the temperature during heat reception from 4_s to 1 is constant and equal to T_A, while during heat rejection the temperature of the refrigerant varies as the state changes from 2_s to 3. On this diagram, the *mean temperature of heat rejection* $\bar{T}_B$ may be defined by writing

$$\bar{T}_B \equiv \frac{\int_3^{2_s} T \, ds}{s_{2_s} - s_3},$$

so that $\bar{T}_B$ is such that the heat rejected, Q_B, is equal both to area $2_s 35_s 62_s$ and to area $EF5_s 6E$. The reciprocal of the coefficient of performance as a refrigerator may then be expressed as

$$\frac{1}{\text{CP}} = \left(\frac{Q_B}{Q_A} - 1\right) = \left[\frac{\bar{T}_B(s_6 - s_{5_s})}{T_A(s_6 - s_{5_s})} - 1\right] = \left(\frac{\bar{T}_B}{T_A} - 1\right). \tag{5.7}$$

This may be compared with the reciprocal of CP for a reversed Carnot cycle operating between the saturation temperatures T_A and T_B in the evaporator and condenser respectively, namely

$$\left(\frac{1}{\text{CP}}\right)_{\text{reversed Carnot}} = \left(\frac{T_B}{T_A} - 1\right). \tag{5.8}$$

Comparing eqns. (5.7) and (5.8), it is easy to see that, since $\bar{T}_B > T_B$, the performance would be improved if superheating of the vapour leaving the compressor could be avoided. Theoretically, this could be done by taking a liquid–vapour mixture from the evaporator in state $1'$, when the values of CP given by eqns. (5.7) and (5.8) would coincide, but this is not practicable. Alternatively, considering different refrigerants, the nearer the slope of the saturated vapour line on the temperature–entropy diagram is to the vertical, the better will be the performance for given values of T_A and T_B. A study of Fig. 5.5(b) and of the answers in line (e) of Problem 5.2 shows that Refrigerant-12 is particularly good in this respect.

5.8. Practical vapour-compression cycles

Practical refrigerating plant operating on the simple vapour-compression cycle differ from the ideal cycle of §5.5 in the following respects:

(1) Unresisted expansion through a simple throttling device invariably takes the place of resisted expansion in the work-producing expander or turbine. Because the specific volume of the fluid expanding from 3 to 4_s in Fig. 5.2 is appreciably smaller than that of the fluid being compressed from 1 to 2_s, the work output of a turbine operating between 3 and 4_s would be small in relation to the work input to the compressor operating between 1 and 2_s. [Note that $(\delta W_x)_{\text{rev}} = -v\,\delta p$.] Although the turbine work output would not always be negligible [cf. the answers in line (4) of Problem 5.3], the amount by which the turbine work output would help to reduce the net work input to the cycle would be, at best, economically marginal. Moreover, the turbine would have to handle a mixture of quite excessive liquid content.

Since the pressure drop from p_B to p_A that would occur in the turbine can be achieved instead by allowing unrestricted expansion of the fluid by passing it through an inexpensive throttling orifice (or, in a small domestic refrigerator, through a capillary tube of small bore), such a device is used in place of the turbine; the plant then takes the form shown in Fig. 5.3(a).

Well upstream and well downstream of the throttle the velocities of the fluid are low, so that there is negligible difference between the upstream and downstream kinetic energies; application of the Steady-flow Energy Equation to the overall throttling process then shows that, if the process is adiabatic, the enthalpy of the fluid downstream of the throttle at 4 is equal to the enthalpy upstream at 3 (though the enthalpy of the fluid changes in passing **through** the throttle as kinetic energy is first generated when the fluid accelerates in flowing through the restriction, and is then later dissipated). Assuming that saturated liquid leaves the condenser at state 3 in Fig. 5.3(b), point 4 will lie on an isenthalp (line of constant enthalpy) passing through 3. Since the state path cannot be traced in this irreversible

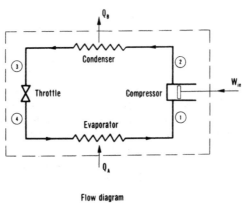

Flow diagram

(a)

FIG. 5.3(a). Practical vapour-compression plant.

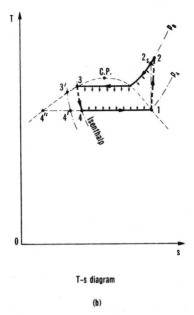

T–s diagram

(b)

FIG. 5.3(b). Practical vapour-compression plant.

process between 3 and 4, the isenthalp is shown as a dotted line. The effect on the plant performance of replacing the turbine process by this throttling process is discussed later in §5.11.

It is seen that at 4 the fluid is a mixture of liquid and vapour, a fraction x_4 of the liquid supplied to the throttle at 3 thus having been "flashed" into vapour in state 1. The energy required for evaporation of this fraction comes from that fraction of the fluid which remains as liquid and which is thereby cooled to state 4″. The liquid–vapour mixture in state 4 then enters the evaporator, in which evaporation is completed as a result of the transfer of heat to the fluid from the refrigeration chamber or, for example, from brine which is circulated through the refrigerating chamber, which is at a **higher** temperature than that of the evaporating fluid.

(2) Instead of all processes being reversible, as are those in the ideal cycle of §5.5, in an actual plant all are, in some measure, irreversible. The throttling process just discussed is, of course, essentially irreversible (note the increase in entropy in this adiabatic process). In Fig. 5.3(b) the only other irreversibility that has been taken into account is that occurring in the compression process from 1 to 2. Frictional pressure drops in the condenser, evaporator and pipework have been ignored. Instead of being isentropic, the compression process results in an increase in entropy, provided that the compression is either adiabatic or that the compressor is

not cooled to such an extent that this more than offsets the increase in entropy due to irreversibility. The required work input to the compressor is increased as a result of the irreversibility, and the coefficient of performance therefore reduced. If the compression were adiabatic, the actual work input could be calculated from a knowledge of the *isentropic efficiency* of the compressor (cf. §3.6). For purposes of calculation and discussion, the compression will always be assumed to be adiabatic.

(3) In an actual plant, there may be some degreee of undercooling of the liquid leaving the condenser and some little superheating of the vapour leaving the evaporator. For the reasons given in §5.5 these will be ignored in the present treatment. Figure 5.3(b) consequently shows saturated liquid leaving the condenser at 3 and saturated vapour leaving the evaporator at 1. If undercooling occurred in the condenser, the state of the liquid leaving the condenser and entering the throttle would be at some such point as 3′, instead of 3, in Fig. 5.3(b). The effect of such undercooling on the plant performance may be learned from Problem 5.5, and the reader is left to assess for himself the effect of any such superheating.

In refrigeration practice, the properties of refrigerants are usually presented on charts in which the logarithm of the pressure is plotted against the enthalpy. On such a chart the cycle would appear as in Fig. 5.3(c), from which the convenience of charts of this kind will be evident.

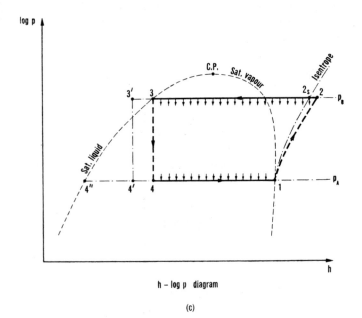

h – log p diagram

(c)

FIG. 5.3(c). Practical vapour-compression plant.

The diagrams in Fig. 5.3 represent conditions for a plant operating at its design conditions. The reader may profitably consider how such a plant would be started up, bearing in mind that a receiver for storage of liquid refrigerant would be provided between the condenser and the throttle valve.

5.9. The quasi-ideal vapour-compression cycle

Having determined the coefficient of performance of a vapour–compression plant from experimental measurements, the engineer needs a *performance criterion* against which to judge the degree of excellence of his plant. The CP of a corresponding fully reversible ideal cycle (§5.5) might seem to provide the appropriate criterion, but since practical refrigerating plant always use unresisted throttle expansion of the fluid instead of resisted expansion in the work-producing expander or turbine of the ideal cycle, a more appropriate criterion is provided by what we shall call the *quasi-ideal cycle*. This incorporates throttle expansion, but all other processes are reversible. It is illustrated in Fig. 5.4. As for the ideal cycle of §5.5, this figure shows saturated liquid leaving the condenser and saturated vapour leaving the evaporator. This is a somewhat arbitrary choice, made for convenience of later study and discussion. When the term *quasi-ideal cycle* is used, these conditions will be implied.

The ratio of the actual CP of a plant to the CP of the corresponding quasi-ideal cycle with the same evaporator and condenser pressures (i.e. saturation temperatures) could be described as the *performance ratio*, analogous to the *efficiency ratio* of §2.5. This must not be confused with

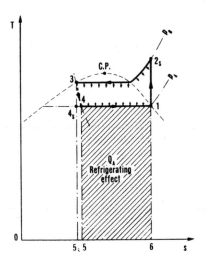

Fig. 5.4. The quasi-ideal vapour-compression cycle.

the term performance energy ratio, sometimes used in place of CP for heat pumps (§5.3).

Not only does the quasi-ideal cycle provide an appropriate performance criterion for vapour-compression plant, but it will be seen in §5.12 to be of use in illustrating the effect of refrigerant properties on plant performance, and so in establishing what are desirable refrigerant properties.

5.10. CP of quasi-ideal vapour-compression cycle

The expressions for the coefficient of performance of the quasi-ideal cycle when operating as a refrigerator or as a heat pump are the same as eqns. (5.5) and (5.6) respectively, with the single exception that h_{4_s} is replaced by h_4. A typical calculation is given in Example 5.1.

5.11. The effect of throttle expansion on refrigerating effect and plant performance

A performance measure which is of practical interest, but which has not hitherto been defined, is the *refrigerating effect*. This is the quantity of heat transferred in the evaporator, expressed either per unit mass of refrigerant circulated or per unit volume of refrigerant vapour drawn into the compressor. When expressed in the latter way, it gives an inverse measure of the swept volume of the compressor, of 100% volumetric efficiency, that would be required for a given refrigerating duty.

When expressed per unit mass, the refrigerating effect is simply Q_A, equal to $(h_1 - h_{4_s})$ in the ideal cycle and to $(h_1 - h_4)$ in the quasi-ideal cycle, namely to the respective shaded areas in Figs. 5.2 and 5.4. It is seen that the introduction of throttle expansion instead of resisted expansion reduces the refrigerating effect and consequently also the CP, the reduction in refrigerating effect being represented by the area $4_s455_s4_s$ in Fig. 5.4. Moreover, since $h_4 = h_3$, so that $(h_4 - h_{4_s}) = (h_3 - h_{4_s})$, this area also represents the work output of the expander or turbine in the ideal cycle, and it therefore further represents the amount by which the net work input in the quasi-ideal cycle is greater than that in the ideal cycle; this increase in net work input further reduces the CP of the former cycle. It has, however, already been noted in §5.8 that the gain resulting from resisted expansion is, at best, economically marginal. The magnitudes of these effects may be seen from a study of lines (3) and (4) in the answers to Problem 5.3, and in a comparison of line (h) in Problem 5.2 with line (5) in Problem 5.3. The reasons why the magnitudes of these effects vary from refrigerant to refrigerant will be discussed in §5.12, after presentation of two typical sets of calculations.

EXAMPLE 5.1. A refrigerating plant operates on the quasi-ideal vapour-compression cycle defined in §5.9. The refrigerant is carbon dioxide, and the saturation temperatures in the evaporator and condenser are respectively $-20\,°C$ and $25\,°C$. The volumetric efficiency of the compressor is 100%.

Calculate (a) the refrigerating effect per unit mass of refrigerant, (b) the refrigerating effect per unit volume of fluid entering the compressor, (c) the mass flow rate of refrigerant per tonne of refrigeration, (d) the compressor displacement per tonne of refrigeration, (e) the superheat at compressor delivery, (f) the compressor work input per unit mass of refrigerant, (g) the power input per tonne of refrigeration, (h) the coefficient of performance.

Using the notation of Fig. 5.4 and the table on page 70:

(a) $h_1 = 323.7\ kJ/kg$, $h_4 = h_3 = 159.7\ kJ/kg$.

Refrigerating effect per unit mass = **164.0 kJ/kg**.

(b) Refrigerating effect per unit volume entering compressor

$$= \frac{164.0}{19.5} = \textbf{8.41 kJ/litre}.$$

(c) Mass flow rate per tonne of refrigeration (see §5.3)

$$= \frac{3.86}{164.0} \times 60 = \textbf{1.41 kg/min}.$$

(d) Compressor displacement per tonne of refrigeration

$$= 1.41 \times 19.5 = \textbf{27.5 litres/min}.$$

(e) $s_{2_s} = s_1 = 1.280\ kJ/kg\ K$.

At p_B and 30 K superheat, $s = 1.241\ kJ/kg\ K$.

At p_B and 60 K superheat, $s = 1.362\ kJ/kg\ K$.

At p_B and $s_{2_s} = 1.280$, superheat $= 30 + \dfrac{0.039}{0.121} \times 30 = \textbf{40 K}$.

(f) $h_{2_s} = 361.5 + \dfrac{0.039}{0.121} \times 41.5 = 374.9\ kJ/kg$.

Compressor work input per unit mass $= (h_{2_s} - h_1)$

$= \textbf{51.2 kJ/kg}$.

(g) Power input per tonne of refrigeration

$$= \frac{1.41}{60} \times 51.2 = \textbf{1.20 kW/t}.$$

The properies of CO_2 are given in the following table:

Saturation			Saturated				Superheated			
Temp. (°C)	Pressure MN/m²	Sp. vol. litre/kg	Enthalpy, h kJ/kg		Entropy, s kJ/kg		By 30 K		By 60 K	
		Vap.	Liq.	Vap.	Liq.	Vap.	h	s	h	s
−20	1.97	19.5	39.7	323.7	0.158	1.280	361.8	1.421	392.5	1.525
25	6.44	4.13	159.7	279.9	0.573	0.976	361.5	1.241	403.0	1.362

(h) Coefficient of performance (CP) $= \dfrac{h_1 - h_4}{h_{2_s} - h_1} = \dfrac{164.0}{51.2} = \textbf{3.20}.$

EXAMPLE 5.2. In the plant of Example 5.1 the throttle valve is replaced by an expander of 100% isentropic efficiency, so that the plant operates on the ideal vapour-compression cycle defined in §5.5.

Calculate (1) the ratio of the condenser pressure to the critical pressure (for CO_2, $p_{crit} = 7.38\,MN/m^2$), (2) the dryness fraction of the fluid entering the evaporator, (3) the percentage increase in refrigerating effect due to replacement of the throttle valve by the expander, (4) the percentage decrease in net work input due to this replacement, (5) the coefficient of performance, (6) the mean temperature of heat rejection in the condenser, (7) the coefficient of performance of a refrigerating plant operating on the reversed Carnot cycle between $-20\,°C$ and $25\,°C$.

(1) $\dfrac{p_{cond}}{p_{crit}} = \dfrac{6.44}{7.38} = \textbf{0.873}.$

(2) $s_{4_s} = s_3 = 0.573\ kJ/kg\ K.$

$$\therefore x_{4_s} = \frac{0.573 - 0.158}{1.280 - 0.158} = \textbf{0.370}.$$

(3) $h_{4_s} = 39.7 + 0.370 \times 284.0 = 144.7.$

Refrigerating effect per unit mass $= 323.7 - 144.7$
    $= 179.0\ kJ/kg.$

Increase in refrigerating effect $= 179.0 - 164.0 = 15.0\ kJ/kg$
    $= \textbf{9.1\%}.$

(4) Work output from expander $= (h_3 - h_{4_s}) = 159.7 - 144.7$
    $= 15.0\ kJ/kg.$

Decrease in net work input $= 15.0\ kJ/kg = \textbf{29.3\%}.$

(5) Net work input $= 51.2 - 15.0 = 36.2\ kJ/kg.$

Coefficient of performance (CP) $= \dfrac{179.0}{36.2} = \textbf{4.94}.$

(6) $Q_{out} = (h_{2_s} - h_3) = 215.2\ kJ/kg,\ (s_{2_s} - s_3) = 0.707\ kJ/kg\ K.$

Mean temperature of heat rejection $(\overline{T}_B) = \dfrac{215.2}{0.707}$

$$= \textbf{304.4 K}.$$

(7) Reversed-Carnot CP $= \dfrac{253}{45} = \textbf{5.62}.$

5.12. The effects of refrigerant properties on plant performance

The tabulated answers to Problems 5.2 and 5.3 provide convenient material for studying the effects of refrigerant properties on plant performance. The reader is recommended to work through these problems before proceeding further.

In these two problems, all four refrigerants are operating between **saturation** temperatures of $-20\,°C$ and $25\,°C$ in the evaporator and condenser respectively. An ideal, reversed-Carnot cycle operating between these temperatures would have a CP of 5.62.

Although the ideal vapour-compression cycles of Problem 5.3 are also internally reversible, and all heat reception in the evaporator is also at $-20\,°C$, heat rejection in the condenser does not take place at a uniform temperature of $25\,°C$. Instead, because the refrigerant entering the condenser is superheated, the mean temperature of heat rejection $\bar{T}_B$ is higher than 298 K (25 °C); this is seen in Fig. 5.2 and in line (6) of Problem 5.3. As explained in §5.7, this higher $\bar{T}_B$ results in a lower value of the CP than the figure of 5.62 for the reversed-Carnot cycle. Thus for Refrigerant-12, with 9 K superheat at condenser inlet, the CP is 5.6, while for CO_2, with 40 K superheat, the CP falls to 4.9. Figure 5.5 shows that the

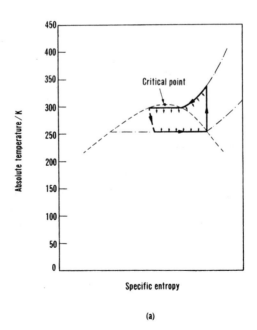

(a)

Fig. 5.5(a). Temperature–entropy diagram for carbon dioxide.

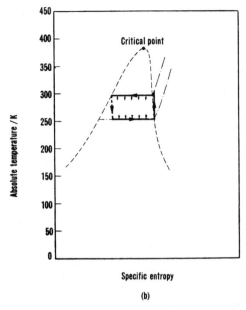

FIG. 5.5(b). Temperature–entropy diagram for Refrigerant-12.

degree of superheat is less for Refrigerant-12 because the saturated-vapour line on the temperature–entropy diagram is steeper, so that in this respect Refrigerant-12 is superior as a refrigerant to CO_2.

Further consideration of the CP figures quoted above reveals another respect in which Refrigerant-12 is superior to CO_2 in the given situation. It is apparent that the magnitude of the reduction in CP below 5.62 does not bear a direct proportional relationship to the degree of superheat at condenser inlet, being only 0.07 for a superheat of 9 K with Refrigerant-12 and as much as 0.68 for a superheat of 40 K with CO_2. In fact, as was seen in §5.7, this reduction in CP is directly related not to the degree of superheat but to the amount by which $\bar{T}_B$ exceeds T_B. It is clear from Fig. 5.2 that this depends not only on the degree of superheat at condenser inlet but also on the relative magnitudes of the respective quantities of thermal energy extracted in desuperheating and in condensing the refrigerant. The greater the relative proportion involved in desuperheating, the greater is the effect of superheat in raising $\bar{T}_B$ above T_B, and therefore in lowering the CP. This proportion is greater the nearer the condenser pressure approaches the critical pressure, since the smaller then becomes the enthalpy of evaporation. It is the fact, illustrated in Fig. 5.5, that the condenser pressure is much closer to the critical pressure for CO_2 than for Refrigerant-12, coupled with the fact that the saturated-vapour line for CO_2 is less steep, that accounts for the noticeably greater reduction in CP when using CO_2.

The influence of these two factors working together is also well illustrated by comparing the figures for CO_2 and ammonia. Although the degree of superheat when using ammonia is greater than when using CO_2, T_B is less because p_{cond}/p_{crit} is much smaller for ammonia than for CO_2, so that the reduction in CP with ammonia is less. We thus conclude that a steep saturated-vapour line and a critical pressure high in relation to the operating condenser pressure are both desirable properties for refrigerants.

In reaching this conclusion we have so far only studied the results of ideal-cycle calculations; in real plant the effects will be qualitatively similar, though different in magnitude. To approach more closely to reality, we now study the results of calculations for the quasi-ideal cycle, in which the isentropic expansion in the work-producing expander of the reversible ideal cycle is replaced by irreversible throttle expansion, without work production. Since this replacement constitutes the only difference between the two cycles, the amounts by which the values of the CP in line (h) of Problem 5.2 are less than those in line (5) of Problem 5.3 are a direct result of this replacement. The CP of the quasi-ideal cycle is less because of the introduction of the irreversible throttling process, the consequences of which are to produce both a smaller refrigerating effect and a greater net work input in the quasi-ideal cycle, both of which result in a smaller value of the CP. The reduction in refrigerating effect is a consequence of the higher specific enthalpy of the fluid entering the evaporator ($h_4 > h_{4_s}$ in Fig. 5.4), while the increase in **net** work input results from the absence of any work output when expanding the fluid through a throttle. The net input is, in fact, increased by the same amount as the refrigerating effect is reduced, since $h_4 = h_3$, so that $(h_3 - h_{4_s}) = (h_4 - h_{4_s})$. It will be seen from the figures set out in the answers to Problems 5.2 and 5.3 that these effects are more pronounced the nearer p_{cond}/p_{crit} approaches to unity; in the light of earlier discussions the reader will readily be able to deduce the reason for this. In particular, it will be noticed that, whereas Refrigerant-12 shows to better advantage than methyl chloride in the ideal cycle, the latter has a slight advantage over the former in the quasi-ideal cycle, which represents a closer approach to reality; this reversal of the situation is a direct consequence of the effect just described, in that p_{cond}/p_{crit} is less for methyl chloride than for Refrigerant-12. Again we see that it is desirable that the critical pressure of a refrigerant should be high in relation to the operating condenser pressure.

Of the four refrigerants, methyl chloride is seen to give the highest CP for the conditions quoted when operating on the quasi-ideal cycle, which, it will be remembered, provides the most suitable criterion of performance against which to judge the excellence of performance of real vapour-compression plant with throttle expansion. However, the margin of advantage of methyl chloride over ammonia and Refrigerant-12 in this respect is not large, and since Refrigerant-12 is negligibly toxic, non-flammable and non-corrosive, while both methyl chloride and ammonia are toxic, flammable and in some circumstances corrosive, Refrigerant-12 has come to find wide usage.

5.13. Desirable refrigerant properties

We may summarise the foregoing discussion by listing the following as the leading properties desirable for refrigerants:

(1) The operating evaporator temperature should be well above the freezing temperature at the operating pressure. (Freezing points at atmospheric pressure: carbon dioxide, $-56.6\,°C$; ammonia, $-77.7\,°C$; methyl chloride, $-97.6\,°C$; Refrigerant-12, $-155.0\,°C$.)
(2) At the desired condenser temperature, p_{cond} should be well below p_{crit}. (Methyl chloride and ammonia good in this respect, Refrigerant-12 quite good and carbon dioxide poor.)
(3) The saturated-vapour line on the temperature–entropy diagram should be as steep as possible (Refrigerant-12 very good in this respect).
(4) The refrigerant should be non-toxic, non-flammable and non-corrosive (Refrigerant-12 very good in this respect, and carbon dioxide also good, although corrosive to copper and iron in the presence of oxygen and water). At the same time, although ammonia is highly toxic, its pungent odour readily reveals the existence of a leak. This ease of detection can be an advantage.

From this list, the reasons for the popularity of Refrigerant-12 as a refrigerant are evident. However, its use is being phased out worldwide, because of its damaging effect on the environment. As a chlorofluorocarbon (CFC), it attacks the ozone layer in the upper atmosphere which protects the Earth from excessive ultra-violet radiation from the sun.

Problems

5.1. Show that the coefficient of performance of a refrigerating plant which operates between two thermal-energy reservoirs, each at uniform temperature, cannot be greater than that of a reversible plant operating between the same two reservoirs.

5.2. Repeat the calculations of Example 5.1 when the refrigerant is (1) ammonia, (2) methyl chloride and (3) Refrigerant-12.

Answer (with results from Example 5.1 included for comparison):

Refrigerant	Carbon dioxide	Ammonia	Methyl chloride	Refrig-erant-12
(a) Refrig. effect/unit mass (kJ/kg)	164	1121	355	119
(b) Refrig. effect/unit vol. (kJ/litre)	8.4	1.80	1.05	1.09
(c) Mass flow rate per tonne of refrigeration (kg/min)	1.41	0.21	0.65	1.95
(d) Compressor displacement per tonne of refrigeration (litre/min)	28	129	221	212
(e) Superheat at comp. delivery (K)	40	73	54	9
(f) Compressor work input per unit mass (kJ/kg)	51	242	74	26
(g) Power input per tonne of refrigeration (kW/t)	1.20	0.83	0.80	0.84
(h) CP	3.2	4.6	4.8	4.6

5.3. Repeat the calculations of Example 5.2 with the same refrigerants as in Problem 5.2.

Answer (with results from Example 5.2 included for comparison):

Refrigerant	Carbon dioxide	Ammonia	Methyl chloride	Refrig-erant-12
(1) p_{cond}/p_{crit}	0.873	0.089	0.085	0.158
(2) Dryness fraction at evaporator inlet	0.370	0.144	0.152	0.237
(3) Increase in refrigerating effect (%)	9.1	1.6	1.6	3.1
(4) Decrease in net work input (%)	29.3	7.4	7.8	14.3
(5) CP	4.94	5.08	5.29	5.55
(6) Mean temperature of heat rejection (K)	304.4	302.9	301.0	298.6
(7) CP of reversed Carnot cycle		5.62		

5.4. Derive the relation between the power input per tonne of refrigeration (in kW/t) and the coefficient of performance of a refrigerating plant. The enthalpy of fusion of ice at 0 °C is 333.5 kJ/kg.

In a test on a vapour-compression refrigerating plant using methyl chloride as the refrigerant, the saturation temperatures in the evaporator and condenser are respectively −5 °C and 40 °C, and the power input per tonne of refrigeration is 0.93 kW/t. Determine the coefficient of performance, and compare it with that of the corresponding quasi-ideal cycle.

Assuming that irreversibilities in the actual plant occur only in the flow through the throttle valve and the compressor, calculate the isentropic efficiency of the compressor.

Answer: CP × (kW/t) = 3.86; 4.15, 5.08; 81.7%.

5.5. What would be the coefficient of performance of a plant in which conditions were the same as for the quasi-ideal cycle of Problem 5.4, with the exception that the liquid leaving the condenser were under-cooled by 5 K?

Answer: 5.20.

5.6. A vapour-compression plant with Refrigerant-12 as the refrigerant is used as a heat pump to supply 30 kW to a building which is maintained at a mean temperature of 20 °C when the mean temperature of the outside air is 0 °C. There is a temperature difference of 5 K between the mean temperature of the outside air and the saturation temperature of the refrigerant in the evaporator, and also between the saturation temperature of the refrigerant in the condenser and the mean temperature of the building.

Saturated liquid enters the throttle valve and saturated vapour enters the compressor, which has an isentropic efficiency of 82%. Calculate the coefficient of performance of the heat pump, and the power input to the compressor.

The compressor is driven by an electric motor, the combined efficiency of the motor and drive being 75%. Express the electrical power input as a fraction of the power that would be required if the same energy input to the room were supplied by direct electrical heating.

Answer: 7.4, 4.0 kW; 0.18.

5.7. In the open-circuit refrigerating plant of Problem 3.5 of Chapter 3, what advantage does this method of aircraft air-conditioning have over one in which the air is throttled at supercharger outlet and supplied direct to an air cooler and thence to the cabin?

5.8. Figure 5.6 gives the flow diagram of a heat-pump installation for the air-conditioning of an underground vault housed in a disused quarry, and used during war-time for the storage of valuable works of art.

Air of relative humidity 90% and temperature 12 °C is drawn into the plant at a pressure of 1 atm through an air-washer and dehumidifier, where it is cooled by a water spray. The air leaves the washer saturated with water vapour at 5 °C and is then heated to 21 °C by being passed over the condenser coil of the heat pump, before being delivered to the vault.

The water is drawn from the air-washer at 5 °C and sprayed over the evaporator coil of the heat pump, where its temperature is reduced to 3 °C before return to the washer.

Calculate, per kg of air circulated:

(a) the amount of moisture extracted from the air in the washer;

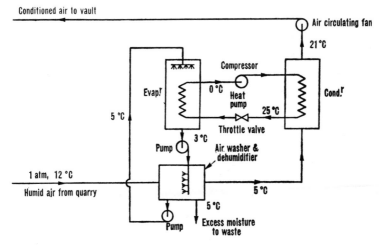

FIG. 5.6.

(b) the amount of spray water;

(c) the coefficient of performance of the heat-pump installation;

(d) the ratio of the actual coefficient of performance to the theoretical coefficient of performance for a quasi-ideal cycle in which dry saturated Refrigerant-12 leaves the evaporator at 0 °C and the Refrigerant-12 is just condensed in the condenser at 25 °C;

(e) the relative humidity at inlet to the vault, taking the pressure there to be 1 atm.

Expain why a reduction in the specific humidity of the air is obtained by spraying water into it in the washer.

Answer: (a) 0.00242 kg; (b) 1.57 kg; (c) 5.23; (d) 0.490; (e) 35%.

PART II

Advanced Power and Refrigerating Plants

CHAPTER 6

Advanced gas-turbine plant

It has already been noted at the beginning of Chapter 3 that the efficiency of a simple gas-turbine plant is often too low to be commercially attractive. This chapter makes a study of the more complex cycles that are necessary if higher efficiencies are to be obtained.

6.1. Limitations of the simple gas-turbine cycle—the importance of the mean temperatures of heat reception and rejection

The temperature–entropy diagram for the simple Joule cycle is given in Fig. 6.1, and on this diagram are shown the mean temperatures of heat reception and rejection, $\bar{T}_B$ and $\bar{T}_A$ respectively. These are defined as follows:

$$\bar{T}_B \equiv \frac{\int_2^3 T \, ds}{s_3 - s_2} \quad \text{and} \quad \bar{T}_A \equiv \frac{\int_1^4 T \, ds}{s_4 - s_1}.$$

Thus $\bar{T}_B$ is such that the heat supplied, Q_B, is equal both to area 23562 and to area $EF56E$, while $\bar{T}_A$ is such that the heat rejected, Q_A, is equal both to area 41654 and to area $GH65G$. An alternative expression for the thermal efficiency of the Joule cycle is consequently given by

$$\eta_{\text{JOULE}} = \left(1 - \frac{Q_A}{Q_B}\right) = \left[1 - \frac{\bar{T}_A(s_5 - s_6)}{\bar{T}_B(s_5 - s_6)}\right] = \left(1 - \frac{\bar{T}_A}{\bar{T}_B}\right). \qquad (6.1)$$

The thermal efficiency of any cyclic device operating between a source at the highest temperature T_b and a sink at the lowest temperature T_a of the Joule cycle could not exceed that of a Carnot cycle operating between these two temperatures. This may therefore be described as the *limiting Carnot efficiency* for these conditions, and is given by

$$\text{Limiting } \eta_{\text{CARNOT}} = \left(1 - \frac{T_a}{T_b}\right). \qquad (6.2)$$

Comparison of expressions (6.1) and (6.2), and inspection of Fig. 6.1,

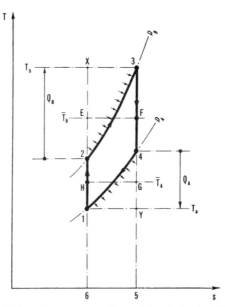

Fɪɢ. 6.1. Temperature–entropy diagram for the Joule (CBT)$_r$ cycle

shows that η_{JOULE} is appreciably less than the limiting η_{CARNOT}, because $\bar{T}_B$ is well below T_b and $\bar{T}_A$ is well above T_a. A visual appreciation of the magnitude of the difference can be obtained by noting that

$$\text{Limiting } \eta_{\text{CARNOT}} = \frac{\text{area } X3Y1X}{\text{area } X356X}$$

and

$$\eta_{\text{JOULE}} = \frac{\text{area } EFGHE}{\text{area } EF56E}.$$

This difference was illustrated in Fig. 3.3, the two efficiencies being equal only at the limiting pressure ratio for which $\rho_p = \theta$; that is, in the impractical case when the compressor outlet temperature was equal to the turbine inlet temperature.

This discussion leads to the important general conclusion that, for reversible cycles, the mean temperatures of heat reception and rejection, rather than the extreme temperatures, govern the value of the cycle efficiency. The same is true of irreversible cycles, though then a gain in efficiency resulting from improvements in these mean temperatures may sometimes be partly or more than offset by the effect of resulting greater internal irreversibility; such is the case, discussed in the next chapter, when feed heating in the steam cycle with a finite number of heaters.

It is clear that, if higher cycle efficiencies in gas-turbine plant are to be obtained, ways must be sought of raising the mean temperature of heat reception and lowering the mean temperature of heat rejection.

6.2. Exhaust-gas heat exchanger—the CBTX cycle

The most direct way of simultaneously raising the mean temperature of heat reception and lowering the mean temperature of heat rejection is to incorporate an *exhaust-gas heat exchanger X* in the cycle, in the manner shown in Fig. 6.2.

When discussing internal-combustion gas-turbine plant in Chapter 4, it was noted that the low efficiency was associated with a high turbine exhaust temperature. That this high exhaust temperature will also be associated with a low efficiency in external-combustion cyclic plant is clear from the fact that it leads to a high mean temperature of heat rejection. An exhaust-gas heat exchanger uses the energy in the high-temperature gases from the turbine to preheat the air supplied by the compressor to the heater in a closed-circuit plant, and to the combustion chamber in an open-circuit plant. The temperature–entropy diagram for an ideal cycle of this kind is shown in Fig. 6.3. Comparison with Fig. 6.1 shows clearly the beneficial effect of the heat exchanger on the mean temperatures of heat reception and rejection. The cycle is known as a CBTX cycle, the *B* symbolising the heater outside the tubes of which the fuel is burnt.[†] The studies in this chapter will be confined to closed-circuit plant and will treat reversible and irreversible cycles in turn; these will be denoted respectively by the subscripts *r* and *i*.

The effect of pressure drops in the heat exchangers and ducting will be neglected initially, but will be considered in a concluding section.

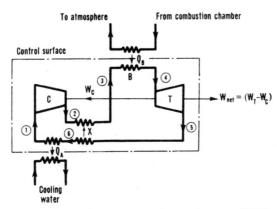

FIG. 6.2. Cycle with exhaust-gas heat exchanger—CBTX.

[†]If a CBTX gas-turbine plant were used in conjunction with a nuclear reactor, so that the working fluid acted also as reactor coolant in the manner described in §8.10 of Chapter 8, Q_B would be the heat taken up by the working fluid in its passage through the reactor.

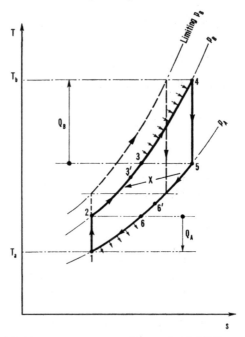

FIG. 6.3. Temperature–entropy diagram for (CBTX)$_r$ cycle.

6.3. Heat-exchanger effectiveness ε

The *effectiveness* ε of a heat exchanger is defined as the ratio of the temperature rise of the cooler fluid to the difference between the temperatures at which the two fluids respectively enter the heat exchanger.

For ideal, reversible heat transfer there must be zero temperature difference between the fluids at all points in the heat exchanger; the cooler fluid will then be raised to the inlet temperature of the hotter fluid, so that the heat-exchanger effectiveness will be unity. To achieve this, the fluids must pass in counterflow through the exchanger, and the surface area provided must be infinite in extent. Such a heat exchanger will be denoted by the subscript r. Real heat exchangers have an effectiveness less than unity, and these will be denoted by the subscript i.

6.4. The (CBTX)$_r$ cycle — $(\eta_T = \eta_C = \varepsilon = 1)$

This is the ideal reversible CBTX cycle, with an exhaust-gas heat exchanger of unity effectiveness. Its temperature–entropy diagram is shown in Fig. 6.3.

Treating the fluid at all points as a perfect gas, the quantities of heat supplied and rejected per unit mass of gas circulated are given by

$$Q_B = c_p(T_4 - T_3) = c_p(T_4 - T_5) = c_p T_b\left(1 - \frac{1}{\rho_p}\right),$$

$$Q_A = c_p(T_6 - T_1) = c_p(T_2 - T_1) = c_p T_a(\rho_p - 1).$$

The thermal efficiency of the cycle is consequently given by

$$\eta_{CY} = \left(1 - \frac{Q_A}{Q_B}\right) = \left(1 - \frac{\rho_p}{\theta}\right), \tag{6.3}$$

where $\theta \equiv T_b/T_a$.

Thus, whereas in the simple Joule cycle $\eta_{CY} \to 0$ as $\rho_p \to 1$, η_{CY} now tends to the limiting Carnot efficiency, $[1 - (1/\theta)]$, as $\rho_p \to 1$. The reader should explain this for himself by sketching the temperature–entropy diagram (Problem 6.2).

Equation (6.3) also shows that η_{CY} falls linearly with increase in ρ_p, and would fall to zero at a value of $\rho_p = \theta$. Such a value of ρ_p is, in practice, unattainable, because an exhaust-gas heat exchanger can only be used when the compressor outlet temperature is less than the turbine outlet temperature. For fixed values of T_a and T_b, a cycle operating at this limiting pressure ratio is indicated by dotted lines in Fig. 6.3, and it is easy to show that ρ_p is then equal to $\sqrt{\theta}$. This is the same value of ρ_p at which W_{net} is a maximum. At this value of ρ_p the cycle is simply a (CBT)$_r$ cycle without exhaust-gas heat exchanger, corresponding to point Y at the intersection of the straight line for the (CBTX)$_r$ cycle with the curve for the (CBT)$_r$ cycle in Fig. 6.4; for values of ρ_p beyond this point the straight line is drawn dotted since such conditions are imaginary. The turbine work

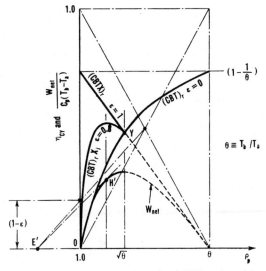

FIG. 6.4. Variation of η_{CY} and W_{net} with ρ_p for CBTX cycles with $\eta_C = \eta_T = 1$.

W_T and compressor work W_C are clearly unaffected by the introduction of the heat exchanger, and the curve for W_{net} is therefore the same for the two cycles.

6.5. The (CBT)$_r$ X$_i$ cycle — ($\eta_T = \eta_C = 1$, $\varepsilon < 1$)

Moving one step nearer to reality, this differs from the previous cycle only in having $\varepsilon < 1$. The states of the fluid at exit from the heat exchanger will be at points such as 3′ and 6′ instead of at 3 and 6 in Fig. 6.3, the mean temperatures of heat reception and rejection will be respectively lower and higher, and the cycle efficiency at a given value of ρ_p consequently less. Moreover, a study of the temperature–entropy diagram shows that as $\rho_p \to 1$, η_{CY} tends to zero instead of to $[1 - (1/\theta)]$, since the heat input now remains finite as the net work output tends to zero. It is clear that there is the same limit to the use of an exhaust-gas heat exchanger as in the previous cycle, so that the efficiency curve again terminates at point Y in Fig. 6.4. For a typical value of $\varepsilon = 0.8$, detailed calculation gives the efficiency curve shown in the figure. The value of ρ_p for maximum efficiency can be found graphically by an extension of the method used for the (CBT)$_i$ cycle in §3.8, and is given by the point at which a straight line from E' in Fig. 6.4 is a tangent to the curve of W_{net}. The reader may either test his ingenuity by trying to prove this or may refer to the original paper.[7] In contrast to the simple (CBT)$_i$ cycle studied in Chapter 3 and Fig. 3.5, the value of ρ_p for maximum η_{CY} is now seen to be less than that for maximum W_{net}.

6.6. The (CBTX)$_i$ cycle — ($\eta_T < 1$, $\eta_C < 1$, $\varepsilon < 1$)

The cycles described in §§6.4 and 6.5 have been of academic interest only, since the turbine and compressor efficiencies were both unity. We now study the practical case in which η_T, η_C and ε are all less than unity. Again, W_{net} is unaffected by the introduction of the heat exchanger so that, for the same values of η_T and η_C, the variation of W_{net} with ρ_p is the same as that shown in Fig. 3.5 for the (CBT)$_i$ cycle, and the maximum value of W_{net} still occurs at a value of $\rho_p = \sqrt{\alpha}$.

Although numerical calculations are not difficult, it is easier and more informative to use the same graphical method as in §6.5 to determine the value of ρ_p for maximum efficiency. This is again given by the point at which a straight line from E' in Fig. 6.5 is a tangent to the curve of W_{net}. Again, the value of ρ_p for maximum η_{CY} is less than that for maximum W_{net}, whereas for the (CBT)$_i$ cycle it was greater.

The variation of η_{CY} with ρ_p is shown in Fig. 6.5. For the special case when $\varepsilon = 1$, namely for the (CBT)$_i$X$_r$ cycle, this again becomes a straight line. As can be seen from Fig. 6.3, $Q_B = W_T$ and $Q_A = W_C$ when $\varepsilon = 1$, so that it is readily shown that the equation for this line is

$$\eta_{CY} = \left(1 - \frac{\rho_p}{\alpha}\right) \qquad (6.4)$$

This compares with eqn. (6.3) for the $(CBTX)_r$ cycle, for which $\eta_T = \eta_C = 1$ so that $\alpha = \theta$.

A comparative study of Figs. 6.4 and 6.5 gives a clear insight into the effects on plant performance of irreversibilities in the turbine, compressor and heat exchanger. It may also be noted from Fig. 6.5 that the optimum pressure ratio for the $(CBTX)_i$ cycle is considerably less than that for the $(CBT)_i$ cycle, being, for the conditions for which this figure is drawn, only 3.7 as against 11.6.

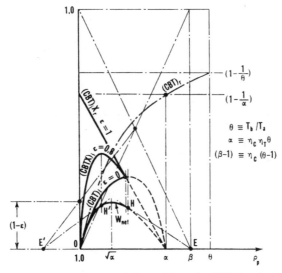

FIG. 6.5. Variation of η_{CY} and W_{net} with ρ_p for CBTX cycles with $\eta_C < 1$, $\eta_T < 1$.

6.7. Reheating and intercooling

It has been seen that the introduction of an exhaust-gas heat exchanger increases the cycle efficiency because it raises the mean temperature of heat reception and lowers the mean temperature of heat rejection. Nevertheless, inspection of Fig. 6.3 reveals that the cycle still fails to achieve the ideal of having all heat reception at the highest temperature T_b and all heat rejection at the lowest temperature T_a. We now consider a way of achieving this ideal in theory and of approximating to it in practice. It involves the use of both reheating and intercooling in conjunction with an exhaust-gas heat exchanger. We shall see that, without the incorporation of the latter, both *reheating* and *intercooling* would decrease the cycle efficiency, though they would increase the net work output per unit mass of gas circulated.

In reheating, the expansion is carried out in two or more turbines in series and the working fluid is reheated between the turbines, usually to approximately the initial turbine inlet temperature. Similarly, in intercooling, the compression is carried out in two or more compressors in series and the working fluid is cooled between the compressors, usually to approximately the initial compressor inlet temperature. Figure 6.6 shows the circuit diagram for a CBTX cycle to which has been added one stage of reheating and one stage of intercooling; this is described as a CICBTRTX cycle. Figure 6.7 shows the corresponding temperature–entropy diagram when the cycle is irreversible.

It is not necessary that intercooling and reheating should be incorporated together; either or both may be added to a CBTX cycle to increase the

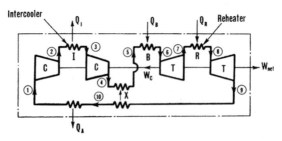

FIG. 6.6. Cycle with intercooler, reheater and exhaust-gas heat exchanger—
CICBTRTX.

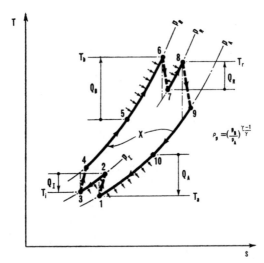

FIG. 6.7. Temperature–entropy diagram for $(CICBTRTX)_I$ cycle.

efficiency. The reasons for this increase will be found by considering each in turn.

6.8. The (CBTRT)$_r$ and (CICBT)$_r$ cycles

Figure 6.8 shows a single stage of reheating added to a simple (CBT)$_r$ cycle. It is not immediately evident from an inspection of the diagram whether the reheat cycle will have a greater or smaller efficiency than the non-reheat cycle.

In the reheat cycle there is clearly both a higher mean temperature of heat reception and a higher mean temperature of heat rejection. The former will tend to increase the efficiency and the latter to decrease it. Which of these two opposing effects is dominant can be decided by noting that the complete reheat cycle may be treated as a compound cycle comprising the non-reheat Joule cycle $N(1234'1)$ of efficiency η_N, to which has been added a hypothetical Joule cycle R (4'4564') of efficiency η_R. The compression ratio for cycle R is less than that for cycle N, so that it is seen from eqn. (3.4) that $\eta_R < \eta_N$. It is easy to show (Problem 6.3) that the thermal efficiency η_{CY} of the complete reheat cycle is given by

$$\left(\frac{\eta_{CY}}{\eta_N} - 1\right) = \frac{[(\eta_R/\eta_N) - 1]}{[(Q_N/Q_R) + 1]}.\qquad(6.5)$$

Hence, since $\eta_R < \eta_N$, η_{CY} will be less than η_N, so that the addition of reheating to a simple (CBT)$_r$ cycle will result in a decrease in cycle efficiency. On this score, reheating alone is not attractive. On the other hand, the work output per unit mass of fluid circulated is increased by the

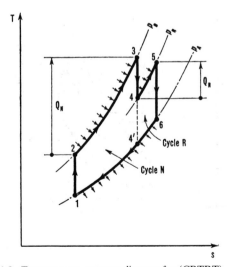

Fɪɢ. 6.8. Temperature–entropy diagram for (CBTRT)$_r$ cycle.

addition of reheating, since the area enclosed by the reheat cycle in Fig. 6.8 is greater than that enclosed by the non-reheat cycle. This results in a smaller volumetric flow rate for a given output and so to a reduction in plant size, though against the resulting reduction in the size and weight of the plant must be offset the introduction of the reheater.

By sketching a temperature–entropy diagram for a $(CICBT)_r$ cycle, the reader may satisfy himself that the addition of intercooling to a simple $(CBT)_r$ cycle will produce effects similar to those of reheating; namely, a decrease in cycle efficiency and increase in work output per unit mass of fluid circulated (Problem 6.4).

Although the discussion has been confined to reversible cycles, the effects will be similar when the cycles are not reversible.

6.9. The $(CBTRTX)_r$ cycle

Continuing the study of reversible cycles, we now consider the effect of the simultaneous introduction of reheating and of an exhaust-gas heat exchanger. Figure 6.9 shows a single stage of reheating added to a $(CBTX)_r$ cycle.

Heat reception in this cycle occurs between T_3 and T_4, and between T_5 and T_6, whereas in the $(CBTX)_r$ cycle it occurred between $T_{3'}$ and T_4, so that the mean temperature of heat reception in the reheat cycle is clearly higher. Heat rejection, however, occurs between T_8 and T_1 in both cycles. Hence, because there is an increase in the mean temperature of heat reception, without change in the mean temperature of heat rejection, the

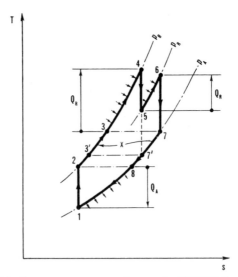

Fig. 6.9. Temperature–entropy diagram for $(CBTRTX)_r$ cycle.

application of reheating to a (CBTX)$_r$ cycle will give an increase in cycle efficiency. Thus reheating results in improved efficiency provided it is accompanied by the introduction of an exhaust-gas heat exchanger.

6.10. The (CICBTX)$_r$ cycle

It may similarly be shown that intercooling will also give an increased efficiency if applied to a cycle with an exhaust-gas heat exchanger. The reader may readily confirm this by sketching a temperature-entropy diagram showing a single stage of intercooling added to a (CBTX)$_r$ cycle.

6.11. Progressive reheating and intercooling to give Carnot efficiency—the (CICI ... BTRTRT ... X)$_r$ cycle

We can now consider how we may, in theory, devise a gas-turbine cycle having a thermal efficiency equal to the limiting Carnot efficiency $[1 - (T_a/T_b)]$. For this to be possible all internal processes must be reversible, while all heat reception must occur at the highest temperature T_b and all heat rejection at the lowest temperature T_a. These conditions are obtained in the cycle shown in Fig. 6.10 when there are an infinite number of stages both of progressive reheating and of progressive intercooling. In these hypothetical circumstances the temperature–entropy diagram takes the simple form 12341. Such a cycle is, of course, of purely academic interest, since it is far from being practicable. It is impracticable not only on account of the infinite number of components, but also because

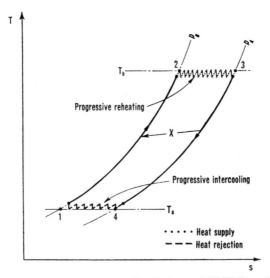

FIG. 6.10. Temperature–entropy diagram for (CICI ... BTRTRT ... X)$_r$ cycle.

all real-life processes are in varying measure irreversible. It may be described as the ideal gas-turbine cycle in "Thermotopia", that idyllic land of the thermodynamicist in which all processes are reversible and all imperfections absent.

Although perfection of the cycle has been achieved in this hypothetical plant, it is necessary to remind ourselves, as we did in §1.4, that high cycle efficiency is not the only factor of importance in the attainment of high overall efficiency of a plant which uses a fossil fuel as its source of energy. In such a plant, the heat supplied to the fluid passing round the closed cycle would be transferred from hot combustion gases coming from a combustion chamber and then discharged to atmosphere, as in Fig. 1.3(a). In these circumstances, the overall plant efficiency is the product of η_{CY} and η_B, the efficiency of the heating device in terms of the calorific value of the fuel. The latter efficiency is directly dependent on the temperature at which the combustion gases are discharged to atmosphere. In the ideal cycle under consideration, this temperature cannot be less than T_b if the gases are discharged direct to atmosphere since the fluid by which they are cooled is all at this temperature, whereas in the simple Joule cycle of Figs. 1.3(a) and 3.1 the combustion gases could theoretically be cooled to T_2. In achieving perfection of the cycle, therefore, a penalty will be paid in a reduction of η_B unless some remedial measure is adopted; this could take the form of an air preheater, in which the combustion gases are cooled before discharge to atmosphere by transferring heat to the incoming air to the combustion chamber. The high temperature of the water fed to the boiler in a feed-heating steam cycle leads to similar introduction of an air preheater on steam boiler plant.

This latter problem would not arise if a nuclear reactor were used as a heat source, since there would then be no combustion gases to be cooled. However, the temperatures required for high efficiency in practical gas-turbine cycles are too high for the materials at present available in nuclear reactors.

Since a large number of stages of reheating and intercooling are impracticable, it is fortunate that even one stage gives a significant gain in efficiency, while the addition of further stages gives a progressively smaller return. The practical, irreversible cycle incorporating only a single stage of both reheating and intercooling will consequently be studied next.

6.12. The practical (CICBTRTX)$_i$ cycle

The temperature–entropy diagram for this cycle has already been given in Fig. 6.7. For given values of T_a, T_b, T_i and T_r, and given values of p_A and p_B, there will clearly be optimum values of p_I and p_R for which the thermal efficiency of the cycle will be a maximum. This will occur when

$$\frac{\partial \eta_{CY}}{\partial \rho_I} = 0 \quad \text{and} \quad \frac{\partial \eta_{CY}}{\partial \rho_R} = 0, \tag{6.6}$$

where $\quad \rho_I = \left(\dfrac{p_I}{p_A}\right)^{(\gamma-1)/\gamma} \quad \text{and} \quad \rho_R = \left(\dfrac{p_R}{p_A}\right)^{(\gamma-1)/\gamma}. \tag{6.7}$

Since $\quad \eta_{CY} \equiv \dfrac{W_{net}}{Q_{in}}, \quad \text{where} \quad Q_{in} = (Q_B + Q_R),$

$$\frac{\partial \eta_{CY}}{\partial p_I} = \frac{1}{Q_{in}} \frac{\partial W_{net}}{\partial p_I} - \frac{W_{net}}{(Q_{in})^2} \frac{\partial Q_{in}}{\partial p_I} = 0;$$

whence, for maximum cycle efficiency,

$$\frac{\partial W_{net}}{\partial \rho_I} = \eta_{CY} \frac{\partial Q_{in}}{\partial \rho_I}. \tag{6.8}$$

By writing down equations for W_{net} and Q_{in}, and eliminating T_3 and T_5, it is only a matter of tedious manipulation[9] to show that eqn. (6.8) gives, for the optimum value of ρ_I:

$$(\rho_I)_{opt.} = \sqrt{\left\{\frac{T_i}{T_a} \rho_p[1 - \eta_{CY}(1 - \varepsilon)]\right\}}. \tag{6.9}$$

Similarly, the optimum value of ρ_R is given by:

$$(\rho_R)_{opt.} = \sqrt{\left[\frac{T_r}{T_b} \rho_p \left(\frac{1 - \varepsilon \eta_{CY}}{1 - \eta_{CY}}\right)\right]}. \tag{6.10}$$

Since these expressions involve η_{CY} it is not just a matter of direct calculation to determine the optimum values. A tedious trial-and-error calculation is still necessary. The following special case is of interest.

If $T_i = T_a$ and $T_r = T_b$, a state of affairs not greatly different from that which usually arises, and if $\varepsilon = 1$, then eqns. (6.9) and (6.10) reduce to

$$(\rho_I)_{opt.} = (\rho_R)_{opt.} = \sqrt{\rho_p}, \tag{6.11}$$

or $\quad p_I = p_R = \sqrt{(p_A p_B)}, \tag{6.12}$

whence $\quad \dfrac{p_I}{p_A} = \dfrac{p_B}{p_I} \quad \text{and} \quad \dfrac{p_R}{p_A} = \dfrac{p_B}{p_R}. \tag{6.13}$

Thus, in this special case, the optimum pressure ratio is the same in both stages of compression and of expansion.

Some idea of the effect of the introduction of reheating and intercooling on the optimum plant performance may be obtained by studying the answers to Problem 6.6.

6.13. Other factors affecting cycle performance

In the foregoing studies, in order to simplify the treatment no account has been taken of pressure drops in the ducting and in all the heat exchangers. In practice these have an effect on the cycle efficiency which is by no means negligible. These parasitic pressure drops are much more serious in gas-turbine than in steam-turbine plant because the overall pressure ratio is always so much less in the gas-turbine cycle. In practical design calculations it is consequently necessary to take proper account of the effects of these pressure drops.[7] It is also necessary to allow for variation in the specific heat capacity of the working fluid with variation in temperature by using tables of the experimentally determined properties of gases.

6.14. Gas-turbine plant for Compressed Air Energy Storage (CAES) systems

Power stations of large output, and particularly nuclear stations, are best suited to continuous base-load operation. With the increasing number of such stations being installed, attention is turning towards means of catering for peak-load demands. Stand-by gas-turbine plant of a conventional type have been quite widely used for this purpose, but another use of the gas turbine in providing a peak-load facility is found in the Compressed Air Energy Storage (CAES) system. Instead of the water reservoirs of a hydroelectric pumped-storage system, making use of the gravitational potential energy of water, a CAES system uses the stored energy of compressed air, with additional energy provided by the burning of a fossil fuel. Until recently, the CAES system has been confined to project studies. However, the first such plant was installed in the late 1970s at Huntorf,[10] near Bremen, in the Federal Republic of Germany. The plant has a full-load output of 290 MW and uses a cavern in an underground salt deposit as the reservoir for air storage. As depicted in Fig. 6.11, it incorporates the elements of a CICBTBTX gas-turbine plant, with the addition of an aftercooler. This avoids the delivery of high-temperature air to the reservoir; maximum storage would be obtained if the air were stored in the reservoir at the ambient temperature.

The air reservoir is charged during the off-peak period by engaging the clutch between the compressors and motor-generator and using the latter to drive the compressors. Two-stage compression is used since the pressure ratio is high, fluctuating between 46:1 and 66:1. Intercooling between the compressors helps to reduce the required work input. After recharging the reservoir, the plant is shut down.

During a period of peak load, the turbines are run up to speed by engaging the clutch between them and the motor-generator, still using the

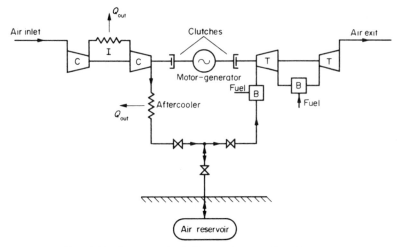

Fig. 6.11. Compressed Air Energy Storage (CAES) system.

latter as a motor. Air is then allowed to flow through the turbines from the air reservoir and, with the ignition of fuel supplied to the combustion chambers, the speed becomes self-sustaining. Thereafter, the turbine plant is run up to load, with the motor-generator now acting as a generator and the air pressure in the reservoir progressively falling. However, in order to maintain constancy of electrical output, the air is throttled between the cavern and the turbine so as to maintain a constant pressure of 46 bar at turbine inlet. By burning fuel in the air, the useful work produced during the peak-load period is between 20% and 40% greater than the work input to the compressors during the off-peak period. Moreover, since the air must be stored in the cavern at around the ambient temperature, the moisture in the air would cause severe icing problems in the turbine if the latter were supplied with unheated air. This may be seen from the answer to part (d) of Problem 6.7.

The ratio of charging time to discharging time is about 4:1, with about two hours of operation at full load and eight hours for recharging the reservoir. The compressors thus need to be designed for only one-quarter of the mass flow rate for which the turbines are designed. The allowed range of pressure fluctuation in the air reservoir is determined from economic considerations of plant size and the required size of reservoir. The mean pressure in the reservoir must be high, otherwise the required size of cavern becomes too large.

Maintenance of a constant pressure in the air reservoir during operation would clearly be more advantageous than a varying pressure. Conceptually, this could be achieved by replacing the outflowing air by water flowing in from a water reservoir situated above the cavern. However, for a pressure of 60 bar in the air reservoir, the water reservoir would need to be situated at a height of about 600 m above the air reservoir. Moreover, this ingress of water would not be permissible in a salt cavern. Constancy of pressure is assumed in Problem 6.7 for ease of calculation.

An *Adiabatic CAES System*[11] is a conceptual alternative to the type of CAES plant just described. It would dispense with the aftercooler and with the burning of fuel. Instead, the air would pass through a thermal-energy reservoir on its way to and from the air reservoir, so that the matrix of this additional reservoir would act as a regenerator, being alternately heated and cooled during the charging and discharging process respectively. Further electrical heat input to the air during the off-peak period could also be arranged. However, such a plant could only be situated where there was a suitable formation of porous rock of an appropriately heat-resistant character.

A detailed application of the concepts and theorems of thermodynamic availability to CAES systems will be found in reference 12. This enables identification to be made of those components whose performance is critical to the efficient performance of the plant.

6.15. Gas-turbine cycles for nuclear power plant

A further discussion of gas-turbine cycles will be found in §8.10 of Chapter 8, in which there is a review of the particular types of cycle that have been considered for use with gas-cooled nuclear reactors.

6.16. Non-cyclic, open-circuit plant

It is advisable to end the chapter with a reminder that all aircraft gas-turbines and many industrial gas turbines are of the internal-combustion, open-circuit type, although large closed-circuit plant have been built. While study of the variation in performance of closed-circuit plant with changes in the design parameters will also give qualitative information about the performance of open-circuit plant, the latter need to be designed with proper consideration of the internal combustion process, and proper use of tables of properties of the combustion products that pass through the turbine. Such studies are outside the scope of the present volume.

Problems

6.1. A CBTX cycle is designed for given values of T_a, T_b, η_C and η_T. The heat exchanger has an effectiveness of unity and causes negligible pressure drop. The plant is designed for the same pressure ratio as that for which a CBT cycle with the same values of the foregoing parameters would produce maximum work output per unit mass flow of circulating fluid, which may be treated as a perfect gas. Show that this CBTX cycle has the same efficiency as an ideal reversible Joule cycle designed for the same pressure ratio.

6.2. The compressor and turbine inlet temperatures in a (CBTX)$_r$ cycle are respectively T_a and T_b. By sketching a temperature–entropy diagram, show that as $\rho_p \to 1$, $\eta_{CY} \to [1 - (1/\theta)]$, where $\theta \equiv T_b/T_a$.

6.3. Prove eqn. (6.5).

6.4. By studying a sketch of a temperature–entropy diagram, show that the addition of a single stage of intercooling to a (CBT)$_r$ cycle will decrease the cycle efficiency.

6.5. For the special case of a (CBTRTX)$_i$ cycle in which $T_r = T_b$ and $\varepsilon = 1$, show, from first principles, that for maximum efficiency $p_R = \sqrt{(p_A p_B)}$. The symbols have the same significance as in §6.12.

6.6. Compare the (CBT)$_i$, (CBTX)$_i$ and (CICBTRTX)$_i$ cycles for the conditions given below, by estimating the values of the maximum efficiency, the pressure ratio for maximum efficiency, and the corresponding net work output per kg of fluid circulated. The fluid may be treated as a perfect gas with $c_p = 1.01$ kJ/kg K and $\gamma = 1.4$.

Compressor inlet temperature	15 °C
Turbine inlet temperature	800 °C
Compressor isentropic efficiency	0.85
Turbine isentropic efficiency	0.88
Heat exchanger effectiveness	0.8
Intercooling to	40 °C
Reheating to	800 °C

Answer (best calculated by computer):

(CBT)$_i$	$\eta_{CY} = 29.9\%$;	$r_p = 11.2$;	$W_{net} = 135.2$ kJ/kg
(CBTX)$_i$	$\eta_{CY} = 38.9\%$;	$r_p = 3.6$;	$W_{net} = 141.1$ kJ/kg
(CICBTRTX)$_i$	$\eta_{CY} = 43.3\%$;	$r_p = 6.5$;	$W_{net} = 224.6$ kJ/kg

$p_I/p_A = 2.51$; $p_R/p_A = 3.26$

6.7. In a conceptual Compressed Air Energy Storage scheme, the air pressure in the cavern is maintained constant at p_c during charging and discharging by means of a head of water applied from a water reservoir at a higher level. As a result of heat exchange with its environment, the temperature of the air within the cavern may also be assumed to remain constant at the ambient temperature T_a. To assist in achieving this, there is an aftercooler between the compressor and the cavern. As a means of reducing the required work input during charging, the compression process incorporates intercooling.

During the charging process, an intercooled compressor of *isothermal* efficiency $\eta_{isoth.}$ is driven by a motor-generator to draw a mass M of air from the atmosphere at p_a, T_a, compress it to p_c and deliver it to the cavern. During the discharging process, the same mass M of air from the cavern is passed through a heat exchanger, where it receives heat from an external source to raise its temperature to T_T at entry to the turbine. The latter exhausts at atmospheric pressure p_a and drives the motor-generator. The isentropic efficiency of the turbine is η_T and the efficiency of the motor-generator is η_{MG} in both the motoring and generating modes. The air may be treated as a perfect gas throughout the process.

Writing $r_p \equiv p_c/p_a$ and $\rho_p \equiv r_p^{(\gamma-1)/\gamma}$, show that the ratio of the electrical output W_G during the discharging process to the electrical input W_M during the charging process is given by the expression

$$\frac{W_G}{W_M} = \eta_{isoth.} \, \eta_T \, \eta_{MG}^2 \left(\frac{\gamma}{\gamma - 1} \right) \left(\frac{\rho_p - 1}{\rho_p} \right) \frac{T_T/T_a}{\ln r_p}.$$

Given that $\eta_{isoth.} = 0.75$, $\eta_T = 0.88$, $\eta_{MG} = 0.97$, $r_p = 45$, $t_T = 700$ °C and $t_a = 10$ °C, calculate the following quantities:

(a) The ratio W_G/W_M.

(b) The ratio of the net electrical output to the heat input from the external source, expressed as a percentage.

(c) The value of W_G/W_M, and the compressor exit temperature, were the compression process to be adiabatic, with an isentropic efficiency of 88%.

(d) The temperature at turbine exhaust were no heat to be added to the air between the cavern and the turbine.

Answer: (a) 1.30; (b) 18.8%; (c) 0.844, 643 °C; (d) −178 °C.

6.8. The compression process in Problem 6.7 is carried out in four adiabatic stages in series, with the same pressure ratio across each and with intercooling between stages. Assuming that the air is cooled to ambient temperature T_a at exit from each intercooler, and that the isentropic efficiency η_C is the same for all four stages, determine the value of η_C corresponding to the given value of $\eta_{\text{isoth.}}$. Neglect parasitic pressure drops.

Answer: 0.862.

Note: The *isothermal efficiency* of a compressor is defined as the ratio of the ideal work input for reversible, isothermal compression over the given pressure ratio to the actual work input to the compressor.

Problems 8.7 and 8.8 in Chapter 8, relating to gas-turbine cycles for nuclear power plant, provide further practice in gas-turbine cycle calculations.

CHAPTER 7

Advanced steam-turbine plant

It was noted in the introduction to Chapter 2 that steam power stations exist to produce electrical power at the least possible cost. This chapter discusses the more complex cycles that have resulted from the consequent drive for higher efficiency.

7.1. Limitations of the simple steam cycle

It was seen in §6.1 that the simple gas-turbine cycle suffers from the defect that the mean temperature of heat reception $\bar{T}_B$ is well below the highest temperature in the cycle T_b, and the mean temperature of heat rejection $\bar{T}_A$ is well above the lowest temperature T_a.

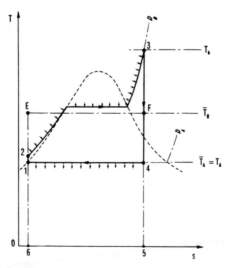

Fig. 7.1. Mean temperature of heat reception $\bar{T}_B$ in the Rankine cycle. Area EF56E = Area 23562 = heat supplied, Q_B.

It is one of the advantages of the simple steam cycle over its gas-turbine counterpart that in the condensing system cycle all the heat is rejected at the lowest temperature, so that $\bar{T}_A = T_A = T_a$. Figure 7.1 shows, however, that $\bar{T}_B$ is much lower than T_b, particularly when the steam is superheated. Consequently, the thermal efficiency of the Rankine cycle is appreciably less than that of a Carnot cycle operating between the same extreme limits of temperature, T_A and T_b, for, by a similar derivation to that given in §6.1, it can be seen that

$$\text{exact } \eta_{\text{RANK}} = \left(1 - \frac{T_A}{\bar{T}_B}\right), \tag{7.1}$$

while

$$\text{limiting } \eta_{\text{CARNOT}} = \left(1 - \frac{T_A}{T_b}\right).$$

This chapter is principally concerned with ways in which $\bar{T}_B$ can be made to approach T_b more closely. As in the case of gas-turbine cycles, this involves a study of more complex regenerative and reheat cycles. Before examining these, however, the effect on the cycle efficiency of advances in the individual terminal steam conditions will be briefly studied, since these advances are themselves responsible for improving the efficiency. At this point the reader would do well to study Appendix B, which sets out the advance in operating conditions in British power stations over the years, and shows the resulting effect on performance.

Since the performance of the steam cycle under different operating conditions is not so readily amenable to direct mathematical analysis as that of the gas-turbine cycle, it must necessarily be discussed in general terms, though some analytical work will be possible. We shall now briefly examine in turn the effect on the cycle efficiency of an advance in any one of the terminal steam conditions when the others are kept constant, at the same time giving an indication of the factors which currently limit further advance.

7.2. The effects of advances in terminal steam conditions

7.2.1. Condenser vacuum

The higher the condenser vacuum (i.e. the lower the turbine exhaust pressure p_A) the lower is the saturation temperature of the condensing steam in the condenser. This is the temperature at which heat is rejected in the cycle, so that the higher the vacuum the lower is the temperature of heat rejection and consequently the greater is the ideal cycle efficiency.

A lower limit to the condensing temperature is set by the inlet temperature of the circulating water and the economic size of the condenser; the design of the last turbine stage may also influence the selection of the operating vacuum at the design load. The condensing temperature may be made to approach more closely to the circulating water inlet temperature by providing greater surface area for condensation and a greater flow rate of circulating water in the condenser; the former will increase the initial capital cost and the latter the running costs. A difference of between 11 K and 14 K may be considered economically suitable. Cooling towers[†] are used to cool the circulating water when supplies are restricted. Since the water is then continuously recirculated, the average inlet temperature over the year is higher than that for river or sea water, being perhaps 21 °C instead of 13 °C. The reader may check that these figures together explain the vacuum figures given in Appendix B.

For the usual range of design steam conditions, a net gain in efficiency of 4–5% would result if the absolute pressure in the condenser were reduced from $6.8 \, \text{kN/m}^2$ to $3.4 \, \text{kN/m}^2$ (corresponding, in British units, to an increase in condenser vacuum from 28–29 inHg). This is a handsome return but, for the reasons given above, little, if any, further improvement in vacuum can be looked for.

†See Problem 16.5 in Refs. 5, C and E.

7.2.2. Initial steam temperature

Figure 7.2(a) shows the effect on the Rankine cycle of an increase in steam temperature T_b at turbine inlet when the initial pressure and the condenser vacuum remain unaltered. It is clear that the mean temperature of heat reception $\bar{T}_B$ is thereby raised, so that the ideal cycle efficiency is

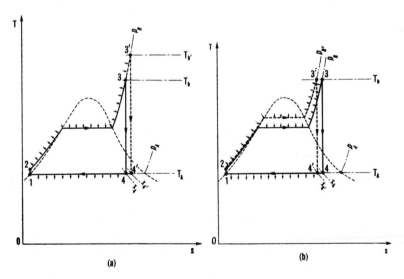

(a)   (b)

FIG. 7.2. Effect of increasing (a) the initial temperature T_b and (b) the initial pressure p_B in the Rankine cycle.

increased; this is achieved, however, by raising T_b without bringing $\bar{T}_b$ any nearer to T_b.

When considering real plant, factors that do not appear in ideal reversible cycles have to be taken into account.[13] One such factor of importance is the effect on the turbine isentropic efficiency (which is always 100% in ideal cycles) of the reduced wetness in the later stages of the turbine when the initial steam temperature is increased. This reduced wetness is indicated in Fig. 7.2(a). In a real plant this would result in a higher isentropic efficiency and so to a gain in cycle efficiency additional to that resulting from the increase in $\bar{T}_B$. Ideal cycle calculations are consequently useless in assessing real-life gains; in this particular case they would seriously underestimate the gain due to a given increase in temperature.

For typical steam conditions, an actual improvement approaching 7% would be obtained by increasing the design initial temperature from 450 °C to 550 °C, and the gain with further increase in temperature would be an approximately linear function of temperature. This is again a handsome return, but in practice the temperature cannot be allowed to exceed about 565 °C without introducing high-cost austenitic steels. At this point, the reader is once more referred to Appendix B.

7.2.3. Initial steam pressure

At first sight there might appear to be no thermodynamic advantage in raising the steam pressure p_B at turbine inlet. However, below the critical pressure, an increase in pressure is accompanied by an increase in saturation temperature. Figure 7.2(b) shows the effect of this on the Rankine cycle when the initial temperature T_b and the condenser vacuum remain unaltered. It is clear that the mean temperature of heat reception $\bar{T}_B$ is thereby increased and so brought nearer to the top temperature T_b. There is a consequent increase in the ideal cycle efficiency. Again, however, other factors have to be taken into account in real-life plant,[13] and Fig. 7.2(b) shows that with an increase in initial pressure there is now an increase in the wetness of the steam in the later stages of the turbine, and this has an adverse effect on the turbine isentropic efficiency and so on the cycle efficiency of an actual plant. For this and other reasons,[13] ideal cycle calculations would seriously overestimate the gain due to a given increase in pressure.

It is not realistic to quote a figure for the gain in cycle efficiency due to a given increase in pressure without noting that such an increase would always be accompanied in practice by an increase in the design final feed temperature, for a reason which will become apparent when regenerative feed heating is considered in the next section. The figures quoted below consequently refer to the effect of an increase in initial steam pressure when accompanied by the appropriate increase in economic final feed temperature. In these circumstances, for typical steam conditions, an actual improvement in cycle efficiency of about 3% could be obtained by increasing the design initial pressure from 5 to $8\,\mathrm{MN/m^2}$. With further increase in pressure, however, the gain would not be a linear function of pressure; indeed, a further increase from 8 to $11\,\mathrm{MN/m^2}$ would provide a gain of something under $1\frac{1}{2}\%$ (Fig. C.1 of Appendix C). The appearance of this law of diminishing returns arises from the way in which the increase in $\bar{T}_B$ for a given increment of pressure is affected both by the smaller enthalpy of evaporation (latent heat) and the smaller rate of rise of saturation temperature with pressure as the pressure rises towards the top of the saturation "dome" in Fig. 7.2(b). Together with other factors, it is this law of diminishing returns that at present makes problematical the economic viability of supercritical pressure plant.

A study of Fig. C.1 also shows that, in a non-reheat cycle, the pressure cannot be increased beyond a certain point, without a simultaneous increase in initial temperature, if the exhaust wetness is not to exceed the value of about 15% beyond which turbine blading erosion becomes excessive. Thus the shaded part of Fig. C.1 is a prohibited area for non-reheat plant.

Since it is important to realise that the conditions for which a power plant is to be designed depend as much on economic as on thermodynamic considerations, Appendix C sets out a simple calculation to illustrate a method by which it is possible to assess the economic limit for the design initial pressure. Similar considerations would determine the economic values for the other design parameters.

FEED HEATING

7.3. Regenerative feed-heating

It has been seen that increasing the initial pressure while keeping the initial temperature constant increases the ideal cycle efficiency by bringing the mean temperature of heat reception $\bar{T}_B$ nearer to the top temperature T_b. A further way of doing this is by the introduction of regenerative feed heating. The Rankine cycle, though reversible internally, suffers from the low temperature at which the returning feed water enters the boiler, for this results in a value of $\bar{T}_B$ much below T_b. This defect can be remedied by using steam to preheat the feed water internally within the cycle by means to be discussed, so raising the mean temperature of heat reception from the external heat source and thus increasing the cycle efficiency. Maximum advantage will be obtained when this is done reversibly. Such ideal, reversible regenerative cycles will first be discussed, and then more realistic practical cycles.

7.4. Reversible feed-heating cycle using dry saturated steam from the boiler

To be reversible, the heat transfer from steam to feed water within the cycle must take place across an infinitesimally small temperature difference. In the hypothetical plant now to be described, this is achieved by passing the feed water coming from the condenser through an infinite number of coils, the coils being placed between successive pairs of an infinite number of turbine stages in the manner indicated diagrammatically in Fig. 7.3. The feed water now enters the boiler at temperature $T_F = T_4$ instead of at T_2, and heat reception from outside the cycle occurs between T_F and T_b. The mean temperature of heat reception is clearly considerably higher and there are no accompanying irreversibilities, so that there is a consequent gain in cycle efficiency.

From a study of Fig. 7.3(b) a simple expression for the cycle efficiency can be written down. Between points 6 and 7 the steam is alternately expanded isentropically in a turbine stage and condensed at constant temperature on the outside of the succeeding heater coil. The steam-expansion line between 6 and 7 consequently takes the staircase form shown in the inset; in the limit, with an infinite number of stages, this staircase becomes the smooth line 67, which is parallel to the feed-water line 23 since, as is shown in the figure, $\delta s_w = -\delta s_s$. Thus $(s_6 - s_7) = (s_3 - s_2)$, so that $(s_7 - s_1) = (s_6 - s_3)$. Hence the heat rejected $Q_A = T_A(s_7 - s_1) = T_A(s_5 - s_4)$ and the exact cycle efficiency is given by

$$\eta_{CY} = \frac{Q_B - Q_A}{Q_B} = \frac{(h_5 - h_4) - T_A(s_5 - s_4)}{(h_5 - h_4)} = \frac{b_5 - b_4}{h_5 - h_4}, \qquad (7.2)$$

where $b \equiv (h - T_A s)$, the specific steady-flow availability function with an environment at T_A.

Equation (7.2) is of precisely the same form as eqn. (2.8) for the exact Rankine cycle efficiency, though of course the efficiency of the regenerative cycle is greater than that of the Rankine cycle for comparable conditions. Equation (7.2) could be written down directly by noting that the plant contained within control surface Y in Fig. 7.3(a) is an open-circuit steady-flow work-producing device which takes in fluid at state 5 and delivers it at state 4, in which all processes are internally reversible and in which the heat transfer to the environment at T_A occurs when the fluid is at T_A, so that the heat exchange with the environment is reversible. There are thus no lost opportunities for producing work, and the net work output of the plant $[W_{\text{net}} = (W_T - W_{P_1} - W_{P_2}) = (Q_B - Q_A)]$ is consequently equal to the available energy, in accordance with the result derived in §§A.9 and A.10 of Appendix A.

It is also instructive to note from Fig. 7.3(b) that, since $(s_7 - s_1) = (s_6 - s_3)$, the entire cycle can be regarded as a hypothetical

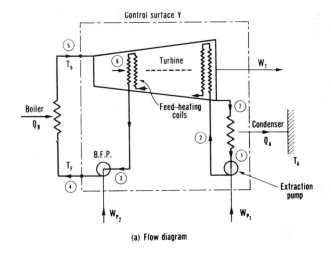

(a) Flow diagram

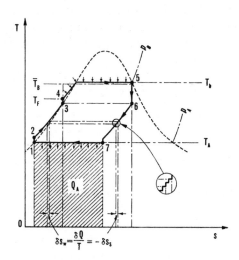

(b) Temperature–entropy diagram

FIG. 7.3. Reversible, regenerative cycle for saturated steam, without steam
extraction.

Carnot cycle 367123 on which is superposed a hypothetical Rankine cycle
34563. It is clear that the greatest advantage of regenerative feed heating in
this type of plant would be obtained if the coils were placed throughout the
turbine to give $T_F = T_b$, since all heat supply would then occur at the top
temperature T_b. The plant would then have the limiting Carnot efficiency

$(1 - T_A/T_b)$. The reader may readily confirm that eqn. (7.2) would give this result in these circumstances.

7.5. Reversible feed-heating cycle using superheated steam from the boiler

If, for the plant illustrated in Fig. 7.3(a), the steam supplied by the boiler is superheated, then there must unavoidably be irreversibility arising from a finite temperature difference between the steam and the feed water passing through the heating coils in that part of the turbine in which the steam remains superheated. Reversibility can be obtained, however, if the feed-water heating takes place in heaters external to the turbine and the superheated steam bled from the turbine is compressed reversibly and isothermally on its way to each heater. A hypothetical plant of this kind is shown in Fig. 7.4(a), and the temperature–entropy diagram in Fig. 7.4(b). The plant incorporates direct-contact (DC) heaters throughout; in these the steam is condensed by direct contact with feed water sprayed into the heater vessel, from which the liquid is pumped forward into the next heater in the train.

In this cycle the steam expands isentropically throughout the turbine, and if bled from the turbine above the point S it must pass through a reversible, multi-stage, intercooled compressor on its way to the heater.

Thus, for the infinitesimal stage shown in Fig. 7.5, the steam must be bled at point M and be brought in the compressor to state N before entering the heater. The process from M to N is represented in Fig. 7.5(b) by the saw-tooth of alternate constant-pressure cooling and isentropic compression in the multi-stage compressor, the cooling being carried out by jacket feed-water as indicated; in the limit, with an infinite number of stages in the compressor, the process from M to N would be isothermal and reversible. In this way, the steam supplied to the heater is at a temperature only an infinitesimal amount higher than that of the incoming feed water, so that in the limit the complete process is reversible. Without isothermal compression of the bled steam, it would have to be bled from the turbine and supplied to the given heater at point M' instead of M: there would then be a large difference between the temperature $T_{M'}$ of the steam and the temperature T_K of the incoming feed water, and resulting irreversibility.

When there are an infinite number of stages, and the exhaust steam is wet, the expression for the cycle efficiency of the plant shown in Fig. 7.4 is again of the same form, since the plant within control surface Y is again a completely reversible open-circuit steady-flow device taking in fluid at state 3, delivering it at state 2 and exchanging heat reversibly with the environment at T_A. Thus, the net work is equal to the available energy, and the efficiency is

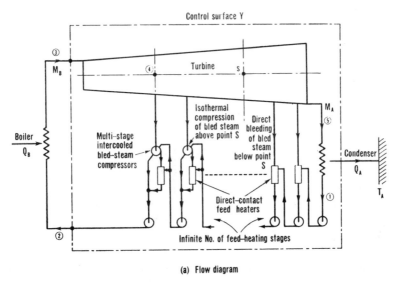

(a) Flow diagram

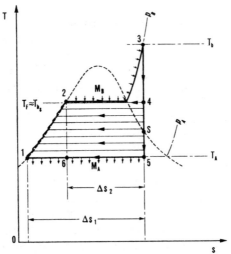

(b) Temperature–entropy diagram

FIG. 7.4. Idealised extraction regenerative cycle using superheated steam and feed-heating to the boiler saturation temperature, T_{b_s}.

$$\eta_{CY} = \frac{b_3 - b_2}{h_3 - h_2}. \tag{7.3}$$

Since $b \equiv (h - T_A s)$, this is equal to the efficiency of a hypothetical reversible cycle 23562 in Fig. 7.4(b). [It is left to the reader to show from

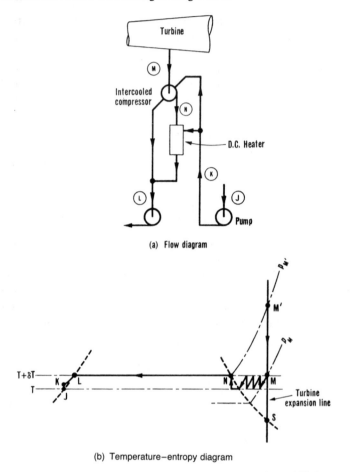

(a) Flow diagram

(b) Temperature–entropy diagram

FIG. 7.5. Infinitesimal ideal feed-heating stage using superheated bled steam.

this that the flow rate through the boiler per unit flow through the condenser is equal to $\Delta s_1/\Delta s_2$ (Problem 7.1).] Although the feed water is raised to the boiler saturation temperature, the efficiency is now less than the limiting Carnot efficiency, since the heat is supplied in the boiler between T_F and T_b, and not all at T_b. A means of carrying the regenerative principle further in order to try and achieve Carnot efficiency for a superheated steam cycle is discussed in §9.3.

7.6. Reversible feed-heating cycles using surface feed heaters

Instead of a feed-heating train of direct-contact heaters with their associated pumps, a train of idealised tubular *surface heaters* could be used, each having negligible terminal temperature differences between the two fluids (i.e. each of infinite heat-exchange area). For complete reversibility, the condensed bled steam, or drain water, from a heater would have to be cascaded down to the next lower-pressure heater in the train via a reversible drain-water turbine, for without this there would be a lost opportunity for producing work. This arrangement is sketched in Fig. 7.6.

The reader may obtain some profit by drawing, for an infinitesimal feed-heating stage of this kind, a temperature–entropy diagram similar to that of Fig. 7.5(b) [Problem 7.2]. The net work from a hypothetical plant such as this would again be equal to the available energy.

7.7. Summary of results for ideal feed-heating cycles

It has been seen that the ideal regenerative cycle is one in which all processes are reversible, requiring an infinite number of feed-heating stages and reversible isothermal compression of the bled steam when it is superheated. Providing this condition of complete reversibility is fulfilled then, whether the system incorporates ideal direct-contact or ideal surface heaters, or a combination of both, the cycle efficiency is given by

$$\eta_{CY} = \frac{\text{available energy}}{\text{heat input}}. \tag{7.4}$$

Furthermore, this efficiency increases continuously with increase in final feed temperature, the limit being reached when the feed water is raised to the saturation temperature T_{bs} corresponding to the boiler pressure.

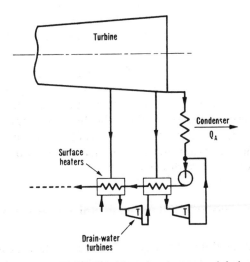

FIG. 7.6. Ideal feed-heating train with surface heaters and drains cascaded.

7.8. Practical feed-heating cycles with a finite number of heaters

A practical regenerative cycle can afford to have only a relatively small number of feed-heating stages. Because the resulting increases in efficiency are not of sufficient magnitude to warrant the extra complication, it is also not an economic proposition to introduce either compressors into the superheated bled steam lines or drain-water turbines into cascaded heater drain lines. Practical cycles therefore have a relatively simple circuit comprising a train either of direct-contact heaters, with their associated pumps, or of surface heaters with orifices in the drain lines in place of drain-water turbines, or, more usually, a combination of both types of heater; any heater placed between the high-pressure boiler feed pump and the boiler must necessarily be of the tubular type. For the sake of simplicity, only a train of direct-contact heaters with their associated pumps will be treated here, but a similar analysis may be made for a train of surface heaters or a combination of both.[14] The enthalpy rise of the feed water in each of the pumps will be neglected, since it is small compared with the enthalpy rise in a heater; this is analogous to neglecting the feed-pump term in the efficiency calculation for the Rankine cycle.

Since practical cycles are being considered, the flow through the turbine is no longer treated as being isentropic, although it is still assumed that stray heat losses are negligible and that there are no pressure drops in the boiler and condenser. We are therefore confronted with the analysis and optimisation of a cycle such as that shown on the temperature–entropy diagram of Fig. 7.7, in which n DC heaters are used to raise the feed water to an arbitrary final feed temperature T_F.

Before commencing the analysis, we may recall the statement in §7.3 that the introduction of regenerative feed heating improves the cycle efficiency because it raises the mean temperature of heat reception $\bar{T}_B$. However, with only one heater, the higher the chosen final feed temperature the greater will be the difference between the temperature of the bled steam at inlet to the heater and the temperature of the entering feed water; hence the greater will be the irreversibility of the process occurring within the heater and so the greater the loss of gross work output[†] due to this irreversibility. This will tend towards a greater reduction in cycle efficiency the higher the final feed temperature. With two such opposing tendencies, there will thus be an optimum final feed temperature somewhere between the condenser temperature and the saturation temperature at boiler pressure. The greater the number of feed heaters, the more will the degree of irreversibility in the heat-exchange

---

[†] This loss may be calculated by evaluating the entropy creation ΔS_c due to irreversibility within the heater, in accordance with Theorem 2 of §A.4 in Appendix A. The reader may test his understanding of these ideas by working through Problems 7.5, 7.6 and 7.7.

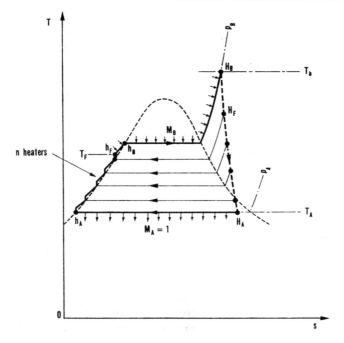

Fig. 7.7. Temperature–entropy diagram for practical feed-heating cycle with
n heaters.

processes in the heaters be reduced. We might therefore expect the
optimum final feed temperature to be higher as the number of feed heaters
is increased. Figure 7.9 shows that this is indeed so. In the hypothetical
case in which there were an infinite number of heaters, it would be
thermodynamically worthwhile to carry the feed heating right up to boiler
saturation temperature. This is confirmed by the analysis that follows.

7.9. Calculation of boiler flow rate per unit flow to the condenser

The first step in the analysis is to find the relation between the flow rates
through the boiler and condenser, since this enables the ratio of the
quantities of heat supplied and rejected to be calculated. To do this an
expression is first found for the ratio of the outflow and inflow rates of feed
water for any heater *j* depicted in Fig. 7.8.

An energy balance for this heater gives

$$m_j(H_j - h_j) = M_i(h_j - h_i),$$

where, for convenience, H in this instance is used to denote the **specific**
enthalpy of steam and h the specific enthalpy of feed water.

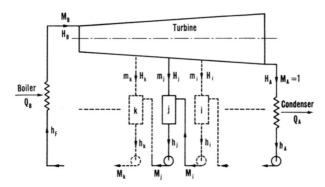

FIG. 7.8. Flow diagram for a plant with a train of DC heaters.

Thus

$$m_j = \frac{r_j}{\beta_j} M_i,$$ (7.5)

where $r_j \equiv (h_j - h_i)$,   the specific enthalpy rise of the feed water in
heater j,

   $\beta_j \equiv (H_j - h_j)$,   the specific enthalpy decrease of the bled
steam in heater j.

But

$$M_j = M_i + m_j,$$

$$\therefore \frac{M_j}{M_i} = \gamma_j,$$ (7.6)

where

$$\gamma \equiv \left(1 + \frac{r}{\beta}\right).$$ (7.7)

Equation (7.6) is a recurrence relation applicable to any two adjacent
heaters, so that the boiler steam flow M_B per unit mass flow to the
condenser is given by

$$M_B = \prod_{j=1}^{j=n} \gamma_j,$$ (7.8)

where

$$\prod_{j=1}^{j=n} \gamma_j$$

signifies the product of the γ terms for all the heaters from 1 to n.

7.10. Calculation of cycle efficiency and heat rate

Having calculated M_B, the cycle efficiency is directly calculated from the relation

$$\eta_{CY} = \left(1 - \frac{Q_A}{Q_B}\right) = 1 - \frac{H_A - h_A}{M_B(H_B - h_F)}. \qquad (7.9)$$

In practice it is more usual to specify the heat consumption rate, briefly described as the *heat rate*; this is the ratio of the heat supplied Q_B to the net work output W_{net}. When both are expressed in the same units we shall denote this by C and then

$$C = \frac{1}{\eta_{CY}}. \qquad (7.10)$$

In the British system of units, it was more usual to express the net work output in kW h, and then

$$\text{Heat rate} = \frac{3412}{\eta_{CY}} \text{ Btu/kW h.} \qquad (7.11)$$

The guaranteed performance given by the turbine manufacturer is usually concerned only with the turbine output W_T and not with W_{net}. The guaranteed figure is then described, somewhat misleadingly, as the *turbine heat rate*. In the foregoing calculation, the heat rate and the so-called turbine heat rate happen to be equal because the work inputs to the pumps have been neglected.

The calculations in this and the previous section can be performed if the final feed temperature T_F, and the allocation amongst the individual heaters of the total rise in temperature or enthalpy of the feed water, have been specified, for when the feed water outlet temperatures from the heaters have been specified the bled steam pressures must be those having saturation temperatures equal to the respective heater outlet temperatures. We shall next be concerned first to find the optimum division of the total enthalpy rise amongst the individual heaters when the final feed temperature is specified, and then to find how to specify the optimum final feed temperature. Lastly, we shall have to consider what determines the choice of the number of heaters.

7.11. Optimum division of the total enthalpy rise amongst the individual heaters

The optimum conditions are those which give maximum cycle efficiency. When the steam conditions and final feed temperature have been specified, the only variable in eqn. (7.9) is M_B. Thus the efficiency will be a maximum when M_B is a maximum and therefore, from eqn. (7.8), when the product of the γ's for all the heaters has its greatest value. The requirements to meet this condition may be deduced by considering **any two** adjacent heaters f and g in a train of heaters such as that of Fig. 7.8, in which the bled steam pressures for all heaters other than for heater f are fixed. Under these conditions, the values of γ for all heaters other than f and g will remain unaltered when the bled steam pressure for heater f is changed by altering the point in the turbine at which this tapping is taken (provided it is assumed that this change does not alter the turbine expansion line). Thus the efficiency will be a maximum when the product $\gamma_f\gamma_g$ is a maximum.

As the position of the tapping point for heater f is changed, the enthalpy rise r_f of the feed water in heater f will change, while the total enthalpy rise R in heaters f and g will remain unaltered since the tapping points for heaters e and g are not varied. Under these conditions

$$\gamma_f = \left(1 + \frac{r_f}{\beta_f}\right)$$

and

$$\gamma_g = \left(1 + \frac{R - r_f}{\beta_g}\right).$$

Thus

$$\gamma_f\gamma_g = \left(1 + \frac{r_f}{\beta_f}\right)\left(1 + \frac{R - r_f}{\beta_g}\right), \tag{7.12}$$

and this will be a maximum when

$$\frac{d(\gamma_f\gamma_g)}{dr_f} = 0. \tag{7.13}$$

This differentiation may be carried out on eqn. (7.12) as it stands to give an expression which may be used in trial-and-error calculations for the optimisation of the system.[15] For the present purpose, however, a more instructive result is obtained by making a simplifying approximation which is valid for non-reheat plant.[14] For the usual range of operating conditions in such plant, the value of β does not vary greatly from tapping point to tapping point along a given turbine expansion line: the reader may check for himself, by drawing lines of constant β on an enthalpy–entropy chart, that this is fortunately more nearly true for an actual turbine expansion line than for an isentropic expansion (Problem 7.3). Thus, to a sufficient degree of approximation,

$$\beta_f = \beta_g \equiv \beta,$$

and it is then easy to show that eqns. (7.12) and (7.13) give the simple result

$$r_f = \frac{R}{2}. \tag{7.14}$$

Thus, to this degree of approximation, the efficiency is a maximum when the feed water enthalpy rises across any two adjacent heaters are equal. It follows that, **for maximum efficiency in a non-reheat plant, the enthalpy rises should, to a first approximation, be the same in all heaters**. Since it is rarely possible to provide tapping points on the turbine which would give exactly the bled steam pressures required to fulfil this condition, it is fortunate for the turbine designer that the cycle efficiency is not greatly sensitive to departures from this optimum division of enthalpy rises.

A similar, though more complicated, analysis may be made for reheat plant.[15] It is then found that, for two heaters whose bled-steam tapping points are respectively upstream and downstream of a reheater, the enthalpy rise of the feed water in the downstream heater **should be greater than that in the upstream heater** (Problem 7.12). Thermodynamically, the need for this arises from the greater superheat at the downstream tapping point which results from reheating.

7.12. Optimum final feed temperature

Referring to Fig. 7.7, the previous section has shown that if the final feed temperature is fixed at some arbitrary value T_F, then, for maximum efficiency in a non-reheat plant, the total enthalpy rise in the feed-heating system from h_A to h_F must, to a first approximation, be divided equally amongst the n heaters; this implies approximately equal temperature rises. The optimum value of T_F can be found by determining the optimum value of $(h_B - h_F)$, the enthalpy rise of the feed water in the economiser section of the boiler. This can be deduced from the foregoing study without further analysis, by noting that the cycle efficiency would be exactly the same if, instead of heating the feed water from h_F to h_B in the economiser, it were raised to h_B in a further DC heater taking its bled steam direct from boiler outlet; that extra bled steam would do no work in the turbine and would require exactly the same amount of heat to produce it in the boiler as would be required to heat the feed water in the economiser of the actual plant. From the foregoing study it is evident that, for maximum efficiency, the enthalpy rise in this imaginary heater, and therefore in the actual economiser, would have to be equal to the enthalpy rises in all the other heaters. Thus, **for maximum efficiency in a non-reheat plant, the enthalpy rises should, to a first approximation, be the same in all heaters and in the economiser**.

An alternative way of specifying the optimum final feed temperature is clearly that it must be such as to satisfy the relation

$$x_n = \frac{n}{n+1},\qquad(7.15)$$

where x is the fraction of the maximum possible enthalpy rise of the feed water $[= (h_F - h_A)/(h_B - h_A)]$, and the subscript n denotes the optimum value of x for n heaters. This agrees with the conclusion stated in §7.7 that the optimum value of x is unity for a system having an infinite number of heaters.

7.13. Gain in efficiency due to feed heating

In §7.12, x expresses the enthalpy range of feed heating non-dimensionally as a fraction of the greatest possible range. It is similarly convenient to express the gain due to feed heating non-dimensionally by expressing it as a fraction of the greatest gain for the given steam conditions. It has already been seen that this is obtained when heating to the boiler saturation temperature with an infinite number of heaters. For purposes of analysis it is found to be more convenient to work in terms of reduction in heat rate rather than increase in cycle efficiency, so that the gain due to feed heating is expressed non-dimensionally by

$$y \equiv \frac{C_0 - C}{C_0 - C_\infty},\qquad(7.16)$$

where, for the given steam conditions, C_0, C and C_∞ are respectively the heat rates when there is no feed heating, when feed heating with a finite number of heaters to any specified temperature, and when feed heating with an infinite number of heaters to the boiler saturation temperature.

The maximum gain when feed heating with n heaters occurs at a value of x equal to x_n. Denoting the corresponding value of y by y_n, it is found that, to a first approximation,[14] this is given by

$$y_n = \frac{n}{n+1},\qquad(7.17)$$

an expression identical to that for the optimum value of x, so that a plot of y against x takes the form of the curves shown in Fig. 7.9.

This is an approximate set of curves which are valid over the full range of steam conditions found in non-reheat plant. More exact calculations, taking account of the variation in β,[16] do not result in marked departures from these curves. In practice, exact calculations would need to be made. These are extremely tedious, but the labour of calculation may be avoided by the use of electronic digital computers. The analysis presented here paints a broad picture which serves to illustrate the kind of results that would come out of such detailed calculations. The optimum gain with a system having five heaters would be about 10%, so that feed heating proves universally attractive in power-station practice.

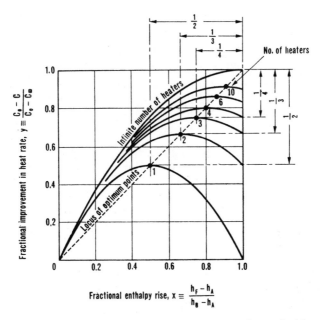

Fig. 7.9. Dimensionless plot of improvement in heat rate due to feed heating in non-reheat plant.

7.14. Choice of the number of feed-heating stages

There remains one important question to be discussed—how to select the number of feed heaters for a given installation. This is a problem similar to that of deciding the economic steam conditions (Appendix C), in that the decision depends on economic as well as on thermodynamic considerations. Such a calculation is outside the scope of the present volume, but attention is drawn to the remarkable fact, illustrated by Fig. 7.9, that the optimum gain due to a single heater is about one-half of that due to an infinite number of heaters. Thus even a few heaters are well worth while. It will also be noted from Fig. 7.9 that, as the number increases, the extra gain due to the addition of a further heater falls off drastically. This accounts for the fact that the number of heaters actually installed does not need to be more than that given in Table B.1 of Appendix B. Furthermore, with too many heaters pipework complications would become prohibitive.

7.15. Subsidiary effects of feed heating

Apart from the effect on the cycle efficiency, the introduction of feed heating has the following advantageous effects on the design of the turbine:

(1) For a given turbine output, the mass flow rate at turbine inlet is increased and the mass flow rate at turbine exhaust is decreased, thus leading both to an increase in height where the blades tend to be excessively short and to a decrease where they tend to be excessively long.

(2) The bled-steam belts in the low-pressure end of the turbine help to act as water-drainage belts, and so to ease erosion problems due to excessive moisture in these stages.

The introduction of feed heating also has an influence on the boiler, for the flue gases passing over the economiser before being discharged from the boiler cannot be cooled to the same extent as in a plant without feed heating. If no corrective step were taken, the increase in η_{CY} would consequently tend to be offset by a decrease in the boiler efficiency η_B. An air preheater is therefore introduced, enabling the flue gases to be further cooled after leaving the economiser by transferring heat to the combustion air fed to the boiler.

REHEATING

7.16. Reheating in the non-regenerative steam cycle

Since feed heaters are incorporated in all large-scale steam power plant, a study of the application of reheating to the non-regenerative cycle is largely of academic interest only. It is easier to begin with this, however, and with a study of ideal, reversible cycles.

When discussing the application of reheating to gas-turbine plant, it was seen in §§6.8 and 6.9 that the cycle efficiency was not increased unless the plant also incorporated an exhaust-gas heat exchanger. This was because, without such a heat exchanger, the adverse effect of the resulting increase in the mean temperature of heat rejection $\bar{T}_A$ more than offset the advantageous effect of the accompanying increase in the mean temperature of heat reception $\bar{T}_B$. In the case of steam plant, the steam entering the condenser at the design load is always wet, so that the introduction of reheating does not affect the temperature of heat rejection, which remains at the saturation temperature corresponding to the condenser vacuum. Hence reheating will increase the ideal cycle efficiency if it results in an increase in $\bar{T}_B$, and this it does provided that the chosen pressure at which the steam is reheated is not too low. The necessity for this proviso is seen from a study of Fig. 7.10.

Cycle 12371 is a simple Rankine cycle. Cycle 1234561 is a reversible, single-reheat cycle in which the steam is expanded in the turbine to point 4, and is then reheated in the boiler from 4 to 5 at a fairly high reheat pressure p_R before returning to the turbine, in which expansion then continues

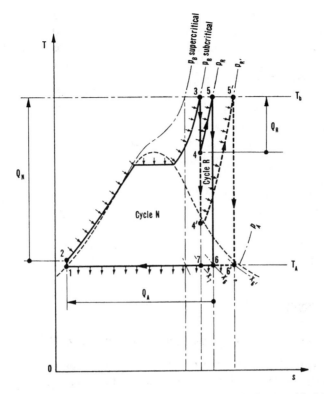

FIG. 7.10. The effects of variation in reheat pressure, and of supercritical initial pressure.

down to the exhaust pressure. Whether or not, in these circumstances, reheating improves the cycle efficiency can best be decided by noting, as in §6.8, that the complete reheat cycle may be treated as a compound cycle comprising the **non-reheat** cycle N (12371) of efficiency η_N, to which has been added a hypothetical cycle R (45674) of efficiency η_R. Again as in §6.8, the thermal efficiency η_{CY} of the complete reheat cycle is given by

$$\left(\frac{\eta_{CY}}{\eta_N} - 1\right) = \frac{\left(\dfrac{\eta_R}{\eta_N} - 1\right)}{\left(\dfrac{Q_N}{Q_R} + 1\right)}. \tag{7.18}$$

From inspection of the areas representing respectively the net work and the heat supplied in cycles N and R, it is evident that for this particular value of p_R, $\eta_R > \eta_N$. It follows from eqn. (7.18) that $\eta_{CY} > \eta_N$, so that reheating at this pressure improves the cycle efficiency. On the other hand, if reheating is carried out at a much lower pressure $p_{R'}$, it is again evident by inspection that $\eta_{R'} < \eta_N$, so that $\eta_{CY} < \eta_N$; thus reheating at this lower pressure gives a decrease in cycle efficiency. It does not follow from this that reheating should be carried out at as high a pressure as possible, for eqn. (7.18) shows that the gain in efficiency, if any, is also dependent on the magnitude of Q_N/Q_R. Inspection of Fig. 7.10 shows that this ratio falls with decrease in reheat pressure. There are thus two opposing tendencies; reduction in reheat pressure leads to a lower value of η_R/η_N and so to a smaller gain in efficiency, while it also leads to a lower value of Q_N/Q_R and so to a greater gain in efficiency. There will consequently be an optimum reheat pressure at some point between p_B and p_A for which η_{CY} will be a maximum.

Detailed calculations show that for reversible, non-regenerative cycles this optimum reheat pressure is about one-quarter of the initial boiler pressure. For irreversible cycles, because of the reduced exhaust wetness with reheating evident in Fig. 7.10 and the resulting improvement in actual turbine efficiencies, there is an advantage in going to a somewhat lower reheat pressure, and the optimum may be as low as one-tenth of the initial pressure.[17] The efficiency, however, is not greatly sensitive to departures from the optimum reheat pressure (Problem 7.11); this is fortunate for the turbine designer, since the reheating must necessarily take place between one turbine cylinder and the next and the pressure there is unlikely to be exactly equal to the optimum reheat pressure.

Calculation of the efficiency of a reversible, non-regenerative reheat cycle is a straightforward matter, the efficiency being given by

$$\eta_{CY} = \left(1 - \frac{Q_A}{Q_N + Q_R}\right). \tag{7.19}$$

7.17. Reheating in regenerative steam cycles

When reheating is added to a feed-heating cycle, it is found that the optimum reheat pressure is higher and the percentage gain due to reheating less than when it is added to a non-regenerative cycle. The reason for this can be seen from eqn. (7.18) by noting that, when there is no reheating, η_N is greater and Q_N is less for the regenerative cycle than for the corresponding non-regenerative cycle. Thus η_R must be greater and Q_R less if maximum advantage is still to be obtained, and it is seen from Fig. 7.10 that both these requirements lead to the need for a higher reheat pressure. Detailed calculations[17] show that the optimum reheat pressure, instead of being about one-tenth, now lies between one-fifth and one-quarter of the initial boiler pressure, the optimum pressure being higher and the percentage gain due to reheating lower the more efficient is the initial non-reheat regenerative cycle. A gain of 4–5% may be expected for a single stage of reheating at normal operating conditions. These figures relate to irreversible cycles and take account of pressure drop in the reheater, the effect of which cannot be neglected.

The adoption of reheating results in a very high degree of superheat at the bled steam points immediately following the reheater. To avoid the resulting irreversibility, less highly superheated steam at the same pressure may be bled from an auxiliary turbine driving the feed pump instead of from the main turbine. Such an auxiliary feed-pump turbine might take its steam from the main HP turbine exhaust line, before the reheater, and itself exhaust into the line between the intermediate and low-pressure cylinders of the main turbine. In that event, two of the heaters in the feed-heating train could use steam bled from the feed-pump turbine, instead of from the main turbine. The optimisation of such cycles is still more involved, and in any case optimisation calculations for reheat cycles are so time-consuming that they are best performed on digital computers.

7.18. Further factors relating to reheating

The improvement in cycle efficiency is not the only advantage of reheating; an equally important advantage is the resulting decrease in wetness at the turbine exhaust which is evident in Fig. 7.10. On both the score of efficiency and of exhaust wetness, it has been found advantageous to have two stages of reheating when the boiler supplies steam at supercritical pressure (Appendix B). Thermodynamically, it would, of course, be most advantageous to use progressive reheating, as in the ideal gas-turbine cycle of Fig. 6.10, but it is found to be uneconomic and impracticable to have more than two stages. With two stages, the optimum ratio of the second reheat pressure to the first is found to be about the same as that of the first reheat pressure to the initial pressure, namely about one-quarter to one-fifth.

Reheating was not greatly favoured when boiler reliability was such as to require the installation of more than one boiler per turbine, since the resulting complication of the interconnecting pipework proved unattractive. When the increased reliability of both boilers and turbines, the unit system of operation (one boiler per turbine) is becoming universal in power-station practice, and reheating is consequently favoured for large machines operating at advanced steam conditions (Appendix B).

It will be seen in Table B.1 in Appendix B that, for plant operating at supercritical pressures, two stages of reheating have been favoured. Turbines in the USA which are not listed in Table B.1 include two double-reheat plants with inlet steam conditions respectively of 31.0 MN/m^2, 621 °C for Philo Unit G and 34.5 MN/m^2, 649 °C for Eddystone Unit 1. However, those turbines, with such high steam conditions, were pioneering plant. A 700 MW double-reheat plant under construction in Japan, for operation at similar advanced supercritical pressure, has reverted to more modest steam temperatures (see Table B.1). The gains in efficiency to be expected from such advanced steam conditions are depicted in Fig. 7.11, taken from a Japanese study.[F]

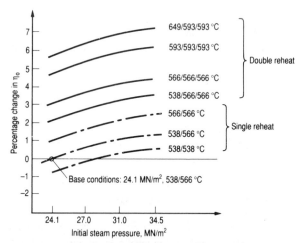

FIG. 7.11. Change in overall efficiency with variation in steam parameters in advanced supercritical, reheat plant. [After A. Suzuki *et al.*[F]]†

The single UK supercritical plant listed in Table B.1, of which two were installed at Drakelow, has not been followed by others. Indeed, at the dawn of the nineties, it was even suggested by the Chairman of the Central Electricity Generating Board (CEGB) in Britain that the era of very large turbines in central power stations might be drawing to a close. No doubt that prospect was influenced by a sudden surge of interest in combined gas/steam plant of the kind discussed in Chapter 9. The primary impulse for that surge of interest was provided by a combination of environmental considerations and the impending privatisation, at the time, of electricity generation and supply in the United Kingdom. That situation, driven by political dogma, made it very difficult to forecast future trends.

COGENERATION (CHP)

7.19. Cogeneration plant

The plants so far considered in this chapter have been of the type used in the large-scale generation of electrical power in stations operating essentially on base load. However, there are many applications in which steam-turbine plant are used for the combined supply of power and process

†References denoted by capital letters are given in the list of **Additional References** appearing at the end of the book.

steam. With the drive towards greater overall efficiency in the cause of energy saving, this type of plant has become of increasing interest. At the same time, such *Combined Heat and Power* (CHP) plant have long featured prominently in continental countries for *district heating*, particularly in the USSR. The remaining sections of this chapter are devoted to a study of this kind of steam plant.

Cogeneration plant may supply:

(a) process steam for industrial processes,
(b) steam to heat water for central or district heating,

though a plant can, of course, be designed to supply steam for both. In both cases, a steam turbine would provide the source of power in large-scale plant, but internal combustion engines, both oil and gas, are also found in smaller plants in conjunction with a *heat recovery steam generator* (HRSG) or "waste-heat" boiler, as it is frequently but misleadingly called. Alternatively, in such situations, industrial gas turbines can provide the source of power, and that type of plant is discussed in Chapter 9. Here we shall confine ourselves to cogeneration plant incorporating steam turbines. For more comprehensive studies of CHP district-heating plant the reader may consult specialist texts.[G,H]

7.20. Performance measures for cogeneration plant

Extraction from the turbine of large quantities of steam for process or district-heating purposes inevitably leads to a decrease in the electrical power produced per unit mass of fuel consumed, though improving the thermal utilisation of the fuel. To take account of the latter benefit, the *total efficiency*, perhaps better called the *energy utilisation factor* (EUF), is a more adequate measure of performance than the simple *work efficiency*. For a cyclic steam plant, these are defined as follows:

$$\text{Work efficiency, } \eta_w \equiv \frac{W_{net}}{Q_{in}}. \tag{7.20}$$

$$\text{Total efficiency (EUF), } \eta_{TOT} \equiv \frac{W_{net} + \text{Useful thermal output}}{Q_{in}}. \tag{7.21}$$

$$\text{Cogeneration ratio, } \lambda \equiv \frac{\text{Useful thermal output}}{W_{net}}. \tag{7.22}$$

$$\text{Hence } \frac{\eta_{TOT}}{\eta_w} = 1 + \lambda. \tag{7.23}$$

The work efficiency η_w is the same as the thermal or cycle efficiency η_{CY} defined in Section 2.1 of Chapter 2. However, it is here called *work efficiency* in order to emphasise the fact that it takes account only of the work produced and ignores the *thermal energy* delivered to the process or district-heating plant.

In a plant burning fossil fuel, Q_{in} would be the product of the boiler efficiency η_B and the calorific value of the fuel, CV.

7.21. Types of steam turbines in cogeneration plant

In steam-turbine cogeneration plant, the turbines may be of the following two types, or a combination thereof:

(1) *Back-pressure*, non-condensing turbines.
(2) *Pass-out* condensing turbines, with auto-extraction of steam. A *single-pass-out* turbine will have auto-extraction of steam at a single pressure intermediate between boiler pressure and condenser pressure. A *double-pass-out* turbine will be used when process or heating steam is required at two different pressures as, for example, when heating separate streams of water to two different temperatures.

A back-pressure turbine will have a pressure-regulator for the automatic regulation of the turbine exhaust pressure to maintain constancy of pressure of the process or heating steam supplied by the plant; this is necessary, since fluctuations in demand for steam will not generally coincide with the fluctuations in demand for power. Similarly, pass-out turbines will have automatic means of controlling the pass-out pressure or pressures; in district-heating schemes, for example, this will be necessary in order that the heating water supplied may be kept at a constant temperature, independently of the electrical demand on the plant. Since the variation of pass-out quantity affects the flow rates passing through the various stages of the turbine, the governor controlling the pass-out pressure will need to be linked (hydraulically and mechanically) to the speed governor of the turbine.

7.22. Double-pass-out turbine providing process steam

Figure 7.14[†] in Problem 7.13 relates to an actual industrial plant[18] in which a double-pass-out turbine supplies process steam at two different pressures, with the condensate from both processes being returned to the single (deaerator) feed heater of the steam plant. The thermal calculations in Problem 7.13 are straight-forward, involving only simple applications of the Energy Conservation Equation for flow processes. It will be seen that, whereas the work efficiency of the plant is only 23.4%, the total efficiency (EUF) is 57.0%.

†Labelled Fig. 7.11 in the 3rd Edition.

7.23. Simple turbine plant providing district heating

In a hot-water district-heating scheme, the heat exchangers or *calorifiers* used to heat the water may be incorporated into the feed-heating system of the steam-turbine plant, or even displace the feed heaters entirely. There are many possible variants. By way of introduction, Fig. 7.12 depicts one of the simplest hypothetical arrangements in which steam is bled from an intermediate stage in the turbine for the purposes of both feed heating and district heating. Real-life plants are more complex, as we shall next see, but an attempt at solution of Problem 7.14 at this stage should prove instructive; it involves the bleeding of steam only for the purposes of district heating. The analysis makes use of the approximating assumption of constancy of β used in Section 7.11. The application of that approximation to cogeneration is due to Horlock.[H,I]

As would be expected, the graph which features in the Solution to Problem 7.14 demonstrates clearly that, as the cogeneration factor λ ($\equiv Q_u/W_{net}$) increases, the total efficiency η_{TOT} (EUF) naturally increases, but the work efficiency η_w decreases. That is the result of the increase in quantity of steam bled from the turbine to heat the water circulating around the district-heating circuit.

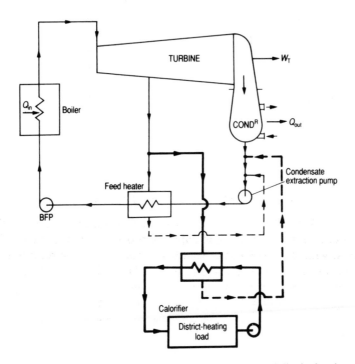

FIG. 7.12. Simple cogeneration plant with feed heating and district heating.

7.24. Complex district-heating (CHP) steam-turbine plant

With its relatively mild climate and high proportion of individual dwellings, in preference to apartment blocks, district-heating CHP plant has featured very little in the United Kingdom, in contrast to continental Europe. Even so, turbine plant for that purpose has been made for export from the UK. The author recalls that, as long ago as 1946, when Chief Turbine Test Engineer with one of the major British companies, he was involved in rectifying unstable behaviour of the complex governing system of a 25 MW double-pass-out turbine while it was on the factory test-bed, prior to export for installation in a power station in Moscow.

With the more severe climate in parts of the USSR, and the high density of apartment blocks in its larger towns, that country has long made use of the principle of cogeneration for district-heating purposes. Fig. 7.13 gives the flow diagram for a double-pass-out plant installed in the USSR in 1972. The plant is designed for a nominal electrical load of 250 MW and a nominal district-heating load of 385 MW.

The diagram has been simplified a little by omitting the drain lines for the condensed bled steam from the feed heaters. The two pass-out lines from the turbine supply the calorifiers and the feed-water heaters numbered 2 and 3 in the figure. Because there are wide variations in the district-heating loads, there is a complex control system for automatic adjustment of the two pass-out pressures. There are no control valves in the extraction piping itself, but a grid-type diaphragm located in the LP cylinder. This acts as a pressure regulator for the two pass-out lines, the pressure of the steam to calorifier 3 being controlled when calorifiers 2 and 3 are both in use, and to calorifier 2 when only that is in use. The peaking boiler comes into operation at times of peak district-heating load. For a fuller description of the system the reader may consult Ref. J.

Problems

7.1. Show that the flow rate through the boiler per unit flow rate through the condenser for the hypothetical ideal plant described in §7.5 is equal to $\Delta s_1/\Delta s_2$, where these quantities are defined as in Fig. 7.4.

Calculate this ratio when the boiler steam conditions are 10 MN/m², 550 °C and the condenser pressure is 7 kN/m².

Answer: 1.825.

7.2. For the hypothetical ideal plant discussed in §7.6, sketch a temperature–entropy diagram, similar to that in Fig. 7.5(b), for an infinitesimal feed-heating stage supplied from a point in the turbine at which the steam is superheated.

7.3. Steam is extracted from a turbine at three points to supply bled steam to a train of feed heaters. The steam conditions at the respective points in the turbine at the design load are as follows:

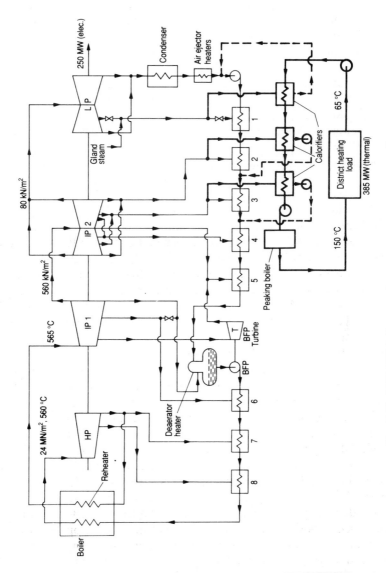

Fig. 7.13. 250 MW(e) Cogeneration steam plant in the USSR.[J]

Position	Inlet	No. 3 heater	No. 2 heater	No. 1 heater	Exhaust
Pressure (kN/m^2)	4500	430	110	34	3.0
Temperature (°C)	440	190	—	—	—
Dryness fraction	—	—	0.978	0.937	0.876

Plot the turbine expansion line on an enthalpy–entropy diagram, and on the same diagram draw a line of constant $\beta = 2180$ kJ/kg (see §7.11).

7.4. Calculate the exact Rankine cycle efficiency (taking due account of the feed pump work) and the corresponding heat rate in Btu/kWh when the steam is supplied to the turbine at 6 MN/m^2 and 500 °C, and exhausts to the condenser at 4 kN/m^2.

In an ideal, reversible regenerative cycle operating with the same steam conditions, the feed water is raised to the boiler saturation temperature in an infinite number of feed-heating stages. Calculate the cycle efficiency and heat rate, and the percentage reduction in heat rate due to feed-heating.

Answer: 40.8%; 8371 Btu/kWh. 47.3%; 7221 Btu/kWh; 13.7%.

7.5. In a hypothetical cyclic steam power plant incorporating a single direct-contact feed heater, the steam leaves the boiler at 1 MN/m^2 and 400 °C and the pressure in the condenser is 3.5 kN/m^2. Expansion in the turbine is reversible and adiabatic. The feed heater takes steam bled from the turbine at a pressure of 70 kN/m^2 and heats the feed water to the corresponding saturation temperature. Temperature and enthalpy changes of the fluid in passing through any pumps may be neglected. Calculate:

(1) the mass of steam bled from the turbine per kilogram of steam leaving the boiler;
(2) the thermal efficiency of the cycle;
(3) the improvement in thermal efficiency due to the introduction of this single stage of feed heating, expressed as a percentage of the Rankine cycle efficiency.

Answer: (1) 0.104 kg; (2) 34.2%; (3) 4.6% ($\eta_{RANK} = 32.7\%$).

7.6. For the same conditions at boiler inlet and exit as in Problem 7.5, and the same condenser pressure, calculate the thermal efficiency of an internally reversible, regenerative steam cycle in which heat is rejected reversibly to the environment, which is at a temperature T_0 equal to the steam saturation temperature in the condenser. Thence determine the resulting percentage improvement in thermal efficiency, expressed as a percentage of the Rankine cycle efficiency.

Show that the thermal efficiency of this internally reversible, regenerative steam cycle is equal to $[1 - (T_0/\bar{T}_B)]$, where $\bar{T}_B$ is the mean temperature of heat reception on the temperature–entropy diagram. Calculate $\bar{T}_B$ for this cycle and for the Rankine cycle. Thence, verify that the improvement in thermal efficiency resulting from the introduction of the reversible feed heating process is due to raising the mean temperature of heat reception.

Answer: 34.8%; 6.4%; 460.1 K, 445.5 K.

7.7. In Problem 7.5, calculate the *entropy creation* due to irreversibility in the feed heater (namely, the *net* entropy increase of the fluid streams in passing through the feed heater). Thence evaluate the loss of gross work output due to the irreversibility in the feed heater, taking the environment temperature T_0 as being equal to the steam saturation temperature in the condenser (see Theorem 2 in §A.4 of Appendix A). Express this lost work as a percentage of the net work output in the cycle and verify that this percentage is approximately equal to the difference between the percentage improvements in thermal efficiency in Problems 7.5 and 7.6.

Answer: 0.064 kJ/K; 19.2 kJ; 1.9% [$\approx$ (6.4% − 4.6%)].

Note: In Problems 7.8 and 7.9 the work input to all pumps may be neglected, the enthalpy of water at any temperature being taken as equal to the saturation enthalpy at that temperature.

7.8. In a regenerative steam plant the turbine inlet conditions and exhaust pressure are as in Problem 7.4, but the feed water is raised to a temperature of 182 °C in four heaters.

(a) If direct-contact heaters were used throughout, each raising the feed water to the saturation temperature of the steam supplied to the heater, and if the overall enthalpy rise of the feed water were divided equally amongst the heaters, what would be the required bled steam pressures?

(b) In these circumstances, the bled steam enthalpies would be respectively 3040, 2868, 2683 and 2490 kJ/kg, and the enthalpy of the steam at turbine exhaust 2285 kJ/kg. Calculate the steam flow rate through the boiler per unit flow rate to the condenser. What would be the calculated value of this ratio if the values of β for all heaters were taken to be equal to the mean value? Determine the cycle efficiency and heat rate.

(c) Calculate the cycle efficiency and heat rate (in Btu/kW h) for a non-regenerative cycle in which the states of the steam at turbine inlet and exhaust are the same as in this regenerative cycle. Thence determine the percentage reduction in heat rate due to feed heating. From this result and that calculated in Problem 7.4 determine the values of x and y, as defined respectively in §7.12 and §7.13, and plot this point on Fig. 7.9.

Answer: (a) 1.05, 0.412, 0.127, 0.0284 MN/m^2.
(b) 1.323; 1.323; 38.3%; 8910 Btu/kW h.
(c) 34.4%, 9904 Btu/kW h; 10.0%; $x = 0.596$, $y = 0.731$.

7.9. In a regenerative steam plant the states of the steam at turbine inlet and exhaust, and at the bled points, are the same as in Problem 7.8, but the feed water from the condenser passes first through two direct-contact heaters and then through two surface heaters. In the surface heaters there is a temperature difference of 5 K between the saturation temperature of the bled steam supplied to a heater and the outlet temperature of the feed water from that heater. The condensed bled steam leaves a surface heater at the saturation temperature of the steam supplied to the heater, and these drains from the surface heaters are cascaded successively from heater to heater, passing finally to the direct-contact heater preceding the surface heaters.

Calculate the flow rate through the boiler per unit flow rate to the condenser, the cycle efficiency and heat rate (in Btu/kW h). (*Hint*: First determine the condenser flow rate per unit flow rate through the boiler, starting the calculation of the bled steam quantities at the heater nearest the boiler.)

Answer: 1.308; 38.1%; 8955 Btu/kW h.

7.10. In Fig. 7.9, it is seen that, for any given number of heaters n, the fractional improvement in heat rate when heating to the boiler saturation temperature (i.e. when $x = 1$) is the same as the optimum fractional improvement for $(n - 1)$ heaters. Explain why this is so.

7.11. In an ideal, reversible non-regenerative steam cycle the initial steam pressure and temperature are respectively 15 MN/m^2 and 500 °C and the condenser pressure is 4 kN/m^2. Calculate the percentage exhaust wetness and the cycle efficiency (1) when there is no reheating, and (2) when the steam is reheated to 500 °C at (a) 6 MN/m^2, (b) 4 MN/m^2, (c) 2 MN/m^2. The work input to the feed pump may not be neglected.

From these results express the optimum reheat pressure as a fraction of the initial pressure, and determine the percentage improvement in efficiency due to reheating at this pressure. Use equation (7.18) to check the calculated improvement.

Answer: (1) 26.4%; 43.6%. (2a) 19.8%; 44.9%. (2b) 17.2%; 45.0%. (2c) 13.0%; 44.8%.
About one-quarter. 3.3%.

Note: In obtaining these answers, use has been made of the Tables listed in reference (2).

7.12. A steam power plant incorporates a homogeneous train of surface feed heaters with the bled-steam drains cascaded. In each heater the outlet feed water is raised to the saturation temperature of the entering bled steam, and the outlet drain water is cooled to the temperature of the entering feed water. Assuming that the enthalpy of water at any temperature is equal to the enthalpy of saturated water at the same temperature and that stray heat losses are negligible, derive an expression for the ratio of the mass flow rates of steam in the turbine upstream and downstream of a bled-steam tapping point.

A multi-reheat plant incorporating such a train of surface feed heaters is to be designed for a specified vacuum and final feed temperature, and for specified steam conditions in the boiler and reheaters. It may be assumed that β, the difference between the specific enthalpies of the bled steam entering a heater and of saturated water at the same pressure, will be the same for all bled-steam tapping points between any two adjacent reheat points, and that the condition line of the steam in the turbine will be unaffected by any changes in the positions of the bled-steam tapping points.

Show that, for maximum cycle efficiency, the enthalpy rises r of the feed water in all heaters between adjacent reheat points must be the same, and that the enthalpy rises in the sets of heaters taking steam from the turbine at points respectively immediately upstream and downstream of a selected reheat point must be such as to satisfy the relation

$$\frac{\alpha_d}{\alpha_u} = 1 + \frac{Q_S}{Q_B},$$

where $\alpha \equiv \beta + r$ and the suffixes u and d refer to conditions immediately upstream and downstream of the reheater respectively, Q_S is the heat supplied to the steam in the selected reheater and Q_B is the heat supplied in the boiler together with all reheaters upstream of the selected reheater.

Answer: $M_u/M_d = \gamma = \alpha/\beta$. [See reference (15)]

Note: Problem 8.6 in Chapter 8, relating to a steam cycle for a nuclear power plant, provides further practice in feed-heating calculations, as also do Problems 9.2 and 9.4 in Chapter 9.

7.13. Figure 7.14 depicts a plant in which a double-pass-out steam turbine is used for the combined supply of power and process steam. All the process steam is returned as

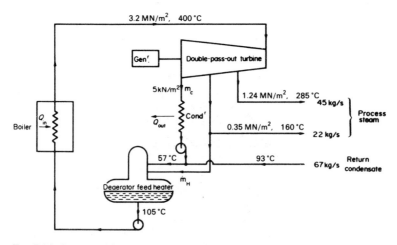

Fig. 7.14. Steam-turbine plant for combined supply of power and process steam.

condensate and this is mixed with the condensate from the condenser before the joint flow enters the deaerator feed heater. The latter is supplied with steam from the second pass-out point and delivers saturated water at the temperature indicated in the figure.

For the conditions depicted in the figure, and taking values of 85% and 97% respectively for the overall isentropic efficiency of the turbine and the efficiency of the generator, calculate the following quantities:

(1) The mass flow rates $\dot{m}_C$ and $\dot{m}_H$.
(2) The electrical output from the plant.
(3) The energy given up by the process steam.
(4) The work efficiency and total efficiency of the plant.

Enthalpy rises in the pumps are to be neglected.

Answer: (1) 100.5 kg/s, 14.5 kg/s; (2) 118.9 MW; (3) 170.6 MW; (4) 23.4%, 57.0%.

Additional problem

7.14. In §7.11 it is stated that, in non-reheat steam plant, the value of β does not vary greatly along a given turbine expansion line, where, at any given pressure at a point on the expansion line, $\beta \equiv (H - h)$; H is the specific enthalpy of the steam at the given pressure and h is the specific enthalpy of saturated water at the same pressure.

(a) Treating β as being constant along the turbine expansion line, and assuming the work input to the feed pump being negligible, the thermal efficiency of a simple steam plant without feed heating is given by

$$(\eta_{CY})_{\text{NON-F.H.}} = \frac{1}{1 + \gamma}, \qquad (7.24)$$

where $\gamma \equiv \beta/\alpha$, and α and β are the areas depicted on the temperature–entropy diagram of Fig. 7.15.

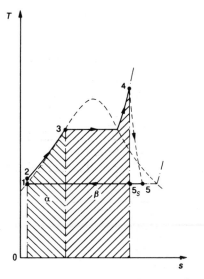

Fig. 7.15. *T–s* diagram for simple steam plant.

(b) In §7.11 it is shown that, for maximum efficiency in a non-reheat plant with one stage of feed heating, the enthalpy rise in the feed heater should, to a first approximation, be equal to $\alpha/2$. For such a plant, sketch the temperature–entropy diagram corresponding to Fig. 7.15, labelling appropriate areas on the diagram. Show that the thermal efficiency is then given by

$$(\eta_{CY})_{F.H.} = \frac{1 + 4\gamma}{(1 + 2\gamma)^2}. \tag{7.26}$$

(c) The single feed heater in (b) above is simply *replaced* by a single heat exchanger (calorifier), in which the bled steam is used to heat the water of a district-heating system. It is to be assumed that the district-heating water enters the calorifier at the same temperature t_1 as that of the condensate leaving the main condenser, and that the condensed bled steam leaves the calorifier at this same temperature t_1.

Show that the performance of the plant is given by the following expressions:

$$\text{Work (thermal) efficiency, } \eta_w = \frac{1 + 2\gamma}{1 + \gamma} \frac{1}{(1 + \lambda) + 2\gamma}, \tag{7.27}$$

$$\text{Total efficiency (EUF), } \eta_{TOT} = \frac{1 + 2\gamma}{1 + \gamma} \frac{1 + \lambda}{(1 + \lambda) + 2\gamma}, \tag{7.28}$$

where $\lambda \equiv Q_u/W_{net}$, the *cogeneration coefficient*, and Q_u is the *useful thermal output* of the cogeneration plant (namely, the heat transferred to the district-heating water passing through the calorifier).

Determine the value of γ for a steam condition at turbine inlet of 2 MN/m² and 353 °C, when the condenser pressure is 7 kN/m². For that value of γ, calculate the values of $(\eta_{CY})_{\text{NON-F.H.}}$ and $(\eta_{CY})_{F.H.}$, and plot graphs of η_w and η_{TOT} against λ for values of λ from 0 to 6.

Taking the value of β at turbine exhaust as the same as that at turbine inlet, estimate the isentropic efficiency, η_T, of the turbine.

Answer: $\gamma = 3.0$; $(\eta_{CY})_{\text{NON-F.H.}} = 25.0\%$; $(\eta_{CY})_{F.H.} = 26.5\%$; $\eta_T = 76.0\%$.

CHAPTER 8

Nuclear power plant

8.1. Introduction

Compared with conventional power plant burning organic fuels, nuclear power plants utilising the *fission reaction* are in a relatively early stage of development. All use steam for the purposes of power production. In the first generation of nuclear power reactors the greatest permissible fuel element temperature was rather low, so that the steam conditions were reminiscent of those that prevailed in conventional plant in the 1920s. By the early 1960s the greatest investment in nuclear power plant on a commercial scale had taken place in the United Kingdom, although that position has since been taken over by the United States. Initially, all commercial nuclear plant in the United Kingdom were of the gas-cooled type, using carbon dioxide as the coolant, natural uranium as the fuel and graphite as the moderator. They had a dual-pressure steam cycle and the first plant of the kind was installed at Calder Hall. The metallic fuel elements of natural uranium were canned in a magnesium alloy which went by the name of Magnox, the the British line of reactors stemming from the Calder Hall design came to be known as *Magnox gas-cooled reactors.*

Although this type of plant has since been superseded by the Advanced Gas-cooled Reactor (AGR), the dual-pressure steam circuit will be discussed in some detail since it is of considerable thermodynamic interest in the means which it uses to reduce the *external* irreversibility involved in the heat transfer process between the CO_2 coolant and the H_2O. Furthermore, as noted in §9.7 of Chapter 9, this type of steam-raising circuit has been adopted for plant which utilise the thermal energy (frequently mis-termed *waste heat*) in the exhaust gases of industrial gas-turbine plant, since there the gas temperatures are also relatively low.

GAS-COOLED REACTORS

8.2. The simple dual-pressure cycle

At Calder Hall, the temperature of the CO_2 supplied to the steam-raising towers which replaced the conventional boiler was considerably lower than the gas temperature in the furnace of a fossil-fuel boiler. It was consequently important to reduce as much as was commercially profitable the degree of irreversibility external to the cycle that resulted from the temperature difference between the CO_2 coolant on the outside of the heat-exchanger tubes and the H_2O fluid circulating through them. This led to the adoption of the dual-pressure cycle, which also facilitated reactor control with varying load. A simplified circuit diagram of the Calder Hall plant is shown in Fig. 8.1.

Figure 8.2 shows the temperatures of the fluids as they pass through the heat exchanger. These temperatures are plotted against the amount of heat exchanged between fluids between the entry point of either fluid and the chosen point, expressed as a percentage of the total quantity of heat transferred in the heat exchanger.

For greatest plant efficiency the CO_2 and H_2O should ideally pass in counter-flow through the heat exchanger with no temperature difference between them at any point, since there would then be no irreversibility in the heat transfer process. However, this condition could only be attained by having a steam cycle in which the steam was raised at an infinite number of successively higher pressures. In practice it was not considered commercially attractive to generate the steam at more than two pressures,

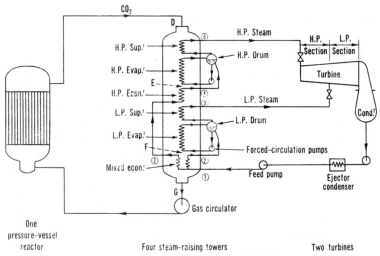

One pressure-vessel reactor   Four steam-raising towers   Two turbines

Fig. 8.1. Gas-cooled reactor with dual-pressure steam cycle (Calder Hall, UK).

though the economics of a triple-pressure cycle were investigated. The greater external irreversibility that would have resulted if a single-pressure cycle had been adopted at Calder Hall is evident from Fig. 8.2; the steam saturation temperature in the single-pressure cycle is well below that of the high-pressure fluid in the dual-pressure cycle, resulting in a much greater temperature difference between the CO_2 and the H_2O over a large proportion of the heat input to the steam cycle. This direct transfer of heat over a finite temperature difference represents a lost opportunity for producing work, the perpetual accompaniment of irreversible processes.

It is seen from Fig. 8.1 that the feed water supplied to the steam generator divides into two parallel circuits; from one, superheated HP steam is supplied to the turbine inlet, and from the other superheated LP steam is supplied to a steam chest located part way down the turbine. The choice of the most economical conditions of steam generation involves a lengthy optimisation calculation which is best done with the aid of a digital computer.

A study of Fig. 8.2 reveals that the design steam conditions are in part governed by the values chosen for the minimum temperature approach

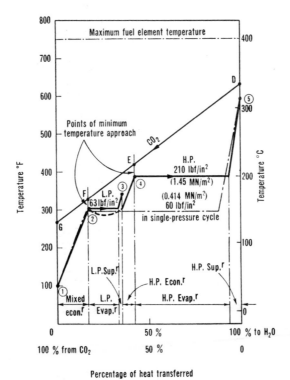

Percentage of heat transferred

Fig. 8.2. Fluid temperatures in the original dual-pressure cycle at Calder Hall.

between the two fluids at the *pinch-points D, E* and *F* in the heat exchanger; they are also dependent on the relative flow rates of HP and LP steam. Not all of these quantities are independent of each other, so that optimisation is a lengthy trial-and-error process. We may, however, show how to calculate the HP and LP steam flow rates given sufficient other data, and we shall then be able to evaluate the efficiency of an ideal, reversible power plant supplied with the same quantities of steam at the given operating conditions. This will provide a criterion against which to judge the excellence of performance of the actual dual-pressure turbine plant. Factors contributing to the poorer performance of the actual plant will include inefficiency of the turbine, and the irreversible mixing of two streams of steam at different temperatures where the LP steam enters the turbine.

8.3. Calculation of HP and LP steam flows, and cycle efficiency

When the CO_2 inlet temperature t_D and the steam conditions have been specified, as in Fig. 8.2, then it is possible to calculate the HP steam flow m_H and the LP steam flow m_L per unit mass of CO_2 circulated, if the minimum temperature-approach values at the two pinch-points E and F have also been specified. The smaller the temperature approach the greater is the required surface area of the heat exchanger, a value of 17 K being chosen as an appropriate economic figure. The calculation then proceeds by considering energy balances for the respective sections of the heat exchanger, treating the CO_2 as a perfect gas having a constant specific heat capacity c_p equal to the mean value between t_D and t_G. In practice, account would, of course, be taken of the variation in specific heat capacity with temperature.

Section DE

$$t_E = t_4 + 17,$$

and $\quad m_H(h_5 - h_4) = c_p(t_D - t_E)$, giving m_H.

Section EF

$$t_F = t_2 + 17,$$

and $\quad m_H(h_4 - h_2) + m_L(h_3 - h_2) = c_p(t_E - t_F)$, giving m_L.

Section FG

$$(m_H + m_L)(h_2 - h_1) = c_p(t_F - t_G), \text{ giving } t_G.$$

In Fig. 8.2 it has been assumed that, in the mixed economiser, both the HP and LP fluid streams are raised to the LP saturation temperature, and that the HP fluid stream leaves the HP economiser at the HP saturation temperature, so that the values of t_2 and t_4 required in the above calculations may be read from the steam tables.

Given the isentropic efficiencies of the HP and LP portions of the turbine, and assuming perfect mixing of the HP exhaust steam and the LP steam at LP turbine inlet, the state of the steam at turbine exhaust may be calculated and thence the heat rejected in the cycle. Finally, calculation of the heat input enables the cycle efficiency to be determined.

8.4. Efficiency of the corresponding ideal dual-pressure cycle

The efficiency of the corresponding ideal, reversible cycle may be calculated from considerations of available energy, in a manner similar to that set out in §2.8 for the single-pressure Rankine cycle. In such an ideal dual-pressure cycle the streams of high and low pressure steam would be fed to the fully reversible, open-circuit steady-flow work-producing device depicted in Fig. 8.3.

As will be seen from a study of §§A.9 and A.10 of Appendix A, the net work output from this ideal device would be equal to the available energy in the given situation, namely

$$\text{Ideal } W_{net} = m_H(b_5 - b_1) + m_L(b_3 - b_1), \tag{8.1}$$

where $b \equiv (h - T_A s)$ and the environment temperature T_A may be taken as the temperature of the cooling water supplied to the condenser. The ideal cycle efficiency is then given by

$$\text{Ideal } \eta_{\text{DUAL}} = \frac{m_H(b_5 - b_1) + m_L(b_3 - b_1)}{m_H(h_5 - h_1) + m_L(h_3 - h_1)}. \tag{8.2}$$

To assess the effect of the irreversible heat transfer in the heat exchanger, the ideal work calculated from eqn. (8.1) may be compared

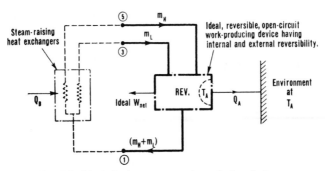

FIG. 8.3. Ideal, dual-pressure work-producing device.

with the energy available for the production of work from the heat transferred from unit mass of CO_2 in its passage through the heat exchanger. This is given by

$$\text{Ideal } W_{max} = -c_p \int_D^G \left(1 - \frac{T_A}{T}\right) dT = c_p \left[(T_D - T_G) - T_A \ln \frac{T_D}{T_G}\right]. \quad (8.3)$$

The corresponding cycle efficiency is then

$$\text{Ideal } \eta_{MAX} = \frac{W_{max}}{c_p(T_D - T_G)} = \left[1 - T_A \left(\frac{\ln T_D/T_G}{T_D - T_G}\right)\right]. \quad (8.4)$$

In Problem 8.1, which is based on the Calder Hall plant,

$$\text{Ideal } \eta_{MAX} = 40.6\%$$

$$\text{Ideal } \eta_{DUAL} = 32.5\%$$

$$\text{Ideal } \eta_{SINGLE} = 28.7\%.$$

In Problem 8.2 it is seen that internal irreversibilities in the steam turbo-alternator plant reduce the efficiency from the ideal of 32.5% to a value of 22.8%.

8.5. The effect of circulator power on the plant efficiency

The power required to drive the CO_2 circulators is by no means negligible and so leads to a significant reduction in overall station efficiency. However, this power is not lost in its entirety, for it is fed into the circulating CO_2, so that the heat transferred in the heat exchanger Q_B is that amount greater than the reactor thermal output Q_R. A certain proportion of the circulator power is consequently recovered in the work output from the cycle, as the following calculation shows.

If Q_R = the reactor thermal output per unit mass of CO_2 circulated, and W_o = the work sent out (neglecting, for the purpose of the calculation, the power required for other auxiliary plant), then, referring to Fig. 8.4, the overall station efficiency is given by

$$\eta_o = \frac{W_o}{Q_R} = \frac{W_{net} - W_C}{Q_R}.$$

But $W_{net} = \eta_{CY} Q_B$ and, neglecting stray heat loss and external losses in the circulators and their driving motors,

$$Q_B = Q_R + W_C.$$

Hence

$$\eta_o = \eta_{CY} - (1 - \eta_{CY}) \frac{W_C}{Q_R}. \quad (8.5)$$

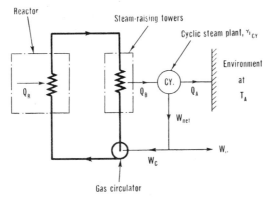

FIG. 8.4. Energy–flow diagram for complete power plant.

It is seen from this equation that a fraction η_{CY} of W_C is recovered, but since W_C is quite large and η_{CY} is relatively low, the power consumed by the blowers is of troublesome magnitude. Problems 8.2 and 8.4 show that the circulator power reduces the efficiency from 22.8% to 21.4%.

8.6. The effects of regenerative feed heating

Feed heating was not employed at Calder Hall because the plant was not designed for maximum efficiency but for a high rate of plutonium production for military purposes. This required a high reactor thermal output and so a high value of θ_{CO_2}, the temperature rise of CO_2 in its passage through the reactor, and thus a low temperature of the gas at reactor inlet. Since this gas comes from the heat exchanger, its temperature is governed by the temperature of the feed water entering the heat exchanger, so that the employment of regenerative feed heating in the steam cycle would have produced the opposite effect to that desired. In later gas-cooled plant built primarily for power production, high efficiency was of greater importance and three or four stages of feed heating were employed.[19]

The purpose of regenerative feed heating is to raise the mean temperature of heat reception in the boiler. It was seen in Chapter 7 that, in a conventional plant using organic fuel, the optimum final feed temperature is a function of the boiler pressure for which the plant is designed, but the choice of boiler pressure is not dictated by the selected final feed temperature—rather the reverse. A difference situation exists in gas-cooled nuclear plant, as may be seen from Fig. 8.5.

We will suppose that the gas inlet and outlet temperatures in the heat exchanger have been fixed, so fixing θ_{CO_2}, and also the minimum temperature approach between the CO_2 gas and the H_2O. It is then seen

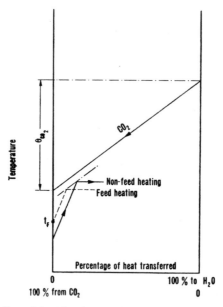

Fig. 8.5. The effect of feed heating on the design steam pressure in gas-cooled nuclear plant.

from Fig. 8.5 that a higher feed temperature t_F at inlet to the steam generator results in a lower steam pressure in a single-cycle plant (or a lower LP pressure in a dual-cycle plant). Thus, not only is the choice of steam pressure no longer independent of final feed temperature, but the gain in cycle efficiency due to an increase in final feed temperature is offset in some measure by a reduction in cycle efficiency due to the resulting lower steam pressure (cf. §7.2.3). Consquently it is not possible directly to apply the type of analysis discussed in Chapter 7, and the choice of appropriate steam conditions and final feed temperature must result from lengthy optimisation calculations which are best made on a digital computer.

8.7. Later developments in Magnox, gas-cooled reactor plant

Considerable developments in Magnox reactor plant took place subsequent to the design of the initial Calder Hall plant in 1952, of which those having the greatest influence on plant efficiency were increases in the gas circulating pressure and in the maximum permissible fuel-element temperature. These advances are typified by the figures given in Table 8.1.

TABLE 8.1. *Advance in operating conditions for Magnox gas-cooled reactor plant with dual-pressure steam cycles*

Station Date of commissioning	Calder Hall 1956–9	Oldbury 1967–8
Net station output (MW)	39	600
No. of reactors	1	2
Thermal output/reactor (MW)	182	893
Circulator power/reactor (MW)	3	20
Gas pressure (MN/m^2)	0.8	2.5
Gas inlet temp. (°C)	140	250
Gas outlet temp. (°C)	336	412
HP Steam		
Pressure (MN/m^2)	1.45	9.45
Initial temp. (°C)	313	390
Reheat temp. (°C)	–	390
Flow (%)	77	60
LP Steam		
Pressure (MN/m^2)	0.435	4.86
Temperature (°C)	177	390
Flow (%)	23	40
No. of feed heaters	–	3
Final feed temp. (°C)	38	141
Net station efficiency (%)	21.5	33.6

8.7.1. Increase in gas pressure

An increase in the pressure level of the CO_2 in the circuit decreases the specific volume of the gas and so reduces both the physical size of the gas ducting and the required work input to the gas circulators for a given pressure rise $[(W_{in})_{REV} = \int v \, dp]$. By increasing the thickness of the steel pressure vessel housing the reactor from 5 cm to 11 cm, it was possible to increase the gas pressure from $0.8 \, MN/m^2$ at Calder Hall to $1.9 \, MN/m^2$ at Sizewell A; while at Oldbury, the last but one of the Magnox stations, a further increase to $2.5 \, MN/m^2$ was made possible by the introduction of a prestressed-concrete pressure vessel which, for the first time, enclosed both the reactor core and the boilers, an innovation described as the *integral-reactor* concept. This step was facilitated by adoption of a once-through design for the boiler circuit.

8.7.2. Results of increase in fuel-element temperature

Improvements in fuel-element design in Magnox reactors permitted increases in the gas temperatures both to and from the reactor. The increase in the temperature at reactor inlet was more marked than that at outlet, so resulting in a lowering of θ_{CO_2}. This raising of the gas-temperature line on the temperature–heat transfer diagram, and particularly its decrease in slope, led to appreciable increases in steam pressures in the dual-pressure cycle and consequent improvement in cycle efficiency. The increase in gas outlet temperature from the reactor was only sufficient, however, to permit modest increases in steam outlet temperatures from the boiler. In the absence of remedial measures, the increase in steam pressures would consequently have resulted in excessive moisture content at turbine exhaust (cf. §7.2.3). The need to adopt remedial measures to avoid turbine blade erosion was enhanced by the simultaneous and very large increase in plant output, leading to turbines with last stages of large tip-diameter. This led to the introduction of reheating, in some cases by the incorporation of a reheater in the gas passes of the boiler, as at Oldbury, and in other cases by the use of both live steam and bled steam for reheating, as at Sizewell.

8.7.3. Other modifications to the steam cycle

The advance in steam pressures made it possible to adopt, in some stations, a direct steam-turbine drive for the gas circulators, possessing the advantage that it eliminated the double inefficiency of generator and driving motor that existed when the circulators were driven by an electric motor. This change led to a dual-pressure cycle at Oldbury in which the entire HP steam supply was fed to turbines driving the gas circulators, the exhaust steam from these turbines being reheated in the boilers before joining the LP steam from the boilers to supply the multi-cylinder main power turbines, from which steam was bled for three or four stages of feed heating.

A further development, resulting from the use of a prestressed-concrete pressure vessel, was the abandonment of the dual-pressure cycle in the Wylfa plant (the last of the Magnox stations) and recourse to a single-pressure cycle with steam-steam reheating. This move towards a simpler cycle resulted partly from the fact that the less steep gas line (smaller θ_{CO_2}) made the addition of a second pressure less attractive economically, and also from the high cost of making penetrations in the prestressed-concrete pressure vessel for the multiplicity of pipes encountered in a dual-pressure cycle.

8.8. Advanced gas-cooled reactor (AGR) plant

By comparing the steam conditions listed in Table 8.1 for the Oldbury nuclear station with those for fossil-fuel stations listed in Table B.1 in Appendix B, it will be seen that the steam conditions in the Magnox stations, and the resulting efficiencies, are well below those in the best conventional fossil-fuel stations. However, considerable advances were made when enriched uranium became available and permitted the Advanced Gas-cooled Reactor (AGR) to be developed for commercial use from a 30 MW(e) prototype built at Windscale. Instead of the metallic uranium used in the Magnox reactors, the AGR uses uranium-dioxide fuel pellets contained in thin stainless-steel (or Zircalloy) cans, but it still has graphite as the moderator and continues to use CO_2 as the coolant. The use of stainless steel, with it higher neutron-absorbing properties, necessitates some enrichment of the fuel with the fissile U^{235}, but it allows the gas outlet temperature from the reactor to be raised to such a level that the plant may be designed for a single-pressure cycle with steam conditions comparable to those in the best conventional fossil-fuel stations operating at sub-critical pressures (Table B.1 of Appendix B). Steam cycle considerations are consequently little different from those discussed in Chapter 7, although, for the reasons discussed in §8.6, the final temperature is somewhat lower than would be selected for a fossil-fuel station having the same boiler pressure.

We may take the conditions in the 660 MW(e) units at Heysham[20] as representative of those in the first generation of AGR nuclear stations. There the CO_2 is under a pressure of about $4 \, MN/m^2$ and the gas temperature of 651 °C at reactor outlet enables the boilers to deliver steam to the turbine at $16 \, MN/m^2$ and 538 °C, with reheating in the boilers to 538 °C. These conditions are nearly on a par with those listed in Table B.1 for fossil-fuel stations, though, for reasons already stated, the final feed temperature is markedly lower at 157 °C. Nevertheless, because there is no waste energy being carried away in flue gases, as in fossil-fuel plant, the overall station efficiency at 41.2% is even a little higher than the figures quoted in Table B.1 for the latter type of plant. A higher thermal efficiency results in less waste energy being carried away in the condenser circulating water. In the context of thermal pollution of the environment, the high efficiency of these gas-cooled reactors is thus a good point in their favour, as against the low thermal efficiency of the water-cooled reactors discussed in later sections. There is a school of thought which also regards gas-cooled reactors as being inherently safer in operation than water-cooled reactors.

A useful comparison of the two types of plant will be found in Ref K, written by a supporter of the AGR. An equally useful counter-comparison, written by supporters of the pressurised-water reactor (PWR), will be found in Ref. L. Some detailed consideration of the latter type of plant is given in §§8.15–8.19.

8.9. High-temperature gas-cooled reactor (HTGR) plant

Although the AGRs just described might be considered to be high-temperature reactors, still higher reactor temperatures can be achieved by using coated-particle, ceramic carbide fuel, together with the use of graphite both as moderator and as structural material for the fuel element, and with the replacement of CO_2 by the chemically inert gas, helium. Rather than leading to higher steam temperatures (which, even in AGRs, are already comparable with those in fossil-fuel stations), the higher gas exit temperature (about 750 °C) achieved in this type of reactor is used to give a more compact boiler through reduction in heat-transfer area. This feature is further enhanced by the fact that the heat-transfer and thermal transport properties of helium are very good and, as a gas, second only to those of hydrogen.

The OECD Dragon Reactor Project at Winfrith in the United Kingdom[21,22] had an experimental reactor of this type, of 20 MW thermal output and zero electrical output. This experience gained with it led to a design study[23] for a commercial reactor of 660 MW(e), but construction of such a plant was not pursued.

From six years of operating experience gained on a prototype power-producing HTGR plant of 40 MW(e) output at Peach Bottom in the eastern United States, a prototype 330 MW(e) uranium-thorium near-breeder reactor at Fort St. Vrain[24] in Colorado was designed for a helium outlet temperature of 777 °C. The steam conditions, however, are very similar to those for the AGR at Heysham, where the CO_2 outlet temperature is 126 K lower. Steam is delivered to the turbine at 16.7 MN/m², 538 °C. The exhaust steam from the HP turbine at 355 °C first passes through the auxiliary turbines driving the helium circulators before being reheated in the reactor to 538 °C and delivered to the IP turbine at 4.1 MN/m². A six-stage feed-heating system raises the feed water to 205 °C, which is again lower than that in a comparable fossil-fuel plant. The overall station efficiency is about 39%.

In the core of the Fort St. Vrain reactor, the graphite block is made up of prismatic elements containing coolant holes in which are mounted the fuel rods. The fuel within these rods is in the form of bonded, coated fissile and fertile particles, fissile uranium-233 being produced in the latter by neutron absorption in fertile thorium-232. This leads to the efficient utilisation of fissionable nuclear fuel as a result of high neutron economy, high conversion ratio and high nuclear fuel burn-up. Operating on this uranium-thorium fuel "cycle", this near-breeder HTGR thus achieves a much better fuel utilisation than does the Magnox type of gas-cooled reactor; the average fuel burn-up is of the order of 10^5 MW days per tonne, as against 4000 in the Magnox reactors. Similar comments apply to an alternative type of reactor core developed in the Federal Republic of Germany, in what is known as a *pebble-bed reactor*. In this reactor, which also operates on the uranium-thorium fuel "cycle", the core consists of a statistical fill of spherical fuel elements. These are made of graphite and contain coated particle fuel. During operation, fresh fuel balls are continuously added at the top of the core and move slowly downward, depleted elements being removed at the bottom. A prototype steam-turbine plant of 300 MW(e) output at Schmehausen,[25] commissioned in 1986, is of this type. In both the prismatic type and the pebble-bed type the coolant gas is helium.

The construction was projected of plants of very much larger output than those at Fort St. Vrain and Schmehausen, with some degree of collaboration between American, German and Swiss constructors. Those plans have not led to firm orders, though there is renewed interest in the USA. Collaboration also extended to project studies for HTGR direct-cycle, gas-turbine plant, such as are next discussed.

8.10. Gas turbine cycles for gas-cooled reactors

From the structural and metallurgical viewpoints, large high-efficiency steam plants have two serious disadvantages; firstly, that the fluid pressure is very high in the high-temperature part of the cycle, and secondly that, with high vacuum at exhaust, the fluid volume there is very great, requiring multiple-exhaust low-pressure cylinders of large diameter. By contrast, closed-circuit gas-turbine plants are designed for a much smaller overall pressure ratio; consequently, the top pressure need not be as high as in steam plant and, with a pressurised circuit in which the exhaust pressure may be well above atmospheric, the very much smaller fluid volume at exhaust reduces the need for multiple exhausts.[26] Moreover, in such closed-circuit plant, variation of output can be achieved by altering the general pressure level in the circuit while keeping the turbine inlet temperature constant, so giving good part-load performance. Even so, the steam plant wins on thermal efficiency on account of its constant and very low temperature of heat rejection, so that a gas-turbine plant must be designed for a considerably higher turbine inlet temperature if it is to compete on equal terms with a steam plant.

With the gas exit temperature of 675 °C that is obtainable from the AGR, consideration has been given[27] to the possibility of using the reactor coolant gas (CO_2) as the working medium for a closed-circuit gas-turbine plant, instead of using it to transfer heat to a steam cycle. With the future prospect of a gas exit temperature of up to 1000 °C from a more advanced type of gas-cooled reactor, increasing attention has been given to the possibilities of gas-turbine plant, but this attention has so far been confined almost entirely to theoretical studies only. There is a wide variety of possible plant arrangements and some discussion of those that appear to have prospects of possible commercial exploitation is not out of place. In all cases, it is one or other variant of the CBTX cycle discussed in Chapter 6 that has come to the fore, with either helium or carbon dioxide as the working fluid.

Helium is particularly attractive as the working fluid for a closed-circuit gas-turbine plant because it is inert and has the further advantages of high specific heat capacity, high thermal conductivity and high acoustic velocity; the latter permits the use of higher blade-tip speeds, and the consequential ability to use longer blades reduces the need for multiple exhausts. [For a perfect gas, the acoustic velocity is given by $a = \sqrt{(\gamma \bar{R} T/M)}$, and for helium γ is high and the molar mass M is very low.] Carbon dioxide is less attractive than helium on all four counts, although, for a given pressure ratio, its lower value of γ leads to a smaller enthalpy drop in the turbine for a given pressure ratio [cf. eqn. (3.6) in Chapter 3] and therefore to fewer turbine and compressor stages.

8.10.1. Helium gas-turbine plant

One of the simplest circuit arrangements for a helium gas-turbine plant is that of Fig. 8.16 relating to Problem 8.7. This figure depicts a *direct-cycle*, closed-circuit, helium CBTX plant in which the helium serves both as reactor coolant and as working fluid for the gas-turbine plant. When provided with three stages of compression, and inter-cooling between the stages, such a plant would be described as a CICICBTX plant in the notation used in Chapter 6. At Geesthacht[28] in the Federal Republic of Germany, a prototype 25 MW(e) unit of this kind is designed for gas conditions at reactor outlet of 750 °C and 25 atm with a pressure ratio of 2.5. The estimated overall efficiency is 38%. This is a high figure for gas-turbine plant, made possible by the high turbine inlet temperature, together with the use of regenerative heat exchange in exchanger X and intercooled compression. In a British study[29] for a CICBTX plant designed for the same pressure ratio, but for gas conditions at reactor outlet of 927 °C and about 50 atm, the estimated overall efficiency is 41% (cf. Problem 8.7).

The temperature at turbine exhaust is considerably higher in a gas-turbine plant than in a condensing steam plant. A number of ways of taking advantage of this otherwise adverse circumstance have been suggested. Considering a CBTX gas-turbine cycle of the type illustrated in Fig. 6.2 of Chapter 6, but with nuclear heating, the turbine exhaust gas is at a relatively high temperature as it leaves the regenerative heat exchanger X. Consequently, instead of the heat Q_A being rejected to cooling water, it could be utilised in the following alternative ways:

(1) The heat could be rejected to a suitable fluid circulating round the *bottoming cycle* of a binary-cycle plant of which the gas-turbine circuit constitutes the *topping cycle*. With helium as the topping fluid, studies have been made for both water substance[30,31], ammonia[32] and isobutane[33] as the bottoming fluid. A plant using the first of these is discussed in more detail in §9.10.2 of the next chapter. Were ammonia or isobutane to be used there would be no high-temperature heat exchanger preceding the regenerative heat exchanger X of Fig. 9.9, while Q_0 would be supplied to the bottoming cycle instead of being dissipated in cooling water. It is estimated that the thermal efficiency of such plant of 1400 MW(e) output would be about 46%.

(2) The heat could be rejected to a calorifier circuit to heat domestic water in a district-heating scheme[32], in the manner depicted in Fig. 8.17 of Problem 8.8.

    In this plant, heat is transferred from the helium to water circulating under pressure in a closed loop. This heated water passes in turn through the calorifier of the district-heating plant and then through the cooling coils of a dry cooling tower. The latter brings the water down to a low temperature in order that it may in turn cool the helium to a low temperature before the latter enters the compressor, thus minimising the work input to the compressor. The water entering the cooling coils of the cooling tower is at a higher temperature than the cooling water from a condensing steam plant, so rendering it economic to use a dry cooling tower. Such a tower dispenses with the need for a large source of cooling water and avoids the discharge of large quantities of heat to natural water sources. It could be used with HTGR plant[34] whether or not the latter were associated with a district-heating scheme.

A comprehensive study of the design and potentialities of HTGRs, including their possible use in providing industrial process heat, will be found in two sets of papers: (a) a series on the Development of Gas-cooled Nuclear Reactors in the USA, which includes reference (34); (b) the Proceedings of a Symposium on Gas-cooled Reactors, in which references (25) and (32) appear.

8.10.2. *Supercritical and hypercritical CO_2 gas-turbine plant*

A form of cycle has been proposed in which, instead of helium, carbon dioxide would be used as the working fluid in two alternative modes of operation, which may most conveniently be classified as *supercritical* and *hypercritical* respectively; in the latter, the CO_2 would be at supercritical pressure throughout the cycle, whilst in the former it would be at subcritical pressure in the low-pressure part of the circuit. Thus, in the supercritical cycle condensation of the CO_2 occurs during the later part of the heat-rejection process, some heat having already been rejected during intercooled compression in the superheated vapour region before condensation. Since the critical temperature of CO_2 is only 31 °C, the condensation necessarily occurs at a pressure not much below the critical (from this description, the reader should sketch a temperature–entropy diagram for the cycle). By contrast, in the hypercritical cycle the CO_2 remains a single-phase fluid at a pressure in excess of the critical throughout the cycle. In both types there is a wide variety of possible circuit layouts, of varying complexity, but although theoretical studies of the condensing supercritical cycle have been made,[35,37] the probability of its practical adoption does not seem high. Being a cycle of greater thermodynamic interest in the present context, only the hypercritical cycle is therefore discussed here; it has been studied by several writers.[35–37] A relatively simple version is illustrated in Fig. 8.6.

The features of particular thermodynamic interest in the cycle of Fig. 8.6 are the existence of supercritical pressure throughout the cycle (hence the name *hypercritical*), with compression starting close to the critical point, and the split-flow compression process.[†] Unlike conventional gas-turbine cycles, in which the pressure at compressor inlet is of the order of not more than a few atmospheres and the consequently large specific volume leads to high compression work input (note that in reversible steady flow, compression work input is equal to $\int v\,dp$), the high density of the fluid during the compression process in the hypercritical cycle is advantageous in reducing the compressor work input and therefore in improving the cycle efficiency (just as liquid compression in the Rankine cycle provides one of the chief attractions of steam plant).

The second feature of particular interest in the cycle of Fig. 8.6, namely split-flow compression, results from the fact that the specific heat capacity of the fluid passing through the higher-pressure side of heat exchanger X_B is appreciably greater than that of the fluid passing through the lower-pressure side [as may readily be seen from the respective slopes of the two isobars on the temperature-entropy diagram, noting that $c_p = T(\partial s/\partial T)_p$]. Thus, in order that the temperature drop on the lower-pressure side should be more nearly equal to the temperature rise on

---

†Feher[38,39] made a study of a straightforward CBTX hypercritical CO_2 cycle without the split-flow compression, giving it the name *Feher cycle*. It seems inappropriate, however, to give such a cycle this distinctive appellation, since these types of cycle have been the subject of independent study by a number of other writers. The term *hypercritical cycle*[37] is more descriptive and is much to be preferred.

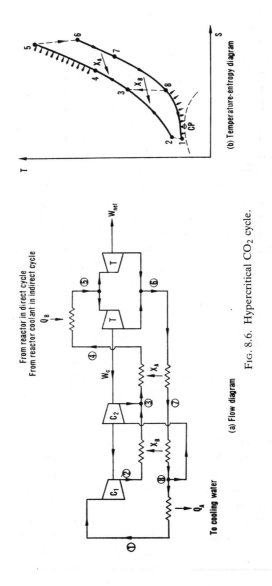

(b) Temperature-entropy diagram

(a) Flow diagram

FIG. 8.6. Hypercritical CO_2 cycle.

the higher-pressure side, a lower mass flow rate must pass through the latter than through the former; hence, about one-third of the flow that passes through the lower-pressure side of heat exchanger X_B is caused to by-pass the higher-pressure side by being diverted at point 8 to compressor C_2. By reducing terminal temperature differences in heat exchanger X_B, this reduces the degree of irreversibility in the heat exchange process and so improves the cycle efficiency.

The hypercritical CO_2 cycle has been studied for both *direct-cycle* and *indirect-cycle* application. In the direct cycle, Q_B in Fig. 8.6 is taken up by the CO_2 as it passes through the reactor, while in the indirect cycle Q_B is taken up by the CO_2 in a heat exchanger from high-temperature helium gas coming from the reactor, through which the helium is circulated as reactor coolant in a non-power-producing closed circuit. It must be stressed that interest in the hypercritical CO_2 cycle has so far been confined almost entirely to theoretical studies.

8.11. Gas-cooled fast-breeder reactor (GCFBR)

Fast reactors operate without the presence of a moderator and have much smaller cores than thermal reactors, so that in them there is a very high energy-release rate per unit volume. Consequently, prototype fast reactors have used a liquid-metal coolant on account of its excellent heat-transfer properties; these reactors are studied briefly in §8.14.4. However, much attention has been given to the possibility of gas cooling of fast reactors. For these, the density of the gas passing through the reactor, and hence the gas pressure, would need to be at least double that in the AGR reactors. The safe containment of such a pressure became a practical possibility with the introduction of the concrete containment pressure vessel.

A helium-steam cycle has been the favoured choice for a possible demonstration plant,[40] but CO_2 has also been considered since it would require a lower gas pressure. Nitrous tetroxide, N_2O_4, is another gas which has been the object of serious study, particularly in the USSR and in the German Democratic Republic.[41] This gas is of special interest because, at reactor temperatures, it dissociates markedly into lower oxides of nitrogen, with considerable absorption of heat. This gives it a high effective specific heat capacity under such conditions, so rendering it ideal for achieving high rates of heat removal in the reactor core. A type of N_2O_4 cycle that has been the subject of some study[41] is the supercritical condensing cycle depicted in Fig. 8.7.

Since the fluid passing through the high-pressure side of the regenerative heat exchanger X in Fig. 8.7 is at supercritical pressure, no boiling takes place in either X or the reactor, so that the latter is cooled by a

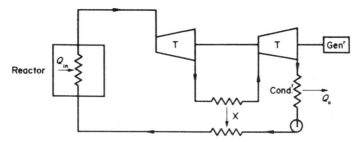

Fig. 8.7. Supercritical condensing N_2O_4 cycle.

supercritical "gas". Cooling a fast reactor in this way would involve only one cooling circuit, instead of the three circuits of the liquid-metal-cooled fast-breeder reactor (LMFBR) discussed in §8.14.4. However, the construction of a demonstration gas-cooled fast reactor still lies in the future,[40,42] whereas prototype LMFBRs have been in operation for one or two decades and full-scale, commercial-size plant have been built in the USSR and France (§8.14.4 hereafter).

In the next section we return to the study of types of plant which have found widespread commercial application.

LIQUID-COOLED REACTORS

8.12. Types of cycle

Liquid-cooled reactors using water either directly or indirectly to generate steam for power production are usually classified as either (a) *pressurised-water reactors* (PWR) or (b) *boiling-water reactors* (BWR). However, since the centre of interest in the present volume lies in the type of cycle rather than in the type of reactor, the classification here will be made in terms of either (a) *direct-cycle* plant, in which the reactor coolant acts also as working fluid and is supplied direct to the power plant, or (b) *indirect-cycle* plant in which, as in the gas-cooled reactor system, the reactor coolant circulates round a closed primary circuit and is used to generate steam as the working fluid in a separate power-producing circuit. Indirect cycles reduce the possibility of the passage of radioactive fluid through the power plant.

Whereas gas-cooled reactors of the Magnox and AGR type need only natural uranium fuel, the liquid-cooled reactors described here require a fuel somewhat enriched in the fissile isotope U^{235}, which comprises only 0.7% of naturally occurring uranium. After the end of the Second World War in 1945, the McMahon Act in the USA denied enriched uranium to that country's wartime partners on the Manhattan atom-bomb project. Consequently, the United Kingdom, France and Canada were obliged to start their development of nuclear-power technology with a neutron moderator which allowed use of natural uranium. The UK and France thus set out on the construction of gas-cooled, graphite moderated reactors, while Canada set out on a successful line of CANDU liquid-cooled reactors, using heavy water (deuterium) as the moderator. In strong contrast to this, having plentiful supplies of enriched uranium from the separation plants installed for the production of nuclear weapons on a truly horrific scale, the USA built up its nuclear power programme on a base of liquid-cooled reactors. As Thomas[M] has pointed out, the roots of that technology were military, liquid-cooled reactors in the USA having been developed initially for submarine propulsion. That, and the withholding of enriched uranium from other countries, gave the USA a great headstart and considerable commercial advantage in the field of civil nuclear power, by the transfer of technology worldwide through licensing and export arrangements. By the 1990s, PWRs in the world outnumbered BWRs in the ratio of about 2:1.

While the BWR bears the disadvantage of delivering lightly radioactive steam to the turbine, containment safety problems are eased in the BWR by the avoidance of the high-pressure water in the primary circuit of the PWR. Thus, comparing Fig. 8.9 for a direct-cycle BWR with Fig. 8.10 for an indirect-cycle PWR, the pressure of the coolant fluid passing through the reactor in the PWR is seen to be double that in the BWR. On the other hand, the PWR has a safety advantage in the fact that the control rods are *lowered* into the reactor, with the assistance of gravity. In the BWR, the control rods are driven *upwards* into the reactor, against the force of gravity, in order to avoid the deposition of minerals on the rods that would occur if they passed through the steam at the top of the reactor.

8.13. Direct-cycle plant

8.13.1. Direct-cycle, dual-pressure, boiling-water reactor (BWR) without superheat

This type of installation is of historical interest, since the first commercial BWR, the Dresden 1 Nuclear Power Station[43] in the USA, was of this kind. It commenced operation in 1960 and was finally shut down in 1978. It is also of interest in bearing a certain similarity to the dual-pressure cycle for gas-cooled reactors. In later plants the external HP steam drum was dispensed with, so that steam separation occurred internally within the reactor vessel. A few other such BWRs were installed, but this type of BWR has given way in popularity to the single-pressure type discussed in §8.13.2.

In the Dreden 1 plant of Fig. 8.8, the light water (H_2O), in which the zircaloy-clad uranium-oxide fuel elements were immersed inside the

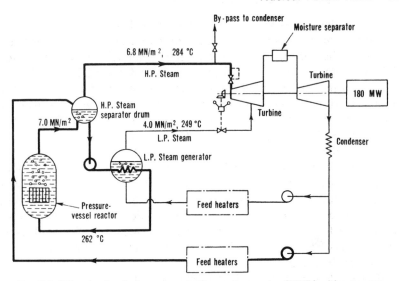

FIG. 8.8. Direct-cycle, dual-pressure, boiling-water reactor (BWR) without superheat (Dresden 1, USA).

reactor pressure-vessel, acted both as coolant and moderator (to slow
down the fission neutrons to the thermal energies at which the chain
reaction can be sustained). The HP steam generated within the reactor
core was separated from the unevaporated water, which was recirculated
through the reactor by a pump, being cooled on its way back to the reactor
by generating LP steam in a heat exchanger. As in the dual-pressure cycle
for gas-cooled plant, that LP steam was supplied to a later stage in the
turbine. Because both the HP and LP steam entered the turbine dry
saturated, a moisture separator was required between turbine cylinders in
order to avoid excessive moisture content in the exhaust stages of the
turbine. At about 29%, the overall station efficiency was low by modern
standards in fossil-fuel plant. This is a feature common to both BWR and
PWR plant, which consequently cause appreciably greater thermal
pollution of the cooling water source than do gas-cooled AGR plant and
fossil-fuel stations.

The direct generation of steam within the reactor has the thermodynamic advantage of eliminating the temperature degradation associated
with heat transfer from a primary to a secondary fluid that occurs in AGR
and PWR reactors. However, it calls for more stringent precautions against
radioactive contamination of the turbine plant in view of the possibility of
carry-over of radioactive salts with the steam. Although this can be
reduced to tolerable proportions by proper treatment of the feed water,
the pressurised-water reactor (in which the steam is generated in a
secondary circuit, as in a gas-cooled reactor) has achieved greater

popularity in the USA and other countries, but the single-pressure BWR type of plant described in §8.13.2 has gained some ground, though to a lesser degree of popularity.

It will be seen that the Dreden 1 plant was really a combination of BWR and PWR, since the LP steam was generated in a secondary circuit, though subsequently mixing with the HP steam. Basically, it was not devised for thermodynamic reasons but to facilitate control of the reactor under varying load. The rate of energy release in the reactor changes with variation in the steam voidage within the core, since steam formation displaces the water moderator and so reduces reactivity. This effect allows the reactor to be self-regulating from 40% to 100% of full load without control-rod adjustment. On increased electrical demand, the turbine speed governor increases the opening of the LP steam inlet valve. The resulting increase in flow rate of secondary steam produces an increased drop in temperature of the HP recirculating water passing through the secondary steam generator, and this results in a lower coolant temperature at reactor inlet. There is consequently an increase in the non-steaming length in the reactor core passages, and the resulting increase in moderator water content raises the reactivity. The greater rate of energy release causes the rate of steam generation in the upper part of the reactor core to be increased to a value which re-establishes equilibrium at the new thermal output required to match the higher electrical load.

Although the dual-pressure cycle was devised to facilitate reactor control, production of steam at two pressures is not essential for that purpose, and the single-pressure BWRs which followed it achieve satisfactory control by adjustment of the rate of forced recirculation of water through the reactor, together with control-rod adjustment.

8.13.2. Direct-cycle, single-pressure, boiling-water reactor (BWR) without superheat

Figure 8.9 gives a simplified circuit diagram showing the initial design conditions for the single-pressure BWR of 515 MW(e) output installed at Oyster Creek, New Jersey,[44,45] in the USA, the net output being later uprated to 650 MW(e). This type has superseded the dual-pressure type described in the previous section and plants of increasing size have since been installed. For example, the reactor and associated turbo-alternator plant at Browns Ferry, Alabama,[46] were of 1100 MW(e) output.

The steam generated in the reactor core passes through steam separators and dryers within the reactor shell before being delivered as saturated steam to the turbine at a pressure which is very low when compared with that in fossil-fuel stations; the steam not being superheated, the tem-

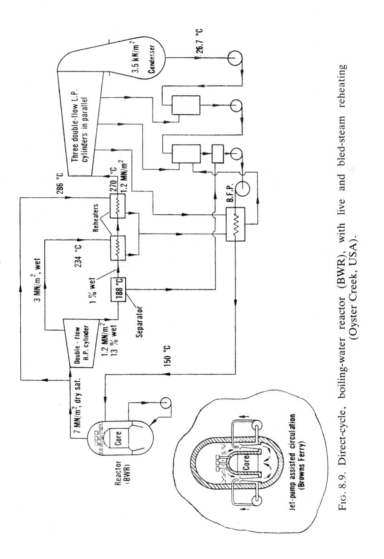

FIG. 8.9. Direct-cycle, boiling-water reactor (BWR), with live and bled-steam reheating (Oyster Creek, USA).

perature is also low, so that steam separators are needed between the HP and LP turbines. Exhaust moisture content is further reduced, and cycle efficiency increased by $1-1\frac{1}{2}\%$,[†] by also reheating the steam at this point with live and bled steam, in two stages, as indicated in the figure. With these relatively low steam conditions, the overall station efficiency is correspondingly low, being only about 33%.

The reactor-boiler combines a system of natural circulation with a system of forced recirculation provided by a number of external centrifugal pumps (assisted, at Browns Ferry, by an interesting system of multiple internal jet pumps, as indicated in the inset of Fig. 8.9, only one-third of the total reactor recirculation water being drawn through the external centrifugal pumps). Variation of reactor output can be brought about by control-rod adjustment, but power control over quite a wide range can also be obtained by varying the rate of forced recirculation of subcooled water through the reactor core, thereby altering the steam voidage in the core and so the reactor reactivity. The provision of adequately subcooled water for recirculation is ensured by designing the plant for a final feed temperature well below the thermodynamic optimum value, as may be noted by comparing the figure quoted in Fig. 8.9 with that listed in Table B.1 in Appendix B for a fossil-fuel plant operating at about the same pressure. This, of course, adversely affects the cycle efficiency. Furthermore, unlike the boiler-turbine plant in a fossil-fuel station, in which the boiler adjusts to a change in steam flow to the turbine, the turbine output here follows a change initiated in the reactor; following an adjustment of reactor output by changing the speed of the recirculating pumps, an automatic pressure regulator adjusts the turbine admission valves to change the steam flow in such a way as to maintain constant pressure in the reactor.

According to Thomas[M], the order for the Oyster Creek BWR was the first to be placed for a nuclear plant by a US utility on the grounds that it was the cheapest option available to it. In the event, the vendor is said to have lost a great deal of money on it, but not before the *apparent* demonstration of the fully commercial status of nuclear power had ended all Government subsidies for liquid-cooled reactors in the USA, Thomas discusses briefly the serious important consequences for the nuclear industry that resulted from the Oyster Creek order.

An earlier prototype plant of 66 MW(e) output, the Pathfinder plant at Sioux Falls[47] in the USA, had an additional superheating section within the reactor core to give a steam outlet temperature of 440 °C, but it was

[†]This gain in cycle efficiency due to reheating is low because it is due solely to the improvement in turbine stage efficiencies resulting from the reduction in moisture content of the steam after the reheaters. There is no increase in mean temperature of heat reception from outside the cycle to give an additional thermodynamic gain, as there is with reheating in the boiler in a fossil-fuel plant.

taken out of commission after a few years and all subsequent BWRs in the USA have delivered only saturated steam.

A British prototype BWR of 100 MW(e) output at Winfrith in Dorset, and known as the Steam Generating Heavy Water Reactor (SGHWR),[48,49] also operated on the direct-cycle, single-pressure system, delivering saturated steam at a pressure of 5.8 MN/m². Unlike the reactors so far described, which are of *pressure-vessel* construction, the SGHWR is of *pressure-tube* construction, and the main moderator is heavy water (deuterium oxide, D_2O), although the light water (H_2O) content of the pressure tubes through which the H_2O coolant pases also contributes significantly to the neutron moderation in the reactor. The zircaloy pressure tubes which contain the slightly enriched uranium-dioxide fuel elements, and in which steam is generated as the H_2O coolant flows through them, are mounted in a calandria vessel containing the D_2O moderator. Since the latter is under low pressure the vessel does not need to be heavy construction. The H_2O is circulated through the pressure tubes by forced circulation pumps, which draw water from the steam-separating drum to which the steam-water mixture from the reactor is delivered. The steam-plant circuit is therefore of traditional layout. As in the Oyster Creek BWR, load adjustment is obtained by causing the turbine output to follow reactor output while keeping constant the steam pressure in the reactor, an arrangement which is permissible for base-load stations.

The pressure-tube construction in the SGHWR not only provides an inherent safety feature, since the need for a large heavy-duty pressure vessel is avoided, but the modulator mode of construction was considered to be readily adaptable to a wide range of reactor sizes. Designs were prepared for commercial plants of 660 MW(e) output but, owing to difficulties in scaling-up to greater outputs and changes in the economic and political outlook on nuclear power, no such plant has unfortunately been ordered.

The SGHWR first went into commercial operation at Winfrith in Dorset in 1967. It is due to be taken out of service in 1992, after 25 years of safe and successful delivery of power to the national Grid.[†]

8.14. Indirect-cycle plant

8.14.1. Indirect-cycle, pressurised-water reactor (PWR) producing saturated steam

The early uncertainties associated with boiling within the reactor, and the possibility of the carry-over of radioactive salts with the steam in the event of inadequate water treatment, led to the development of the *pressurised-water* reactor (PWR) system (also sometimes described as a

---

†For reasons which may not have been solely technological, the SGHWR was taken out of service peremptorily in late 1990.

closed-cycle system). The PWR, stemming from naval designs, became the principal reactor type in the initial nuclear power programmes of both the USA and USSR. In this system, boiling within the reactor is suppressed by maintaining the coolant light water (H_2O) in the closed primary circuit at a pressure in excess of the saturation pressure corresponding to the coolant temperature at reactor outlet (although some localised nucleate boiling within the reactor can be tolerated).

Figure 8.10 gives a simplified circuit diagram showing the initial design conditions for a typical PWR plant, the 460 MW(e) unit at Haddam Neck[50,51] in Connecticut, USA, the net electrical output being later uprated to 575 MW(e). It will be seen from the diagram that, by a process of forced circulation, the primary coolant passes through a heat exchanger on the secondary side of which saturated steam is generated at a much lower pressure; thus, the system is similar to that of a gas-cooled reactor, with the gas coolant replaced by pressurised water, which serves as the reactor moderator in place of the graphite moderator in gas-cooled reactors. A pressuriser vessel connected to the primary circuit controls the primary pressure and provides surge capacity.

Since there is no bulk boiling within the reactor core, it is not possible to adopt the method of reactor control used in the BWR plant described in §8.13.2, namely adjustment of void fraction by forced recirculation of subcooled water, so that variation of reactor output is achieved solely by control-rod adjustment. The conditions that dictated a low final feed temperature in the BWR plant of Fig. 8.9 are consequently not present in the PWR, and it will be seen that the final feed temperature in Fig. 8.10 [†] is much higher than that in Fig. 8.9, in spite of the lower pressure of the steam supplied to the turbine. However, the improvement in cycle efficiency resulting from the higher feed temperature is offset by the adverse effect of the lower steam pressure in an indirect cycle, so that the overall station efficiency (31.5%) of the PWR plant at Haddam Neck is a little lower than that of the BWR plant at Oyster Creek (33%). Thus, as in BWR plants, the overall station efficiency of PWR plants is low in relation to that of fossil-fuel stations because of the low steam pressure and the absence of steam superheating. Again, steam separators and live-steam reheating together help to maintain an acceptable moisture content at turbine exhaust and to boost the efficiency.

A similar type of indirect steam cycle is associated with the CANDU type of reactor, typified by the 752 MW(e) units at Bruce[52] and the 600 MW(e) standard units[53] at Point Lepreau in Canada. However, these use pressurised heavy water (D_2O) as both moderator and coolant in a *pressure-tube* reactor having a pressure of about 10 MN/m² in the primary

---

†To facilitate calculation in Problem 8.6, Fig. 8.10 depicts surface feed heaters throughout the feed heating system, but it will be appreciated that, in practice, the LP feed heaters before the boiler feed pump are more usually of the direct-contact type.

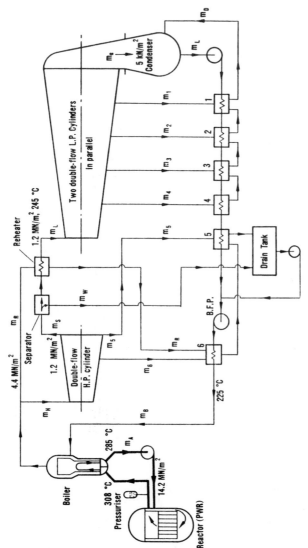

Fig. 8.10. Indirect-cycle, pressurised-water reactor (PWR), with live-steam reheating (Haddam Neck, USA).

circuit. In the secondary circuit, the steam (H_2O) supplied to the turbine at about $4.5 \, MN/m^2$ is only dry saturated, so that again steam separation and live-steam reheating between turbine cylinders are used and the overall station efficiency is only about 29%. These reactors use natural uranium (UO_2) and have good neutron economy but require a source of heavy water.

In spite of their low thermal efficiency, PWR plants of 1100 MW(e) unit capacity and upward have been built. France, with its lack of indigenous fossil-fuel resources, and its more *dirigiste* political system, has latterly concentrated solely on *standardised* lines of PWR plant. In consequence, by the 1990s about 75% of France's electrical power had become nuclear generated. The resulting array of nuclear stations lining the Channel coast of France, and the existence of the world's largest nuclear reprocessing plant at Cap de la Haye, near Cherbourg, is viewed with a certain apprehension by those living on the south coast of England.

A somewhat more detailed study of PWR plant of the latest designs, both under construction and proposed, is given in §§8.15–8.19 at the end of this chapter, but brief studies of other types of indirect-cycle plant are made first.

8.14.2. Indirect-cycle, boiling-water reactor (BWR) with graphite moderator

In the direct-cycle BWRs described in §8.13, direct generation of steam within the reactor can lead to the carry-over of radioactive salts with the steam if water treatment is inadequate. This possibility was avoided in the indirect-cycle BWR shown in Fig. 8.11, which gives a simplified circuit diagram showing the initial design conditions for a 100 MW(e) prototype Beloyarsk plant in the USSR.[54]

Like the British SGHWR, the Beloyarsk reactor was of *pressure-tube*, instead of pressure-vessel, construction. However, unlike the SGHWR, in which the main neutron moderator is heavy water (D_2O), the Beloyarsk reactor had a graphite core to act as main moderator, with further moderation provided by the water content. The enriched-uranium fuel was contained within a multiplicity of tubular elements, of double-tube construction, mounted vertically in the graphite block, only a single evaporating element and a single superheating element being shown in Fig. 8.11 for the sake of simplicity. Each tubular element comprised two concentric tubes, the high-pressure H_2O coolant flowing through an inner pressure tube and the fuel being contained in the annular space between the two tubes, both of which were of stainless steel. With this mode of construction a high primary steam pressure could safely be used, and that in turn allowed the secondary steam to be generated at a reasonably high pressure. Consequently, with the addition of appreciable superheat, the

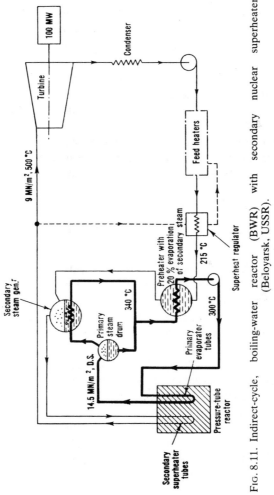

Fig. 8.11. Indirect-cycle, boiling-water reactor (BWR) with secondary nuclear superheater (Beloyarsk, USSR).

condition of the steam supplied to the turbine, and the overall station efficiency of about 33%, approached the values obtaining in fossil-fuel plant of similar output. Nitrogen filling of the graphite core enabled the reactor to operate at the temperature required to give the high degree of superheat.

The primary steam generated in the reactor at a pressure of $14.5 \, \text{MN/m}^2$ was not supplied to the turbine but was used to complete the evaporation of the secondary steam, about one-fifth of which was produced in the preheater through which the recirculated primary water and the condensed primary steam were passed on their way back to the reactor. Only the non-radioactive secondary steam at a pressure of $9 \, \text{MN/m}^2$ was supplied to the turbine and it was superheated in the reactor to 500 °C before passing to the turbine.

In a second plant at Beloyarsk,[55,56] of twice the electrical output of the first, the graphite-moderated, pressure-tube reactor was retained, but the indirect cycle was replaced by a simple forced-circulation, single-pressure, direct cycle with nuclear superheating. The cycle thus became similar to that of a fossil-fuel plant, the reactor simply replacing the fossil-fuel-fired boiler. Use of the indirect cycle in the first plant arose from the fear of carry-over of radioactive salts with the steam passing to the turbine. Close attention to water treatment and control in the second plant permitted change to a direct cycle.

In later reactors of this type,[57] giving outputs of 1000 MW(e) and 1500 MW(e), each with four steam drums and two turbo-generators, superheating of steam in the reactor was abandoned, so that they produced only saturated steam at a pressure of about $6.4 \, \text{MN/m}^2$. It was a reactor of this type, known as RBK-1000, which in 1986 suffered a truly catastrophic explosion at Chernobyl, north of Kiev, in the Ukraine. Further reference to that disaster is made in §8.19.

8.14.3 Indirect-cycle, boiling-water reactor (BWR) with superheating by fossil fuel

Practically all BWR plant deliver saturated steam to the turbine. An exception was the 260 MW(e) unit near Lingen,[58] in the Federal Republic of Germany, but there the superheating was provided by a separate oil-fired superheater. The plant was also unusual in that the saturated steam generated in the reactor at a pressure of about $7 \, \text{MN/m}^2$ was confined within a primary circuit in which it was simply condensed in a heat exchanger and then subcooled by secondary-circuit feed water before being returned as feed water to the reactor; in this respect, there was some resemblance to the Beloyarsk circuit described in §8.14.2. The condensation of the primary steam was brought about by the generation of saturated steam at a lower pressure on the secondary side of the heat exchanger, and this secondary steam was superheated in an oil-fired superheater before being delivered to the turbine at about 4.4MN/m^2 and $530\,°\text{C}$. The condensate from the condenser was fed back to the secondary side of the heat exchanger after being first heated in a bled-steam feed-heating system and then further heated by sub-cooling the high-pressure primary fluid on its return from the heat exchanger to the reactor. From this description the reader will find it instructive to sketch a circuit diagram for the plant. The provision of superheating in the oil-fired superheater enabled the plant to achieve an overall station efficiency of about 33%, with the superheating accounting for around one-third of the total electrical output of 260 MW. However, a metallurgical problem known as *intergranular stress-corrosion cracking* seriously affected early BWRs, and the Lingen plant had an operating life of only about 10 years. No further plants with fossil-fuel-fired superheating were built.

8.14.4. Indirect-cycle, liquid-metal fast-breeder reactor (LMFBR)

Because the boiling-point of sodium at atmospheric pressure is about $880\,°\text{C}$, its use as a reactor coolant permits the attainment of high temperatures without subjecting the reactor to the high pressures associated with water cooling. Furthermore, a liquid-metal coolant with its excellent heat-transfer properties is ideally suited to the cooling of *fast reactors*, which operate without the presence of a neutron moderator, for these have small cores in which there is a very high energy-release rate per unit volume. The use of sodium as a coolant also has advantages from the point of view of reactor physics.

The first experimental plant of this type to be constructed in the United Kingdom was that at Dounreay,[59] in the north of Scotland. This plant was followed at Dounreay by the full-scale commercial prototype fast reactor (PFR) of 254 MW(e) output,[60] of which a simplified circuit diagram is given in Fig. 8.12. The PFR has three seperate closed circuits in sequence, namely a primary and a secondary circuit round each of which liquid sodium is circulated, and a tertiary power-producing circuit with H_2O as the working fluid. The degree of radioactivity induced in the primary sodium coolant in its passage through the reactor necessitates the use of a secondary sodium circuit as a buffer between the primary coolant and the H_2O in the tertiary steam-generating circuit.

The entire primary circuit of the PFR is contained within a stainless-steel vessel of about 12 m diameter and 15 mm wall thickness. The sodium circulating round the closed secondary circuit first passes in parallel through similar shell-type heat exchangers, one of which serves as a steam superheater and the other as a steam reheater. The two sodium streams emerging from there reunite to pass through a third shell-type heat

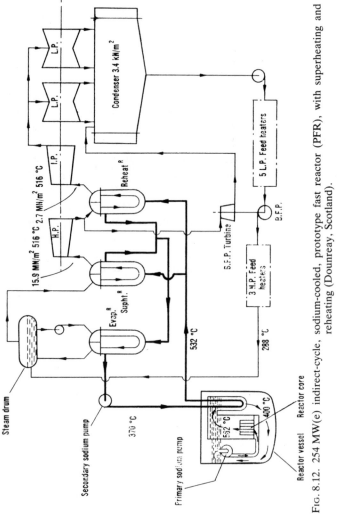

FIG. 8.12. 254 MW(e) indirect-cycle, sodium-cooled, prototype fast reactor (PFR), with superheating and reheating (Dounreay, Scotland).

exchanger which serves as an evaporator in which steam is generated from H_2O passing through the tubes. From the evaporator the sodium returns to the secondary side of the sodium–sodium heat exchangers, which are mounted within the reactor containment vessel. The superheated steam is delivered to the turbine at $15.9\,MN/m^2$ and $516\,°C$, with reheating to $516\,°C$ and bled-steam feed heating to $288\,°C$.

Although the steam conditions in the PFR are comparable to those in the AGR station at Heysham quoted in §8.8, the final feed temperature is appreciably higher, since the design final feed temperature in a sodium-cooled plant is free from the restrictions in a gas-cooled plant that are discussed in §8.6; this freedom arises from the fact that the thermal capacity of a liquid-metal coolant is much greater than that of a gas, so that the temperature range of the coolant is much smaller. At 42.3%, the overall station efficiency of the PFR plant is thus a little higher than that of the AGR at Heysham; thus both steam conditions and station efficiency are well on a par with those of the most advanced fossil-fuel stations.

Design studies[O] were made for a 1320 MW(e) Commercial Demonstration Fast Reactor (CDFR) to follow the PFR. That would have had steam-steam reheating at the turbine,[61] instead of by a reheater in the secondary sodium circuit. However, the economic and political outlook on nuclear power, and on fast reactors in particular, prevented realisation of these plans. In France, on the other hand, a full-scale commercial plant of 1200 MW(e) output has been built at Creys-Malville; this is the Super Phénix.[62] It is similar in concept to the Dounreay PFR, but without reheating in the steam cycle, in which steam is supplied to the two turbines at $17.7\,MN/m^2$ and $487\,°C$.

At the end of the Eighties, sodium-cooled fast-breeder reactor plant (LMFBRs) of 350 MW and 600 MW electrical output had been in successful operation in the USSR for respectively 16 and 9 years. Prior to the catastrophic destruction of the thermal reactor at Chernobyl (§8.19), the Soviet intention was to build a series of 800 MW LMFBR stations, followed by twin-turbine 1 600 MW stations. The plant parameters for those, not greatly different from those of the PFR at Dounreay are given in Ref. P. However, the following year, the upbeat tone of that report was damped down in a short report by Tryanov[Q], when he stated that, following increased attention in the USSR to safety of nuclear plant following the Chernobyl catastrophe, the beginning of large-scale deployment of commercial LMFBRs was being postponed. In that context, it is of interest to note that both those reports draw attention to the importance of the *natural circulation process* (Appendix D) in helping to ensure safe operation, particularly in the disposal of decay heat in the reactor. That process also features prominently in the design of "safer" pressurised-water reactors (PWRs) discussed in §8.15 *et seq.*

An LMFBR enables plutonium to be bred from uranium-238 at a rate greater than the primary fuel is consumed in the reactor core. It is also a very efficient user of plutonium. Such reactors would therefore be able to make good use of the increasing quantities of this fissile material that will result from the operation of the large number of thermal reactors that are being installed throughout the world. They could therefore make vastly better use of the world's limited uranium supplies than can thermal reactors alone. It remains to be seen, however, whether society will accept the grave environmental hazards that might be associated with their widespread installation and operation.

PWR PLANT–FURTHER STUDIES

8.15. Introduction

Except in the United Kingdom, currently building its first pressurised-water reactor at Sizewell B, the PWR has become the most widely adopted type of nuclear power plant worldwide, outnumbering the rival boiling-water reactor (BWR) by a factor of about 2:1. The discussion of PWR plant in §8.14.1 centred on the associated steam cycle. It is therefore appropriate here to give more detailed attention to the design of the steam generators (boilers) themselves. This type of PWR plant, which may now be classified as "conventional", has been criticised as lacking inherent safety because of the numerous, very large interconnecting pipes that carry the primary hot water under high pressure between the reactor and its encircling steam generators. Because of this, some interesting attempts have been made to design PWR plant with a greater degree of ultimate "safety". Two which are discussed below are the *Safe Integral Reactor* (SIR) and the PIUS (*Process Inherent Ultimate Safety*) design. The former is of British origin and the latter Swedish.

In discussing these three types of PWR plant, the different modes of circulation of the primary and secondary fluids through the steam generators are of particular interest. All involve, in varying degrees, the process of *natural circulation*, which is treated theoretically in some detail in Appendix D.

8.16. Sizewell B steam generator

The Sizewell B power plant will have two 600 MW units and a thermal efficiency of 32.5%, low by the standards of modern fossil-fuel plant. The reactor/boiler tie-up will be basically the same as that depicted in Fig. 8.10 *ante*. Each reactor will supply hot pressurised water to four steam generators (boilers) in parallel, each with its own primary pump for circulating the pressurised-water reactor coolant through the heat-exchanger tubes, by a process of *forced circulation*. This arrangement of multiple steam generators involves eight runs of connecting pipework of large diameter. For reasons of safety, these pipes must be of very high integrity, since they carry hot water under high pressure and complete rupture could be catastrophic.

A somewhat simplified diagram of a Sizewell B steam generator is given in Fig. 8.13, which is based on the information given in Ref. R. It will be seen that the primary pressurised water from the reactor passes through the tubes of an inverted-U heat exchanger. The secondary feed water from the turbine feed-heating system is fed into a circular distributor ring near the top of the boiler. From this ring, the feed water drops through nozzles in to the recirculating-water plenum, where it is joined by unevaporated water coming from the steam/water separators. The combined flow of fresh feed water and recirculated water then passes downwards through the annular *downcomer* passage, on the periphery of the tube bundle. The direction of the water flow is reversed at the bottom of the annulus, just above the tube-plate, the water then travelling upwards over the tubes in the central *riser* leg, where a steam/water mixture is generated. This flow, downwards and then upwards, takes place by the gravitational process of *natural circulation*. (See Appendix D.)

The mixture at the top of the riser leg comprises approximately 25% by weight of steam, and 75% water. Being a very wet mixture, a considerable array of water separators has to be provided in the upper part of the vessel. These separators reduce the water content at exit to less than 0.25%, so that the steam delivered by the generator is virtually dry saturated. As noted above, the separated water is recirculated to join the incoming feed water.

8.17. SIR – Safe Integral Reactor

SIR is an advanced collaborative design project of the United Kingdom Atomic Energy Authority as a member of a consortium of four UK/US organisations. Unlike the Sizewell B plant, the SIR design is for an integral PWR, in which all major primary circuit components are contained within a single pressure vessel. The projected plant is designed as a standard unit of 300 MW electrical output, but would be suitable for a 600 MW station with twin units, or a 1200 MW station with four units.

While Fig. 8.13 depicts just one of the four steam generators per reactor in the Sizewell B plant, Fig. 8.14 depicts the entire assembly of the SIR plant, in which the reactor core, pressuriser, 12 modular steam generators and 6 primary coolant pumps are all contained within a single pressure vessel. As a particular safety feature, it is intended that the main part of this primary containment vessel would be installed below ground level. It may be noted that, unlike the conventional PWR plant depicted in

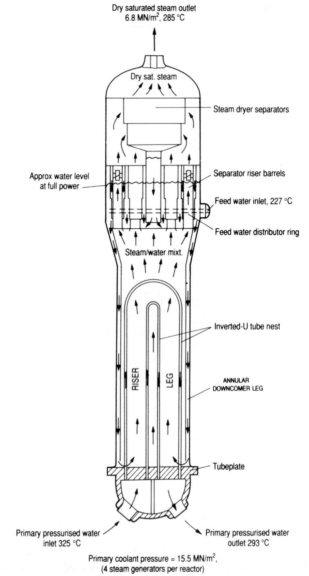

FIG. 8.13. Simplified diagram of Sizewell B steam generator. [After Beckett and Clarke[R]]

Fig. 8.10 *ante*, and the Sizewell B plant, the pressuriser is not a separate vessel, but is formed by the dome of the pressure vessel itself.

From Fig. 8.14 it will be seen that, in SIR, circulation of the primary coolant through the reactor core is provided by a process of *assisted natural circulation*. In this, the process of *natural circulation* (Appendix D), with

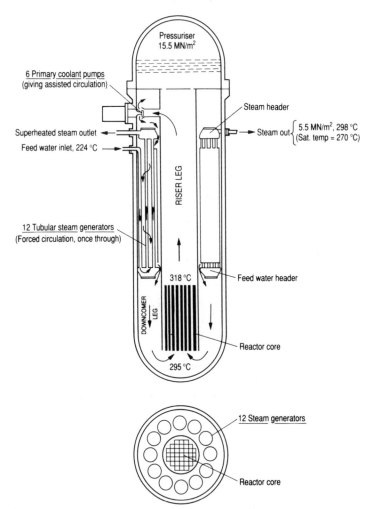

FIG. 8.14. Layout of Safe Integral Reactor (SIR) (depicting circulation of primary and secondary fluids). [After Hayns[S]]

the pressurised water being cooled as it flows down past the twelve heat exchangers in the annular *downcomer* leg and heated in the reactor core as it flows up the central *riser* leg, is assisted by a degree of *forced circulation* provided by the six pumps spaced at equal circumferential intervals around the upper part of the vessel. This process of *assisted natural circulation* contrasts with the process of purely forced circulation of the primary pressurised water through the inverted-U tubes of the Sizewell B heat exchanger, and it gives a distinct advantage in respect of safety of reactor operation. That is because the plant could operate at 20% load even when the six primary pumps were out of action, sufficient cooling flow through the reactor core being then provided by the process of gravitational natural circulation alone. This would enable the plant to cope with the emission of "decay heat" from the reactor that continues for some time after reactor shut-down or "scram", even when the primary coolant pumps were not in operation. There would also be an external closed-cycle system relying solely on natural circulation, and so able to function in the total absence of A.C. power.

The flow of the steam-generating secondary fluid in SIR also contrasts with that in Sizewell B. While, in the latter, the secondary fluid flows upwards over the *outside* of the heat-exchanger tubes by a process of natural circulation, in SIR the secondary fluid flows *through* the tubes of the heat exchangers. Feed water from the boiler feed pump of a conventional turbine/condenser system is fed to feed water headers through the cool *downcomer* legs, and steam generated in the tube nests of the *riser* legs is discharged to steam headers at the top. It is anticipated that restricting orifices may be required at feed-water inlet to each steam generator, in order to ensure stability between the twelve boilers operating in parallel.

The heat-exchangers (boilers) are of the *once-through* type, there being a complete change of phase from water to steam as the secondary fluid is heated in passing through the vertical riser tubes. Moreover, the design legislates for the steam at exit to be superheated by about 28 K. Thus, in SIR, there is no need at all for a complex assembly of steam/water separators, such as are required within the steam generators of Sizewell B.

The consortium responsible for SIR has set as its goal the achievement of a licensing certificate by the mid 1990s. If that were achieved, a pilot Safe Integral Reactor might be in operation around the turn of the century, possibly at Winfrith in Dorset, where the SGHWR discussed in §8.13.2 was taken out of service peremptorily in late 1990. For an informative and much fuller description of this interesting project, the reader may consult Ref. S.

8.18. PIUS reactor – Process Inherent Ultimate Safety

This novel PWR concept, conceived in Sweden, is designed to ensure that the process which would follow in the event of a major malfunction of the reactor would be inherently safe; hence the acronym *PIUS – Process Inherent Ultimate Safety*. Figure 8.15, which is highly diagrammatic, is based on one given by Rowland in an excellent article entitled *The design of safe reactors*.[T]

The unique feature of PIUS is the immersion of the entire reator within a prestressed-concrete pressure vessel full of a solution of borate in water under essentially the same pressure as that of the pressurised water in the

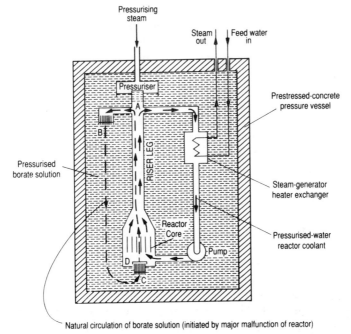

FIG. 8.15. Conceptual PIUS pressurised-water reactor (PWR)
[After Rowland.[T]].

main reactor circuit. Borate is a very effective neutron absorber, so that passage of that solution through the reactor would automatically close down the nuclear reaction.

The borate solution would not normally be allowed to enter the main reactor circuit containing the pressurised-water coolant, in spite of the fact that that circuit is open to the immersant borate solution at B and C. Such flow of the immersant solution into or out of the main circuit would be avoided by ensuring, by careful control, that $p_A = p_B$ and $p_C = p_D$ under normal operating conditions. This would mean that the following condition would have to be satisfied:

$$(p_C - p_B) = (p_D - p_A). \tag{8.6}$$

This can only be achieved by control of the flow rate of the pressurised-water coolant up the riser leg. That fact can best be understood by reference to the following equation from Appendix D:

$$-\int dp = g \int \frac{dz}{v} + \int \frac{V \, dV}{v} + \int \frac{dW_f}{v}. \tag{D.19}$$

The terms on the right-hand side of this equation represent respectively hydrostatic, accelerational and frictional pressure drops. With no flow from B to C under normal conditions, and since the density of the coolant fluid in the heated riser leg DA will be less than the density of the cooler borate solution between B and C, equation (8.6) can only be satisfied by so setting the flow rate up the riser leg DA that the sum of the hydrostatic, accelerational and frictional pressure drops from D to A is equal to the hydrostatic pressure difference alone between C and B.

In the event of a major malfunction of the reactor tending towards gross overheating, the resulting rapid decrease in density of the circulating fluid in the riser leg DA would initiate, without human intervention, the *natural circulation* of borate solution through the reactor, as indicated by the dotted line in Fig. 8.15. This would automatically close down the nuclear reactor, so scramming the reactor; hence the claim of "inherent ultimate safety".

Such control of flow rate would clearly imply sophisticated design and computerised operation. While computer modelling at the laboratory stage has been carried out, it is believed that the design has not yet been taken up commercially. The chances of that happening could not have been improved by a decision of the Swedish Government, post-Chernobyl, to start phasing out nuclear power in Sweden by the year 1995,[†] with completion by the year 2010.

8.19. Conclusion

The need to build 'safe' reactors has been more than amply demonstrated by four known disasters that have already occurred, *viz*:

(1) In 1957, in the UK, a calamitous fire in the gas-cooled, graphite-moderated reactor at Windscale (Sellafield). That arose from the *Wigner effect* in the graphite which was imperfectly understood at the time. Information on the resulting widespread dispersion of radioactive fallout over Britain and into Europe was withheld for many years, under orders coming from the highest political level. The reactor was built for the production of bomb material.

(2) In 1975, in the USA, a near disaster in the PWR plant at Browns Ferry, when the main control panel was put out of action by a fire. The fire was caused by a technician who was following the tradiational practice of using a candle flame to locate air leakage into piping under vacuum. The candle set alight the *flammable* insulation of the electrical wiring under the control panel. A *nuclear* power station nearly brought to disaster by a *candle* and a careless technician!

(3) In 1979, in the USA, a hair-raising near melt-down of the reactor core of a PWR at Three Mile Island in the Susquehanna river, ten miles from the town of Harrisburg. This resulted from a combination of equipment malfunction, human error, misjudgement and inade-

†In February 1991, postponement of such action was announced.

quate design. The disasterous destruction of the reactor core, and the release of considerable quantities of radioactivity into the atmosphere, effectively ended the ordering of nuclear power plants for installation in the USA.

(4) In 1986, in the USSR, a truly catastophic destruction of a water-cooled, graphite-moderated reactor at Chernobyl on the Pripyat river north of Kiev (see §8.14.2). There was widespread dispersion of long-lived nuclear isotopes throughout Europe. Serious long-lasting effects on farms and agriculture in parts of Wales, England and Scotland, some 1500 miles from Chernobyl, still persisted four years later. The catastophe occurred during an operational experiment which went dreadfully wrong, due to operator errors and misjudgements, on a reactor which had itself been seriously criticised on safety grounds.

Chilling descriptions of the avoidable operational and design faults that led to the last three of these alarming accidents will be found in References N and U, while Ref. T also discusses in some detail those at Three Mile Island and Chernobyl.

Since these disasters struck as a result of both operator error and design fault, it should be self-evident that, if the building of nuclear power plants is to continue, it is an absolute necessity that they should possess the property of inherent ultimate safety. In the real world, however, technological considerations may well continue to be downgraded, and indeed sometimes overruled, as a result of military and commercial pressures, and of political dogma and ideology. The student could therefore profit from reading two books published in 1988, *The Realities of Nuclear Power* by Thomas[M], and *Technocrats and Nuclear Politics* by Massey.[U]

Nuclear power has sadly lost its original "shine". This is not only because of what Thomas has fittingly called its "military contamination", but also on grounds of safety, cost and failure to tackle with sufficient urgency the serious problems of nuclear waste disposal. With respect to cost, "creative accountancy" has indeed for many years appeared to work miracles, but all resulting delusions were swept away when the British Government was obliged to prepare *honest* prospectuses for the sale of the nuclear generating industry into private hands. As a result, the plans had to be abandoned. The student of such matters might possibly be excused for coming to the conclusion that information on the subject had suffered more than its fair share of military, political and commercial obfuscation, if not outright mendacity. That situation has certainly inhibited the taking of rational decisions. Contrariwise, in the matter of safety, the hard-pressed British nuclear industry might well be tempted to ask the student to compare its own safety record with the 5000 or so deaths *every year* on

Britain's roads alone which result from the massive use of that other technical invention, the internal-combustion engine; to say nothing of the resulting and appalling atmospheric pollution.

There is no doubt that, in the face of the so-called *greenhouse effect*, mankind faces agonisingly difficult decisions on how to satisfy its voracious appetite for energy. It used to be thought that the *fission* reaction would ultimately yield place to the *fusion* reaction[64] as a source of large-scale power, but realisation of that hope still appears to be a far-off prospect. However, it is of interest to note that a conceptual form of such a plant, put forward in 1971,[63] proposed primary, secondary and tertiary circuits, just as with the sodium-cooled fast reactor depicted in Fig. 8.12 *ante*. That proposal envisaged molten lithium as primary coolant, molten potassium in the secondary circuit, and H_2O in the tertiary circuit, which would be that of a conventional steam-turbine plant. Thus it was conjectured that, even in that distant future, mankind would still be living in the Age of Steam for the large-scale generation of electrical power, as has always been so from the earliest days of the Industrial Revolution.

Problems

For the solution of Problems **8.1** to **8.6**, use is recommended of the **UK Steam Tables in SI Units 1970.**[2]

Problems **8.1** to **8.5** relate to the Calder Hall type of plant illustrated in Figs. **8.1** and **8.2**.

8.1. At the design load the feed water is heated just to the saturation temperature of the LP steam in each section of the mixed economiser, the minimum temperature approach between the two fluids is 17 K in both the HP and LP heat exchangers, and the conditions in the steam-raising towers are as follows:

Temperature of entering CO_2	= 337 °C
Temperature of entering feed water	= 38 °C
HP steam conditions at exit	= 1.45 MN/m², 316 °C
LP steam conditions at exit	= 0.435 MN/m², 177 °C

Treating CO_2 as a perfect gas with $c_p = 1.017$ kJ/kg K, and neglecting pressure drops and external heat losses in the heat exchangers, make the following calculations for the plant per kg of CO_2 circulated:

(1) Calculate the masses of HP and LP steam produced, the temperature of the CO_2 at exit from the towers, and the heat transferred in the steam generators.

(2) Taking as the environment temperature that of the circulating water at inlet to the condenser, which is at 24 °C, calculate the energy available for the production of work from (a) the heat transferred from the CO_2 in its passage through the steam generators, and (b) the H_2O. In each case express the available energy as a percentage of the heat transferred in the steam generators.

(3) Calculate the reduction in the available energy of the steam if the environment temperature is taken as being the saturation temperature of the steam in the condenser, in which the pressure is 6 kN/m². Express the new available energy as a percentage of the heat transferred in the steam generators.

(4) If the dual-pressure cycle were to be replaced by a single-pressure cycle, producing steam at 316 °C from feed water at 38 °C, with the inlet and exit temperatures of the CO_2 and the minimum temperature approach between the two fluids the same in the dual-pressure cycle, determine (a) the steam pressure, and the mass of steam

produced, (b) the available energy for an environment temperature of 24 °C, again expressing it as a percentage of the heat transferred in the steam generators.

Answer: (1) 0.0560 kg; 0.0176 kg; 130.5 °C; 210.0 kJ.
            (2) 85.1 kJ; 40.5%; 68.7 kJ; 32.7%.
            (3) 5.8 kJ; 29.7%.
            (4) 0.418 MN/m^2; 0.0714 kg; 60.2 kJ; 30.0%

8.2. There is a pressure drop of 70 kN/m^2 and a temperature drop of 6 K in each of the steam mains between the steam generators and the turbine, and the condenser pressure is 6 kN/m^2. The internal isentropic efficiency of the HP section of the turbine is 85% and of the LP section is 80%.

(1) Determine the specific enthalpy of the steam at the end of the HP section of the turbine, and the shaft work obtained from this section per kg of CO_2 circulated through the reactor.
(2) Assuming perfect mixing of the HP exhaust steam and LP inlet steam, determine the specific enthalpy of the steam entering the first stage of the LP section of the turbine, the specific enthalpy of the steam at LP turbine exhaust, and the shaft work obtained from this section of the turbine per kg of CO_2 circulated through the reactor.
(3) The combined efficiency factor for the turbine external losses and alternator losses is 94%. Calculate the output from the alternator terminals per kg of CO_2 circulated through the reactor, and express this output as a percentage of the heat transferred in the steam generators.

Answer: (1) 2811 kJ/kg; 14.2 kJ.
            (2) 2808 kJ/kg; 2311 kJ/kg; 36.6 kJ.
            (3) 47.8 kJ; 22.8%

8.3. The pressure of the CO_2 at inlet to the gas circulators is 0.8 MN/m^2; they provide a pressure rise of 28 kN/m^2 and their internal isentropic efficiency is 80%. Estimate the temperature rise of the CO_2 in its passage through the circulators, and calculate the thermal output of the reactor per kg of CO_2 circulated.

Answer: 3.2 K; 206.7 kJ.

8.4. The combined efficiency factor for the external losses in the circulators and the losses in their driving motors is 93%. At the design load the mass flow rate of CO_2 is 880 kg/s. Calculate the following quantities at this load, all expressed in MW:

(1) The thermal output of the reactor.
(2) The electrical output at the alternator terminals.
(3) The total input to the circulator motors. Express this as a percentage of (a) the electrical output at the alternator terminals, and (b) the reactor thermal output.
(4) The net electrical output of the station, and the overall station efficiency, neglecting the power required for other auxiiary plant. For comparison, calculate the ideal thermal efficiency of a cyclic heat power plant working between a source temperature equal to the maximum permissible fuel-element temperature of 400 °C and a sink temperature equal to that of the environment at 24 °C.

Answer: (1) 181.9 MW. (2) 42.1 MW. (3) 3.1 MW; 7.4%; 1.7%.
            (4) 39.0 MW; 21.4%; 55.9%.

8.5. The reactor uses natural uranium fuel, each tonne (10^3 kg) of which contains 7 kg of the fissile U^{235}, of which half can be consumed before the fuel elements have to be withdrawn from the reactor. The U^{235} is consumed at the rate of 1.3 g per megawatt-day of reactor thermal output. What steady rate of fuel replacement would be required if the reactor operated continuously at the design load?
If the heat transferred in the steam generators were supplied in a boiler of 85%

efficiency and burning coal having a calorific value of 25.6 MJ.kg, what would be the rate of coal consumption at the design load?

Answer: 67.6 kg/day; 734 tonne/day.

8.6. For the operating pressures and temperatures in the PWR plant illustrated in Fig. 8.10. perform the following calculations. Frictional pressure drops and stray heat losses are to be neglected, and the boiler may be assumed to deliver dry saturated steam at the pressure indicated.

(1) Determine the steam wetness at HP turbine exhaust, given that the isentropic efficiency of the HP turbine is 81%.
(2) Determine the fraction of the wet steam mixture supplied to the separator that is drained off as saturated water to the drain tank, given that the steam wetness at separator exit is 1%.
(3) Determine the steam wetness at LP turbine exhaust, given that the isentropic efficiency of the LP turbine is 81%.

It may be assumed that the heating steam is just condensed in the reheater and in all six feed heaters, and that the feed water leaving each heater is raised to the saturation temperature of the bled steam supplied to the heater. The work input to all pumps may be neglected, the specific enthalpy of water at any temperature being taken as equal to the saturation specific enthalpy at that temperature.

(4) From energy balances on the reheater, on feed heater 6 and on feed heater 5, calculate the values of the following ratios of the flow rates indicated in Fig. 8.10: (a) m_H/m_B, (b) m_R/m_B, (c) m_L/m_B.

The bled steam supplied to feed heater 6 is at a pressure of 2.55 MN/m^2 and its specific enthalpy is 2715 kJ/kg.
(5) From energy balances on feed heaters 4, 3, 2 and 1 in turn, calculate the bled steam quantities supplied to these heaters, each expressed as a fraction of m_B, and thence evaluate m_D/m_B and m_C/m_B.

The following table gives the condition of the bled steam supplied to each of these feed heaters:

Heater no.	4	3	2	1
Pressure, MN/m^2	0.585	0.248	0.088	0.0245
Specific enthalpy, kJ/kg	2803	2674	2538	2392

(6) Determine the thermal efficiency of the steam cycle.
(7) From an energy balance for the boiler, determine the value of m_A/m_B.
(8) Taking the environment temperature as being equal to the saturation temperature of the steam in the condenser, calculate the energy available for the production of work from (a) the high-pressure fluid passing through the reactor, and (b) the low-pressure fluid passing through the boiler. In each case, express the available energy as a percentage of the thermal output of the reactor.
(9) Neglecting pressure drops on both sides of the boiler, determine the minimum temperature approach between the two fluids in their passage through the boiler (cf. the *pinch points* discussed in §8.2 for gas-cooled reactors).

Answer: (1) 9.1%. (2) 0.0817. (3) 13.4%.
(4) (a) 0.930, (b) 0.070, (c) 0.735.
(5) 0.166, 0.569. (6) 33.6%. (7) 14.6.
(8) (a) 46.3%, (b) 42.0%. (9) 31.0 K.

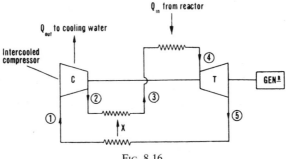

FIG. 8.16.

8.7. Figure 8.16 shows a circuit diagram for a closed-circuit gas-turbine plant to be used in conjunction with a projected nuclear power plant incorporating a high-temperature, gas-cooled, fast reactor.

Sketch the temperature–entropy diagram for the corresponding ideal cycle in which a perfect gas flows round the circuit, there are no stray heat losses, no mechanical losses and no pressure drops in the heat exchangers and ducting, the expansion in the turbine is reversible and adiabatic, the compression in the compressor is reversible and *isothermal*, and heat exchanger X has zero terminal temperature differences, so that $T_1 = T_2$ and $T_3 = T_5$. Derive an expression for the thermal efficiency of this ideal cycle in terms of T_1, T_4, the pressure ratio of compression, r_p, and the ratio of specific heat capacities, γ.

It is estimated that, when the temperatures at points 1 and 4 are respectively 25 °C and 660 °C and $r_p = 3$, the actual thermal efficiency of a plant of this kind using helium as the circulating gas will be 35%. Calculate the ratio of this estimated efficiency to the ideal cycle efficiency for these conditions when helium is the circulating gas.

Answer: $\eta_{CY} = \left[1 - \dfrac{T_1}{T_4} \dfrac{(\gamma - 1)/\gamma}{(\rho_p - 1)/\rho_p} \ln r_p\right]$ where $\rho_p = r_p^{(\gamma-1)/\gamma}$; 0.58.

8.8. Figure 8.17 shows the circuit diagram for the HTGR plant discussed in §8.10.1, in which the reject heat from the gas-turbine cycle is utilised to provide a heat supply for a district-heating scheme. Helium, for which $\gamma = 1.67$, circulates round the closed gas-turbine cycle, while water under pressure circulates round the closed calorifier circuit. In addition to the temperatures shown on the diagram, the following data are given:

Turbine: Pressure ratio   = 2.4; isentropic efficiency   = 90%.
Compressor: Pressure ratio   = 2,6; isentropic efficiency   = 90%.

The *effectiveness* of both the regenerative heat exchanger X and the calorifier heat exchanger is 0.87, where the effectiveness is defined as the ratio of the temperature rise of the cooler fluid to the difference between the entry temperatures of the two fluids.

Assuming that the specific heat capacities of both the helium in the gas-turbine circuit and the water in the calorifier circuit are constant, calculate the following quantities: (1) Q_P/Q_{in}, (2) the work efficiency, W_{net}/Q_{in}, (3) Q_C/Q_{in}, (4) the total efficiency, $(W_{net} + Q_C)/Q_{in}$, (5) Q_0/Q_{in}, (6) the effectiveness for which the precooler heat exchanger must be designed.

Answer: (1) 0.578; (2) 0.422; (3) 0.378; (4) 0.800; (5) 0.200; (6) 0.560.

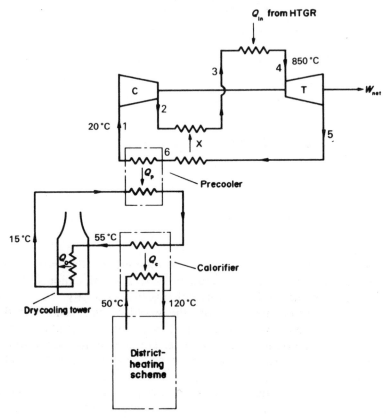

Fig. 8.17. HTGR gas-turbine, direct-cycle plant with reject heat utilised for district heating.

CHAPTER 9

Combined and binary power plant

9.1. Introduction

In this chapter a study is made of plant in which a combination of working substances is used, dealing first with combined gas and steam plant, and then with binary cycles. In binary plant a *topping* cycle is superposed directly on another cycle, the former using a working substance other than steam and the latter, in practice, using steam, although this is a matter of convenience and suitability rather than of necessity.

Here we are sometimes far from the world of established practice. Although there are in existence an increasing number of combined steam and gas-turbine plants, and one or two installations operating on the binary vapour cycle, this is the limit of present application of the ideas discussed in this chapter. Some of the ideas are purely conceptual, some at present partly so. Nevertheless, the economics of power generation are continually changing with advances of technology, and the studies made in this chapter help to point the way to possible future progress.

9.2. Combined gas–steam plant

In a gas-turbine cycle the fluid remains gaseous throughout the cycle, but the condensible fluid of a steam plant is gaseous at some points in the cycle and liquid at others. The two kinds of plant are similar in many ways, but they each have their own distinctive features. The exhaust-gas heat exchanger is a characteristic feature of all but the simplest gas-turbine plant, while the advanced steam-turbine cycle is characterised by its system of feed-water heating. These features are both examples of the regenerative principle, and both raise the mean temperature of heat reception. While the exhaust-gas heat exchanger also simultaneously lowers the mean temperature of heat rejection in the gas-turbine cycle, the steam plant has the advantage over the gas-turbine plant in rejecting all its heat at the lowest temperature in the cycle. The steam plant thus shows to particular advantage at the "bottom" end of the cycle. At the same time, we have seen in §3.6 that a high "top" temperature is particularly necessary for high efficiency in gas-turbine installations. It is this combination of circumstances that accounts for the fact that in all the combined gas–steam plant discussed in this chapter the gas-turbine section acts effectively as a topping unit to the steam section.

9.3. The ideal, super-regenerative steam cycle

This is not a combined gas and steam cycle as commonly understood, in that the fluid which circulates round the "gas-turbine" section of the plant is neither a single "permanent" gas nor a mixture of combustion gases, but superheated steam. Nevertheless, it has the characteristics of a gas-turbine plant superposed on a condensing steam plant and so serves as a good introduction to the study of combined cycles. It is a purely hypothetical, ideal cycle such as might be found in Thermotopia, that idyllic land where all is thermodynamically perfect and there are no lost opportunities for producing work, and its complexity ensures that it will never be found in all its purity in any earthly power station. Its inclusion is justified by the fact that it represents a theoretical means of attaining Carnot efficiency in a steam cycle even when the steam is superheated. The conception of the super-regenerative steam cycle is due to Field.[65,66]

It was shown in §7.5 how to devise a reversible, regenerative feed-heating cycle using superheated steam, but it was pointed out that, although the feed water was raised reversibly to the boiler saturation temperature, the efficiency was still less than the limiting Carnot efficiency. The super-regenerative cycle carries the regenerative principle into the superheat zone by a method normally associated with gas-turbine cycles, namely the use of an exhaust-gas heat exchanger.

The cycle utilizes the technique described in §7.5 for the isothermal compression of superheated bled steam, as well as the technique of progressive reheating described in §§6.11 and 7.18, and depicted in Fig. 6.10. A flow diagram for one possible version of this cycle is shown in Fig. 9.1, and a temperature-entropy diagram in Fig. 9.2.

It will be seen from these diagrams that the complete plant is equivalent to a CICIC ... BTRT ... RTX "gas-turbine" plant (using superheated steam as the "gas") effectively superposed on a fully regenerative condensing steam plant. Progressive reheating between 4 and 5 allows all the heat supplied to the cycle to be transferred at the top temperature T_b. Instead of the steam being superheated in a boiler from 3 to 4 it is passed through a regenerative heat exchanger X, in which the steam flow M_G exhausted from the last "reheat turbine" is simultaneously cooled from 5 to 6. At 6 the steam flow divides, a quantity M_B passing to a fully regenerative condensing steam plant employing the technique already illustrated in Fig. 7.4 and raising the feed water to point 2 at boiler saturation temperature. The remainder of the steam ($M_G - M_B$) coming

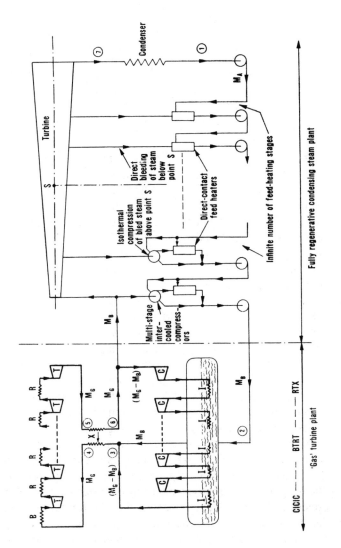

FIG. 9.1. Ideal super-regenerative cycle.

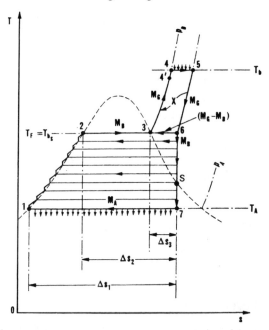

FIG. 9.2. Temperature–entropy diagram for ideal super-regenerative cycle.

from the heat exchanger X is compressed reversibly and isothermally from 6 to 3 by multi-stage compression with intercooling coils placed in the saturated steam drum, the heat transferred in the intercoolers serving to evaporate the feed water from 2 to 3 without the supply of any heat from an external source. The fireless saturated-steam drum thus takes the place of an externally fired boiler, and no heat from an external source is supplied at any temperature below the top temperature T_b. All heat rejection occurs at the bottom temperature T_A in the condenser.

In the limit, with infinite numbers of stages of reheating, compression and feed heating, the cycle would be completely reversible if superheated steam behaved as a perfect or semi-perfect gas and, since all heat reception would occur at T_b and all heat rejection at T_A, the cycle would have Carnot efficiency $(1 - T_A/T_b)$. The reader may thus find it instructive to derive the expressions given in Problem 9.1 for the respective steam flow ratios. In point of fact, the properties of superheated steam depart somewhat from those of a perfect gas, the specific heat capacity increasing with pressure instead of being independent thereof. Hence, even in an infinite heat exchanger, steam cooled on one side from 5 to 6 would only raise the steam passing through the other side to point 4′ instead of to point 4, so that the overall heat transfer process in X would not be reversible and the cycle efficiency would consequently be something less than the limiting Carnot efficiency. Nevertheless, the cycle serves to illustrate the theoretical requirements that would have to be met if any close approach to the limiting Carnot efficiency in a superheated steam cycle were to be obtained. The reader will gain a better understanding of this hypothetical ideal cycle by working through Problem 9.1.

It is clear that the cycle is highly impracticable. An alternative method of constructing an ideal super-regenerative cycle which would be fully reversible, and so would have Carnot efficiency, was suggested by Field[65]; in that cycle, the same regenerative heating process was applied to the "feed steam" from 3 to 4 as was applied to the feedwater from 1 to 2 in Fig. 9.1. The steam would then have been expanded isentropically all the way from 5 to 7, the extraction of bled steam for subsequent isothermal compression also taking place all the way from 5 to 7. This was equally impracticable, and a much simpler, though much less efficient, variant was consequently proposed by Field[67,68] and has come to be known by his name.

9.4. The Field cycle

It is a feature of the ideal, super-regenerative cycle described in §9.3, that the mixing of the two fluid streams at point 3 in Figs. 9.1 and 9.2 is reversible, since they are at the same pressure and temperature before mixing. This requires, however, the entirely impracticable process of multi-stage isothermal compression from 6 to 3, a process avoided in the simpler Field cycle illustrated in Figs. 9.3 and 9.4. Here the reversible mixing process is replaced by irreversible mixing of the steam and the return feed water in a spray desuperheater preceding a single stage of steam compression. The impracticable process of progressive reheating is also replaced by a single reheat stage.

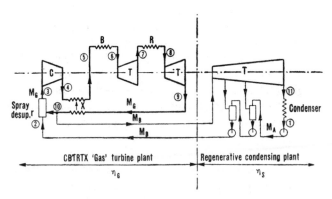

Fig. 9.3. Field super-regenerative cycle.

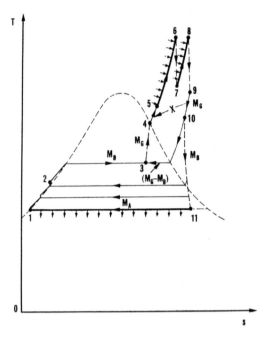

Fig. 9.4. Temperature–entropy diagram for Field super-regenerative cycle.

Evaporation of the feed water occurs in the spray desuperheater, so that the cycle retains the advantage of avoiding any heat reception at the steam saturation temperature and thus achieves a high mean temperature of heat reception. The gain due to the latter, however, is now rather seriously offset by the reduction in cycle efficiency resulting from the irreversibility of the mixing process.

The complete cycle is again equivalent to a regenerative "gas-turbine" cycle effectively superposed on a regenerative condensing steam plant, and a simple expression for the thermal efficiency η_F of the Field cycle in terms of the thermal efficiencies η_G and η_S of these hypothetical gas and steam cycles may readily be devised. By writing down the steady-flow energy equation for the mixing process in the spray desuperheater, the ratio of the flow rates M_B and M_G is seen to be given by

$$\frac{M_B}{M_G} = \frac{h_{10} - h_3}{h_{10} - h_2}. \tag{9.1}$$

Referring to Fig. 9.4, the thermal efficiency η_G of a hypothetical CBTRTX "gas-turbine" cycle 3456789–10–3, in which heat rejection Q_{10-3} occurred between 10 and 3, would be given by

$$(1 - \eta_G) = \frac{Q_{10-3}}{Q_{in}}, \tag{9.2}$$

where

$$Q_{10-3} = M_G(h_{10} - h_3) \tag{9.3}$$

and Q_{in} is equal to the heat reception in the actual Field cycle.

Similarly, the thermal efficiency η_S of a hyothetical, regenerative condensing steam cycle 123–10–11–1 in which heat reception Q_{2-10} occurred between 2 and 10 would be given by

$$(1 - \eta_S) = \frac{Q_{out}}{Q_{2-10}}, \tag{9.4}$$

where

$$Q_{2-10} = M_B(h_{10} - h_2), \tag{9.5}$$

and Q_{out} is equal to the heat rejection in the actual Field cycle.

From eqns. (9.1), (9.3) and (9.5),

$$Q_{10-3} = Q_{2-10},$$

so that

$$(1 - \eta_G)(1 - \eta_S) = \frac{Q_{out}}{Q_{in}}.$$

But the thermal efficiency of the Field cycle is given by

$$(1 - \eta_F) = \frac{Q_{out}}{Q_{in}},$$

so that

$$\boxed{(1 - \eta_F) = (1 - \eta_G)(1 - \eta_S)} \tag{9.6}$$

This relation enables the variation in efficiency of the Field cycle with varying steam and gas-turbine cycle parameters[69] to be related to the variations in efficiency of these cycles already studied in previous chapters. Other variants of the Field cycle are possible but their discussion is beyond the scope of the present volume.

The attraction of the Field cycle lies in its ability to give reasonably high cycle efficiency without necessitating the use of the very high steam pressures usually associated with steam plant of high efficiency. However, the original performance estimates given by Field were over-optimistic, and developments that have since taken place in more conventional plant have rendered it less easy to make a case for the competitiveness of the

Field cycle, particularly as it presents technical difficulties of its own. The cycle has consequently not found practical application, although some experimental work on certain associated problems was done. High-temperature compression is one of the less attractive features of the cycle. Field originally envisaged wet compression of the steam as in Fig. 9.4 and Problem 9.2 but this presents its own special difficulties, while if the steam is superheated the work input required to compress it is much greater. Other problems arise from the fact that all heat reception occurs at a high temperature, an aspect which is discussed in the next section.

9.5. The effect on plant overall efficiency of high heat-reception temperature in super-regenerative cycles

In §1.4 it was shown that in a plant burning fossil fuels such as coal, oil or natural gas, the plant overall efficiency η_o is a product of the cycle efficiency η_{CY} and the boiler or heating-device efficiency η_B. It has already been pointed out in §§1.4 and 7.15 that, if the temperature at which heat is transferred from the combustion gases to the working fluid is uniformly high, then unless remedial action is taken an improvement in η_{CY} due to increase in mean temperature of heat reception will not be reflected in an equal improvement in η_o because of the adverse effect on η_B. It was noted in §7.15 that the remedial action that followed the adoption of feed heating was the introduction of the air preheater, by means of which the flue gases could be further cooled before discharge from the boiler. The even higher temperature at which heat reception commences in the Field cycle aggravates this problem, and leads to the need for an air preheater capable of giving an unusually high degree of air preheat; this presents one of the practical difficulties associated with the application of the Field cycle. An alternative method of reducing the temperature of the flue gases before discharging them to the atmosphere would be to use them in a combined cycle of the kind discussed in the next section, the superheater-reheater of the Field cycle taking the place of the boiler in Fig. 9.5. It is also of interest to note that this problem would not arise if a boiler fired by fossil fuel were replaced by a nuclear reactor as the source of heat. Reactor technology has not, however, reached a stage of commercial development which would permit of temperatures sufficiently high to provide a suitable heat source for a super-regenerative cycle.

9.6. Combined gas–steam plant incorporating gas turbines

We now turn to a consideration of true gas–steam combined plant. Though usually described as operating on a combined gas–steam cycle, in the type of plant more generally envisaged the gas-turbine section is a non-cyclic (open-circuit) internal-combustion plant of the kind discussed in

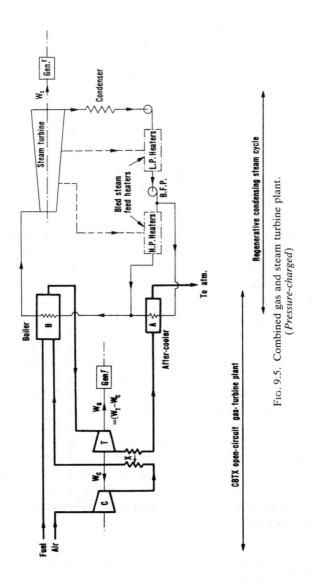

Fig. 9.5. Combined gas and steam turbine plant.
(*Pressure-charged*)

§§1.2 and 6.16. An example of such a plant is illustrated in Fig. 9.5. Many variants are possible.[70–72] In that shown, an open-circuit gas-turbine plant is linked with the steam generator supplying the steam cycle.

The air supply drawn in by the compressor is used to pressurise or supercharge the combustion chamber of the steam generator, the flue gases from which supply the working fluid for the gas turbine. The latter exhausts into a regenerative heat exchanger X and so preheats the incoming combustion air, but since this comes from the compressor at a temperature in the neighbourhood of 250 °C the exhaust gases leaving the heat exchanger are at a prohibitively high temperature to be discharged direct to atmosphere. They are consequently passed through an after-cooler fed with low-temperature feed water tapped from an intermediate point in the feed-heater train of the steam plant; this, of course, has a slightly adverse effect on the steam cycle efficiency.

The attraction of the combined cycle lies in the increase in station efficiency that it can give when applied to a steam station operating under even the most advanced conditions. This is discussed in the next section.

The plant illustrated, which is described as a *pressure-charged*, combined gas–steam plant, suffers from the drawback when burning coal that the dirty products of combustion pass through the gas turbine. To use a closed circuit on the gas side, with external combustion, would require a large, high-temperature heat exchanger at a cost which would be uneconomic even if adoption of the combined cycle could itself be otherwise justified economically. An alternative solution is provided by the *exhaust-heated* combined gas–steam plant. In one example of this,[72,73] a simple CBT gas-turbine unit has its own combustion chamber burning clean natural gas or distillate oil, and the gas turbine exhausts into the unpressurised combustion chamber of the steam generator, in which more fuel (which may be either residual fuel oil or coal) is burnt in the excess of air coming in with the exhaust gases from the gas turbine. This procedure is known as *supplementary firing* and it enables plants of substantially greater output to be built.

This type of exhaust-heated plant with supplementary firing is discussed further in §§9.16 and 9.17, and the reader would profit from working through Problems 9.6 and 9.9 at this stage. The latter relates to a *cogeneration* plant for the simultaneous delivery of power and process steam in a factory. Unlike the cogeneration plant discussed in Chapter 7, in the plant of Problem 9.9 the power is not provided by a steam turbine but by a simple CBT gas-turbine plant.

As a means of reducing possible adverse effects on the blading of the gas turbine when the products which pass through it come from the burning of coal, the use of *fluidised-bed combustion* (FBC) of coal has become of increasing interest in combined gas/steam plant. Interest in FBC is also being shown for conventional steam plant, with the aim of reducing atmospheric pollution. Even cleaner products of combustion result if the reactants supplied to the gas-turbine combustion chamber come from a *coal-gasification* plant. These matters are discussed in greater detail in §§9.20–9.22.

9.7. The overall efficiency of a combined gas–steam turbine plant

A simplified diagram of the plant shown in Fig. 9.5 is given in Fig. 9.6. This is a combined fluid-flow and energy-flow diagram.

In the following analysis all energy quantities are expressed **per unit mass of fuel** supplied to the plant, and the symbols are defined in the figure.

We first define the efficiency (perhaps better called effectiveness) of energy extraction from the open-circuit gas stream as

$$\eta_B \equiv \frac{H_R - H_P}{CV}, \qquad (9.7)$$

where H_R and H_P are the enthalpies of the reactants and products respectively entering and leaving the plant, and CV is the calorific value per unit mass of fuel burnt. η_B is thus analogous to the boiler or heating-device efficiency defined in §1.4. If the flue gases left the plant at the incoming temperature of the reactants, η_B would have a value of 100%.

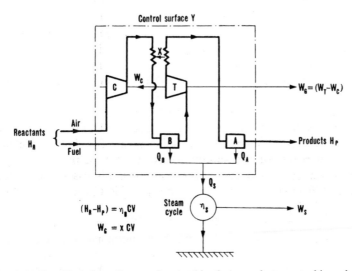

FIG. 9.6. Simplified flow diagram for combined gas and steam turbine plant. (*Pressure-charged*)

We next express the net work output from the gas-turbine plant as a fraction x of the calorific value of the fuel. Thus

$$x \equiv \frac{W_G}{CV}. \tag{9.8}$$

In steady flow, the steady-flow energy equation for control surface Y is

$$Q_S + W_G = (H_R - H_P) = \eta_B \, CV. \tag{9.9}$$

Hence

$$Q_S = (\eta_B - x) \, CV, \tag{9.10}$$

while

$$W_S = \eta_S Q_S. \tag{9.11}$$

The overall efficiency of the complete plant, neglecting the power required for auxiliaries, is defined by

$$\eta_o \equiv \frac{W_G + W_S}{CV}. \tag{9.12}$$

For the reasons given in §4.4, since the gas-turbine plant is not cyclic, this is an arbitrary but convenient measure of performance.

From eqns. (9.10), (9.11) and (9.12)

$$\boxed{\eta_o = x + \eta_S(\eta_B - x) = \eta_B\eta_S + x(1 - \eta_S)} \tag{9.13}$$

By writing

$$\eta_o' \equiv \eta_B\eta_S, \tag{9.14}$$

eqn. (9.13) becomes

$$\boxed{\eta_o = \eta_o' + x(1 - \eta_S)} \tag{9.15}$$

η_o' would be the overall efficiency of a conventional steam plant having a boiler efficiency equal to η_B and a cycle efficiency equal to η_S. Thus eqn. (9.15) shows that the overall efficiency of the combined gas–steam plant is greater than that of a conventional steam plant having the same values of η_B and η_S, by an amount equal to $x(1 - \eta_S)$. This may be simply explained by the fact that an extra quantity of work equal to x CV is obtained from the gas stream by direct generation in the gas-turbine plant, causing a reduction of this amount in the quantity of heat transferred to the steam cycle and a resulting reduction of $\eta_S.x$ CV in the net work output from the steam cycle. The percentage increase in efficiency of the combined plant over that of a conventional steam plant with the same values of η_B and η_S is therefore given by

$$\% \text{ gain} = \frac{x(1 - \eta_S)}{\eta_o'} \times 100. \qquad (9.16)$$

The reader is encouraged to check that eqn. (9.15) also holds for the *exhaust-heated* (or *supplementary-fired*) combined gas–steam plant described in §9.6 if CV is interpreted as the combined heating value of the fuel burnt in the two combustion chambers (Problem 9.3).

Since $W_G = x$ CV and the total work output $W_{TOT} = \eta_o$ CV, the fraction of the total plant output that is produced by the gas-turbine plant is given by

$$\frac{W_G}{W_{TOT}} = \frac{x}{\eta_o}. \qquad (9.17)$$

Optimisation calculations[70,71,74] for combined cycles with high-efficiency steam plant incorporating reheating and a high degree of regenerative feed heating show an optimum value of x of about 0.05. With this value of x, and taking $\eta_S = 0.44$ and $\eta_B = 0.9$ as appropriate values for this type of plant, so that $\eta_o' = 0.396$, we have:

$$\text{Gain} = \frac{0.05 \times 0.56}{0.396} \times 100 = 7.0\%,$$

$$\eta_o = (0.396 + 0.05 \times 0.56) \times 100 = 42.4\%,$$

$$\frac{W_G}{W_{TOT}} = \frac{0.05}{0.424} = 0.118.$$

This relatively low percentage of work output from the gas-turbine part of the plant is a reflection of the fact that the plant is of the exhaust-heated type with supplementary firing, so enabling a high output to be obtained from the steam plant as a result of the further burning of fuel in the steam generator. In the past decade, considerable progress has been made in the installation in Europe[75] of high-efficiency combined-cycle plant of this type and of high output. Typical of these are a number of plants of 417 MW total output, with an output from the gas-turbine plant of about 52 MW, so that $W_G/W_{TOT} = 0.125$. The quoted overall efficiency is 43.6%. It will be seen that these figures are of the same order as the values quoted above.

Combined gas–steam plant of the pressure-charged type described in §9.6 and depicted in Fig. 9.6 (namely, plant without supplementary firing in the steam generator) are inevitably of smaller output than the exhaust-heated type. Again, in the pressure-charged type, the fraction of the total work output that is produced by the gas-turbine plant is, of course, appreciably greater than in the exhaust-heated type without supplementary firing and may be as high as $\frac{1}{2}$ to $\frac{2}{3}$. Since we shall find that η_o will not be greatly different in the two types of plant, it will be seen from equation (9.17) that the value of x will be appreciably greater. Equation (9.16) thus shows that, in the pressure-charged type, there will be a greater percentage gain due to topping the steam plant with a gas-turbine plant. In the exhaust-heated plant without supplementary firing, the thermal efficiency η_S of the steam plant itself will, however, be markedly lower. This is because steam will then only be generated at a relatively low pressure and temperature by heat transfer solely from the exhaust gases from the gas turbine. The temperature of these gases will be something less than 500 °C at entry to the steam generator. Furthermore, their temperature will fall as they pass through the steam generator, just as was shown in Fig. 8.2 for the passage of the CO_2 through the steam generators of gas-cooled nuclear reactors. Indeed, as for the gas-cooled Magnox reactors, if we want to obtain as high a value as possible for the thermal efficiency of the steam cycle, it will be advantageous and economic to use a *dual-pressure* steam cycle, as in Fig. 8.2, or even a *triple-pressure* cycle. This practice has, in fact, been adopted for a 118 MW plant of the exhaust-heated type, without supplementary firing, at Donge in Holland,[75,76] the overall efficiency being reported[75] as 44.4% with $W_G/W_{TOT} = 0.64$. The latter corresponds to a value of x of about 0.28, as compared with the value of 0.05 for the exhaust-heated type of plant with supplementary firing. More detailed discussion of a triple-pressure steam cycle will be found in §9.18.

We may now apply the same analysis to this type of plant as we did to the previous type. Taking $\eta_S = 0.26$ as an appropriate value for the thermal efficiency of a dual-pressure steam cycle in the above circumstances and again taking $\eta_B = 0.9$, so that $\eta_o' = 0.234$, we shall have the following results for a value of $x = 0.28$:

$$\eta_o = (0.234 + 0.28 \times 0.74) \times 100 = 44.1\%,$$

$$\frac{W_G}{W_{TOT}} = \frac{0.28}{0.441} = 0.635.$$

These results are in line with the quoted values for the Donge plant. The percentage gain calculated from equation (9.16) would be 89%. However, in this case, this is not a very meaningful comparison because, in the absence of the gas turbine, one would not install a steam plant of such low thermal efficiency. In the previous case, on the other hand, one was comparing two plants in each of which the steam plant would have been of high efficiency, with reheating and regenerative feed heating in both. The percentage gain was thus essentially that due to topping such a steam plant with a gas-turbine plant. What is notable in both the pressure-charged and exhaust-heated types, however, is the high overall efficiency achieved by the combined gas–steam plant. Still higher values will be possible as improvements in gas-turbine materials and design allow for the use of higher inlet temperatures. Proven gas turbines of modern design can now operate with an inlet gas temperature of 900–1000 °C. With an advance to 1200 °C in prospect,[76] overall efficiencies of 50% or higher will be possible.

The advantages of combined-cycle plants with respect to high thermal efficiency and low specific capital cost can also be applied in plants providing both electrical power and thermal energy;[77] in these, steam may be supplied, for example, for industrial processes, for hot-water district heating or for the desalination of sea-water. Such a plant is studied later in §9.16 and Problem 9.9.

An alternative form of pressure-charged combined gas–steam plant is one in which the gas turbine is replaced by a *magnetohydrodynamic generator*. This type of plant is discussed in the next section, but its successful exploitation is a more distant prospect.

9.8. Combined gas–steam plant with magnetohydrodynamic (MHD) generation

MHD generation is at a very early stage of development,[78,79] though some notable efforts have been made, particularly in the USSR, to develop the system for commercial use.

In the projected type of combined gas–steam plant discussed in this section the MHD generator takes the place of the gas turbine as the workproducer in the gas circuit. Its principle of operation is illustrated in Fig. 9.7. An electrically conducting ionised gas or plasma at elevated temperature enters the flow passage of the generator at high but probably subsonic velocity. An intense magnetic field passes across the flow passage in a direction perpendicular to the direction of flow, so that an electric field is created in a direction normal to both the magnetic field and the duct axis. A voltage is consequently generated across electrodes suitably placed on

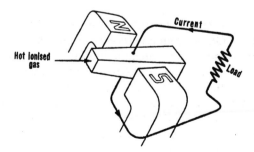

FIG. 9.7. Principle of operation of magnetohydrodynamic (MHD) generator.

the duct walls in contact with the gas stream, and energy can be extracted from the stream by connecting the electrodes to an external load. Thus electromagnetic braking of the fluid stream takes the place of the mechanical braking by the turbine blades in a gas turbine. The velocity is maintained by allowing the pressure of the subsonic gas stream to fall as it passes through the divergent duct, so that there is a drop in temperature and pressure as the gas expands and delivers work, just as in a gas turbine. Consequently, from the point of view of cycle analysis, the plant presents no basically new problem. A simplified diagrammatic arrangement is shown in Fig. 9.8. It will be seen that this is a non-cyclic, open-circuit plant on the gas side.

In the MHD generator the electrically conducting properties of the plasma result from ionisation of the gas when it is raised to an elevated temperature. This ionisation can be assisted by seeding the gas with atoms of an alkali metal such as caesium, but even so the temperature required is of the order of 3000 K. This requires an exceedingly high degree of air preheat, so that the air supplied by the compressor to the combustion chamber is passed first through a regenerative heat exchanger X, which also serves to reduce the temperature of the exhaust gases from the MHD generator to a suitable value before they enter the steam generator. Alternatively, oxygen enrichment of the combustion air may be used instead of a high degree of air preheating. Only because an MHD generator has none of the highly stressed rotating parts present in a gas turbine is it thought to be feasible to handle gas at the temperature required. Before they enter the MHD duct, the combustion gases pass through a nozzle in which the necessary velocity is generated. A diffuser at generator outlet gives some recovery of pressure and temperature by reducing the velocity of the stream before it enters the steam generator.

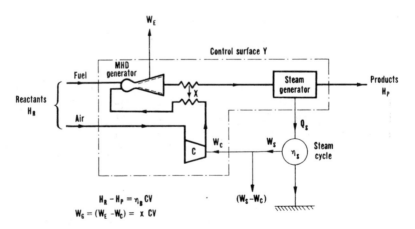

FIG. 9.8. Simplified flow diagram for combined MHD and steam plant.

The compressor provides the necessary pressure difference across the nozzle and duct of the MHD generator, and since there is no gas turbine it must either be driven electrically or by a steam-turbine unit of the steam cycle.

9.9. The overall efficiency of an open-circuit MHD combined plant

In deriving an expression for the overall efficiency of the plant, neglecting the power required for auxiliaries, the *net* work output from the gas circuit (**not** the gross electrical work output W_E from the MHD generator itself) is again expressed as a fraction of the calorific value of the fuel, all quantities again being expressed per unit mass of fuel supplied. Thus

$$x \equiv \frac{W_G}{CV} = \frac{W_E - W_C}{CV}. \tag{9.18}$$

The overall efficiency is

$$\eta_0 = \frac{W_E + (W_S - W_C)}{CV} = \frac{W_G + W_S}{CV}. \tag{9.19}$$

This is the same as eqn. (9.12), and since eqns. (9.9), (9.10) and (9.11) are equally applicable to control surface Y and the steam cycle of Fig. 9.8, the overall efficiency is again given by eqn. (9.15), so that

$$\boxed{\eta_0 = \eta_0' + x(1 - \eta_S)} \tag{9.20}$$

where $\eta_0' = \eta_B \eta_S$.

With the very high temperature of the order of 3000 K at inlet to the MHD duct, the exit temperature from the duct will also be high, so that supplementary firing in the following steam generator will not be needed to ensure high steam parameters and correspondingly high efficiency for the steam cycle. The plant will thus correspond to the pressure-fired type of gas-turbine plant, rather than the exhaust-heated type. Hence the value of x will be of the same order as that for the former type, though possibly somewhat higher in view of the high inlet temperature. Thus, taking a value of $x = 0.3$, a steam cycle efficiency of 40% and $\eta_B = 90\%$, we would have:

$$\eta_0 = (0.36 + 0.3 \times 0.60) \times 100 = 54\%,$$

$$\frac{W_G}{W_{TOT}} = \frac{0.3}{0.54} = 0.556.$$

In the early stages of development, the overall efficiency would probably be nearer to 50%.

When x is large, eqn. (9.20) might better be written as

$$\eta_o = x + \eta_S(\eta_B - x) \qquad (9.21)$$

An increase in overall efficiency from 36% to 54%, much greater than that obtainable from a combined gas-turbine installation, is clearly an attractive proposition, but the development of an MHD generator to operate at elevated temperatures presents formidable difficulties. Because the dissociation and ionisation of the gas which occur at these temperatures are endothermic reactions, a given increase in preheat of the air supplied to the combustion chamber is not matched by an equal increase in combustion temperature, so that the air must be preheated to over 2000 K to achieve a combustion temperature of 3000 K. The provision of a suitable air preheater thus presents a much more serious problem than was posed in the Field cycle. Since it is the high proportion of nitrogen in the air which keeps the temperature of the combustion products down, a means of avoiding the high degree of air preheat is to supply the MHD installation with an oxygen-enriched mixture.

In a pilot plant in the USSR, up to 40% oxygen enrichment of the air was employed, together with seeding with potassium carbonate in solution for enhancing the electrical conductivity. The plant formed part of a collaborative programme on MHD development between the USSR and the USA.[80,81,82] That collaboration looked towards the construction of large-scale, commercial MHD-steam plant, but that appears to be a long-term prospect.

For a more detailed study of the problems associated with MHD power generation, the reader may consult a more specialised text.[83]

9.10. Closed-circuit gas–steam binary cycles for nuclear power plant

In the combined gas–steam plant discussed in §§9.6–9.9, only the steam plant is cyclic, the gas-turbine and MHD plant being of the open-circuit variety using fossil fuel. By contrast, in a combined gas–steam plant using a nuclear reactor as the source of thermal energy, both the gas and steam plant are cyclic. This type of installation, in which a gas cycle is superposed on a steam cycle, is described as a *binary-cycle* plant.

9.10.1. MHD/steam-turbine binary cycle

A closed-circuit MHD plant having a nuclear reactor as the heat source, with cooling by an inert monatomic gas such as helium or argon, is an alternative to the open-circuit plant discussed in §9.8, but one which is likely to remain speculative for some time to come. The higher electrical conductivity of monatomic gases at elevated temperatures due to non-equilibrium electron temperatures[84] might allow the use of a lower temperature than that required for molecular gases, but the gas supplied to the MHD generator from the nuclear reactor would still have to be at a temperature in the region of 2000 K.

In the event of its development such a combined MHD-steam plant would be cyclic in its entirety, the helium passing round a closed cycle through the compressor, reactor, MHD generator and steam generator in turn, the heat rejected in the steam generator by the MHD gas cycle serving as the heat supply to the steam cycle. It is easy to show that, neglecting stray heat losses, the thermal efficiency η_{CY} of the binary cycle would be given by

$$\boxed{(1 - \eta_{CY}) = (1 - \eta_G)(1 - \eta_S)} \qquad (9.22)$$

where

$$\eta_G \equiv \frac{W_G}{Q_R}, \qquad (9.23)$$

Q_R being the thermal output of the reactor. Neglecting the power required for auxiliaries, η_{CY} is also the overall efficiency of the plant since η_B does not enter into the calculations for a plant in which there are no flue gases exhausted to atmosphere.

It may be recalled that the Field cycle was equivalent to one cycle superposed directly on another, and the similarity of eqns. (9.6) and (9.22) may be noted. The expression for the thermal efficiency of any type of cyclic binary plant in which one cycle is superposed directly on another, so that the heat rejected by the upper cycle is equal to the heat received by the lower, is of this form.

9.10.2. Gas-turbine/steam-turbine binary cycle

The problems associated with the development of a successful MHD generator, and the difficulty of finding materials for both the MHD duct and the nuclear reactor able to withstand the very high temperatures required for the type of plant described in the preceding section, make it very questionable as to whether such a plant will ever reach the stage of commercial development. A more conventional type of gas–steam binary cycle is that mentioned briefly in §8.10.1 of the previous chapter, in which a modified CBTX gas-turbine cycle is superposed on a conventional steam cycle. Again, no such nuclear plant has been built, but that indicated in highly diagrammatic form in Fig. 9.9 has been the subject of theoretical study.[85]

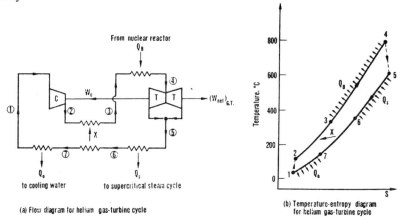

(a) Flow diagram for helium gas-turbine cycle

(b) Temperature-entropy diagram for helium gas-turbine cycle

FIG. 9.9. Gas-turbine/steam-turbine binary cycle.

It will be noted that the CBTX helium cycle is somewhat unconventional, in that heat rejection from it occurs at both a high-temperature level (in the steam generator) and at a low-temperature level (to cooling water), with the regenerative heat exchange X taking place between these two heat rejection processes. The high-temperature heat exchange in the steam generator provides the heat supply for a supercritical, reheat, regenerative steam cycle operating at steam parameters comparable to those in supercritical, fossil-fuel stations. It is left to the reader to show that, in this case, the thermal efficiency η_{CY} of the binary cycle is given by the expression

$$(1 - \eta_{CY}) = (1 - \eta_G)(1 - y\eta_S), \qquad (9.24)$$

where η_G and η_S are the thermal efficiencies of the gas and steam cycles respectively and $y \equiv Q_s/(Q_s + Q_o)$, the fraction of the total heat rejection from the CBTX helium cycle that is supplied to the steam cycle. η_G, η_S and y are, of course, all interdependent and are also related to the relative net outputs of the gas and steam cycles, so that optimisation calculations would be quite complex. For the temperatures indicated in Fig. 9.9, the overall station efficiency might be of the order of 48%.

A more conventional gas/steam binary cyclic plant is studied in more detail in §9.15.

9.11. Binary vapour cycles

In the types of binary cycle described in the previous section the gas cycle acts as a *topping plant* to a conventional steam cycle. A gas is not an ideal working fluid for the topping plant of a binary cycle because, in particular, the mean temperature of heat reception is inevitably well below the top temperature, although this is compensated in an MHD plant by the much higher temperatures which can be permitted if the working fluid does not have to pass through the highly stressed blades of a rotating turbine (the problem of finding suitable materials for the reactor and the MHD duct nevertheless remains). If the topping plant in a binary cycle is to incorporate turbo-machinery, to the search for superior materials is added a search for a working fluid with more suitable properties than those of a gas. In particular, a condensible fluid is needed with a much higher critical temperature than that of water in order that it can be evaporated at the maximum allowable cycle temperature and so give a cycle with a much greater fraction of the total heat supply occurring near or at the top temperature than is possible with gas. Liquid metals can meet this requirement and plants using mercury as the working fluid in the topping cycle have been built in the USA, though in very small numbers. The critical temperature of mercury is about $1500 \,°C$, and its accompanying high critical pressure of $106 \, MN/m^2$ is an additional advantage, in that it results in the saturation pressure at a working temperature of $550 \,°C$ being only about $1.4 \, MN/m^2$. The extremely low vapour pressure at room temperature would be a disadvantage if the mercury were used in a simple single-vapour cycle, as it is only about $0.3 \, N/m^2$ at $30 \,°C$, so that both the condenser vacuum and the specific volume of the vapour at turbine exhaust would be excessive. This disadvantage is avoided in a binary mercury–steam cycle, in which the mercury is condensed at about $260 \,°C$ in a *condenser–boiler* by evaporating the steam for the steam cycle; at this temperature, the mercury vapour pressure is about $14 \, kN/m^2$.

To illustrate the theoretical advantage of a binary vapour cycle, Fig. 9.10 shows a combined temperature–entropy diagram for an ideal, reversible Rankine cycle using dry saturated mercury, superposed on an ideal, reversible fully regenerative steam cycle using dry saturated steam and feed heating to the steam saturation temperature with an infinite number of feed heaters. The entropy scales for the mercury and steam cycles are different, being in the ratio of the mass flow rates of mercury and steam through the condenser–boiler; the area on the diagram representing the heat rejected by the mercury cycle is then the same as the area representing the heat received by the steam cycle. The diagram also compares the mean temperature of heat reception for the binary cycle $\bar{T}_{binary}$ with the value $\bar{T}_{steam}$ for the single superheated steam cycle with the same steam pressure and the same top temperature as in the binary cycle; this provides clear evidence for the thermodynamic superiority of the binary cycle. Furthermore, if fully regenerative feed heating were also used in the mercury cycle, all heat addition in this ideal binary cycle would take place from B to C at the top temperature and the thermal efficiency would equal the

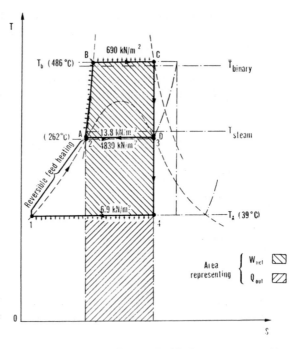

FIG. 9.10. Temperature–entropy diagram for ideal mercury–steam binary vapour cycle

limiting Carnot efficiency $(1 - T_A/T_b)$. Even without mercury feed heating, $\bar{T}_{\text{binary}}$ is very close to T_b, so that it is evident that there is no incentive to incorporate feed heating in the mercury cycle. This is because the specific heat capacity of liquid mercury is very much smaller than that of water, so that the saturated-liquid line for mercury on the temperature–entropy diagram is much nearer to the vertical.

It is readily shown that the thermal efficiency η_{CY} of this ideal cycle is identical in form to eqn. (9.22), with the thermal efficiency η_M of the mercury cycle simply replacing η_G, so that

$$\boxed{(1 - \eta_{\text{CY}}) = (1 - \eta_M)(1 - \eta_S)} \qquad (9.25)$$

In practical binary–vapour plant, the complication of feed heating in the mercury cycle is certainly not warranted. To increase η_S, however, it pays not only to have feed heating in the steam cycle, but also to use superheated steam. In these circumstances the steam superheater is placed in the gas passes of the fossil-fuel fired mercury boiler. Further economy is achieved by passing the boiler flue gases over an economiser through which the feed water from the steam plant is passed on its way to the condenser–boiler. (The reader should, at this point, sketch the flow diagram for the plant described in Problem 9.4.) Since heat is then supplied from the combustion products to both the mercury and steam, eqn. (9.25) requires modification.

Referring to Fig. 9.11, in which all energy quantities are expressed **per unit mass of fuel supplied**, since the thermal efficiency of the binary cycle (within control surface S) is given by

$$(1 - \eta_{CY}) = \frac{Q_{out}}{Q_{in}},$$

inspection of the figure shows that

$$(1 - \eta_{CY}) = (1 - q\eta_M)(1 - \eta_S), \qquad (9.26)$$

where q is the fraction of the total heat supply to the binary cycle that is absorbed by the mercury cycle, namely

$$q \equiv \frac{Q_1}{Q_{in}}. \qquad (9.27)$$

We see from eqn. (9.26) that, if η_M and η_S were independent of q, then η_{CY} would tend towards η_S as q tended towards zero; while if q were unity,

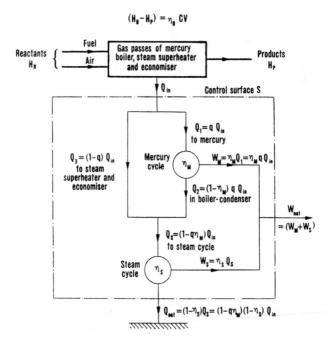

FIG. 9.11. Simplified energy-flow diagram for mercury–steam binary vapour plant.

eqn. (9.26) would become the same as eqn. (9.25). Although η_S could not be fully independent of q, this rather self-evident result at least points to the fact that, other things being equal, it is best to transfer to the steam cycle as little of the total heat supply as possible.

The overall efficiency of the binary plant, neglecting the power required for auxiliaries, is given by

$$\eta_o = \eta_B \eta_{CY}, \tag{9.28}$$

where η_B is the efficiency of the mercury boiler (*cum* steam economiser and superheater) as a heating device. Hence, from eqns. (9.26) and (9.28),

$$\boxed{\eta_o = \eta_o' + x(1 - \eta_S)} \tag{9.29}$$

where

$$\eta_o' \equiv \eta_B \eta_S \tag{9.30}$$

and x is the ratio of the net work output from the mercury cycle to the calorific value of the fuel, namely

$$x \equiv \frac{W_M}{CV} = q\eta_M\eta_B. \tag{9.31}$$

Equation (9.29) is identical to eqns. (9.15) and (9.20).

The design conditions for the plant described in Problem 9.4 resemble those in the Schiller station, New Hampshire, one of the last of the mercury–steam binary-cycle installations to be installed in the USA.[86,87] In that Problem, $q = 0.723$, $\eta_M = 0.220$ and $\eta_B = 0.85$, so that x is 0.135 and $\eta_o = 38.3\%$. This value of x may be compared with the figures of 0.05 and 0.28 quoted in §9.7 for exhaust-heated and pressure-charged gas–steam plant respectively, and the value of 0.30 for MHD–steam plant. The Schiller station achieved an overall efficiency of just over 37%.

The costliness of the working fluid in the mercury cycle, its toxicity and the technical problems associated with its use have limited such installations to outputs which are relatively small compared with those of the latest conventional plant. For these reasons, together with the fact demonstrated in Table B.1 of Appendix B that the efficiencies of conventional plant have since passed that of the Schiller station, no mercury–steam binary plant have been built since 1950. Nevertheless, the increased experience in the handling of liquid metals which has come from their use in experimental nuclear plant has led to a certain revival of interest in the binary cycle, and, indeed, in the possibility of using fluids other than mercury and water. In a proposed Alkali Metal Topping Cycle,[88] potassium has been put forward as a suitable working fluid, having a critical temperature somewhat higher than that of mercury. The potassium cycle would be associated with a conventional supercritical steam cycle to give an estimated overall station efficiency of about 50%. In another proposal, a still more advanced concept is put forward.[89] This would comprise a *ternary system* in which a gas-turbine plant would act as a topping plant to a potassium–steam binary cycle. The compressor of the gas-turbine plant would pressurise the combustion chamber, which would serve as both the combustor for the gas turbine and the heat source for the potassium boiler. There would thus be incorporated in the one plant a pressure-charged gas-turbine/potassium-turbine combined plant, with the potassium cycle acting as the topping plant in a potassium–steam binary cycle. It is estimated that the overall efficiency would be about 53%, but it must be doubted whether such a concept will ever prove more than an academic exercise. An excellent review of the potentialities and problems of liquid-metal binary-cycle plant will be found in reference 90.

By contrast to these *topping cycles*, the use of ammonia in a *bottoming cycle*, or in a ternary mercury–water–ammonia cycle has been mooted. The exhaust volumetric flow rates to be handled in the very large steam units being built are so great that a single unit has as many as four large low-pressure turbines operating in parallel and discharging to a single condenser. Ammonia has the advantage of a much higher vapour pressure than water at room temperature ($728 \, kN/m^2$ at $15 \, °C$ compared with $1.7 \, kN/m^2$), so that if ammonia were used in a bottoming cycle the resulting reduction in specific volume would allow the four low-pressure steam turbines to be replaced by a single, relatively small medium-pressure ammonia turbine. A water–ammonia boiler would be required, however, and since the combined cycle would offer no improvement in thermal efficiency, the economic attractions of bottoming plant are much more open to question than are those of topping plant. Nevertheless, the search has continued rather spasmodically for possible suitable fluids, amongst which some of the freons have appeared quite attractive.

9.12. Binary cycle with thermionic generation

It was noted in §9.11 that a binary cycle that relies entirely on highly stressed turbo-machinery is limted in the top temperature that can be utilised. Replacement of the mercury vapour cycle by a thermionic generating plant as the topping device is a concept which might make possible the use of elevated temperatures in a plant which, in fact, also operates on a binary cycle, although it is not usually described as such.

The thermionic diode generator[83] would operate on the same principle as the radio diode. It makes use of the Edison effect, in which electrons are emitted from a hot cathode to a cooler anode.

Figure 9.12 gives a diagrammatic representation of a possible method of

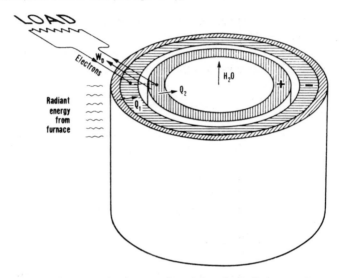

Fig. 9.12. Diagrammatic representation of thermionic diode generator.

construction of a thermionic generator, showing its essential mode of operation. It indicates the futuristic concept of thermionic diodes superposed on the boiler tubes of the steam circuit. The boiler tube, suitably coated on its outer surface, forms the anode and is surrounded by a concentric tubular cathode of appropriate material housed within a ceramic sheath which receives radiant energy from the boiler furnace. The heat quantity Q_2 transferred from the cooler anode to the fluid passing through the tube is less than the quantity Q_1 transferred from the sheath to the cathode, by the amount of the electrical work delivered to the external circuit. The device is thus a *cyclic heat power plant*, in which the electron "gas" passing round the electrical circuit is the working fluid, the sheath is the heat source and the H_2O fluid passing through the tube the heat sink. The thermal efficiency of the diode generator is thus

$$\eta_D \equiv \frac{W_D}{Q_1} = \frac{Q_1 - Q_2}{Q_1}. \tag{9.32}$$

The boiler tubes, which thus also serve to provide the cathodes for the thermionic generator, would line the boiler furnace and receive radiant energy from the flame resulting from the combustion of fossil fuel. In addition, further plain boiler tubes would receive a quantity of heat Q_3 from the combustion products in the convection passes of the boiler. Thus the fluid passing round the steam cycle would pick up heat Q_2 rejected by the diode cycle and heat Q_3 direct from the gases. The diode cycle thus simply takes the place of the mercury cycle in Fig. 9.11, so that eqns.

(9.26), (9.29) and (9.31) are equally applicable to this binary cycle if η_M is replaced by η_D, and W_M by W_D.

Being a cyclic heat power plant, the upper limit for the thermal efficiency of a diode generator is the Carnot efficiency

$$\eta_{\text{CARNOT}} = \left(1 - \frac{T_{\text{sink}}}{T_{\text{source}}}\right).$$

For $T_{\text{source}} = 1400$ K and $T_{\text{sink}} = 700$ K, this would be 50%, but a practical efficiency might be only between one-fifth and two-fifths of this value. Taking $\eta_D = 0.15$, $\eta_B = 0.9$, $q = 0.6$ and $\eta_S = 0.42$, we have:

$$x = q\eta_D\eta_B = 0.6 \times 0.15 \times 0.9 = 0.081,$$

$$\eta_o = (0.378 + 0.081 \times 0.58) \times 100 = 42.5\%,$$

$$\frac{W_D}{W_{\text{TOT}}} = \frac{0.081}{0.425} = 0.191.$$

These figures are roughly in line with those quoted in §9.7 for the exhaust-heated gas-turbine topping plant, which uses well-established techniques, while the diode generator is still in the development stage. However, the assumed diode efficiency of 15% has been achieved[91] with cermet electrodes (a sintered mixture of ceramic and metal) having caesium vapour in the electrode gap, and a value of up to 30% is thought to be possible. Nevertheless, the construction of a plant of any size must be considered to be a distant project.

More specialist texts[83,92] should be consulted for further information on thermionic energy conversion.

9.13. Summary

For all the combined and binary plant which derive their energy from the combustion of fossil fuel, whether the topping plant comprises a gas-turbine installation, MHD generator, mercury vapour or thermionic cycle, the overall efficiency has been seen to be given by

$$\eta_o = \eta_o' + x(1 - \eta_S),$$

where $\eta_o' = \eta_B\eta_S$ and x is the ratio of the net work output of the topping plant per unit mass of fuel burnt to the calorific value of the fuel. A direct explanation of the applicability of a common formula to these different kinds of plant is provided by a study of the simplified block diagrams of Fig. 9.13, noting the identity of the energy quantities crossing control surface Y in (a) and (b). In the gas-turbine plant the total net work output $W_{\text{TOT}} = W_G + W_S$, and in the MHD plant $W_{\text{TOT}} = W_E + (W_S - W_C) = W_G + W_S$, since $W_G = (W_E - W_C)$, so that Fig. 9.13(a) applies equally to both of these installations.

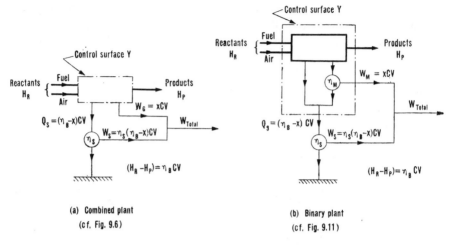

(a) Combined plant
(cf. Fig. 9.6)

(b) Binary plant
(cf. Fig. 9.11)

FIG. 9.13. Simplified energy-flow diagrams for combined and binary plants.

The values which we have assumed and calculated for the various parameters in each of these kinds of topping plant are gathered together in Table 9.1.

TABLE 9.1. *Approximate performance parameters for combined and binary plant, using steam in the bottoming cycle.*

Performance parameter	Gas turbine		MHD	Mercury	Thermionic
	Pressure-charged	Exhaust-heated			
$x \equiv W_{TOP}/CV$	0.28	0.05	0.3	0.135	0.081
η_S	0.26‡	0.44	0.40	0.347	0.42
η_B	0.9	0.9	0.9	0.85	0.9
η_o	0.234	0.396	0.360	0.295	0.378
η_o	44.1%	42.4%	54.0%	38.3%	42.5%
W_{TOP}/W_{TOT}	0.635	0.118	0.556	0.352	0.191

‡Dual-pressure steam cycle.

The values in this table are approximate and to some extent conjectural. It is evident from the table that a high value for the fraction of the total work output that is produced in the topping plant is conducive to high overall efficiency. In this, the MHD plant and the pressure-charged gas-turbine plant lead the field. However, in the latter case the high value of x is offset by the low value of η_S resulting from the relatively low temperature of the gases entering the steam generator; hence the use of a dual-pressure steam cycle. Because of the supplementary firing in the steam generator, the exhaust-heated gas-turbine plant suffers no such offsetting effect and so follows close behind in overall efficiency. Well leading the field in overall efficiency is the MHD topping plant as a result of the very high temperature at which the MHD duct would operate. However, whereas gas-turbine topping plant use well-established techniques and are being built in increasing numbers, the building of a commercial MHD topping plant is a distinctly long-term prospect. The same may be said of a thermionic topping plant.

Not least interesting in this study of present and future prospects is the fact that steam features in the application of all these methods of power generation at high efficiency. No single known working substance appears to offer any possibility of replacing steam, a fluid despised only by the ignorant.

We next turn to a more detailed study of combined gas/steam plant.

9.14. Simple recuperative plant, without supplementary firing

The type of plant that was depicted in Figs. 9.5 and 9.6 is a little more complex than that now to be studied in this section. In the simpler plant shown in Fig. 9.14, the exhaust products leaving the gas turbine are still at a sufficiently high temperature to be used for generating steam in what has commonly been called, somewhat inaccurately, as a "waste-heat" boiler. This utilises what would otherwise have been waste thermal *energy* (not heat) in the exhaust products at a fairly high temperature. Such a boiler is more generally described as a *heat recovery steam generator*, or HRSG for short.

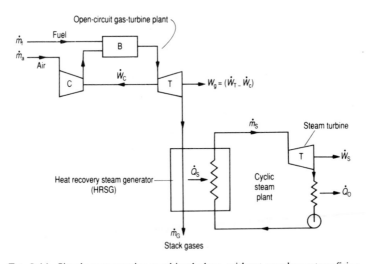

Fig. 9.14. Simple recuperative combined plant, without supplementary firing.

Since steam is generated through the transfer of heat from *gaseous* exhaust products from the gas turbine, the situation in the HRSG is not unlike that in the steam-raising towers of the gas-cooled nuclear power plant discussed in Chapter 8. However, in the *simplest* recuperative plant presently under consideration, there is no call on economic grounds for a dual-pressure steam cycle such as that depicted in Figs. 8.1 and 8.2. Thus here a single-pressure steam cycle suffices. In such a case, without supplementary firing, the *pinch point* (i.e. the point of minimum temperature approach between gas and steam) becomes a controlling factor. That is, for a given design steam pressure, the mass ratio of steam to gas, m_S/m_G, is controlled by the economic minimum temperature difference at the pinch point in Fig. 9.15. The steam pressure is inevitably fairly low.

Problem 9.6 provides an example of this type of plant, in which the open-circuit (internal combustion) CBT gas-turbine plant is superposed on a simple single-pressure steam plant. The problem provides an opportunity for evaluation of the various measures of performance defined in §9.7.

It is important to note that, in this single-pressure type of combined plant, there would be no advantage from bled-steam feed heating, other than to ensure proper deaeration of the feed water supplied to the boiler through the provision of a single *deaerator* heater.[77] Although feed heating would improve the thermal efficiency of the steam cycle, it would *reduce* the *overall* efficiency of the combined plant. That is a consequence

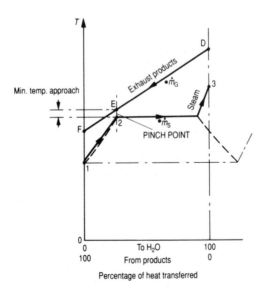

FIG. 9.15. Fluid temperatures with single-pressure steam cycle.

of the controlling influence of the pinch point in single-pressure plant, as may be seen from the following analysis.

With reference to Fig. 9.15, T_D is fixed by the temperature of the exhaust products from the gas turbine. With the steam specified, T_2 is known. Hence, with the temperature approach at the pinch point also specified, the value of T_E is also fixed. The ratio $\dot{m}_S/\dot{m}_G$ of the mass flow rates is then calculable from the Energy Conservation Equation for the heat exchange process from DE to 23. Thus:

$$\frac{\dot{m}_S}{\dot{m}_G} = \frac{h_D - h_E}{h_3 - h_2}. \tag{9.33}$$

For the specified steam parameters, the specific enthalpies on the right-hand side of this equation would be the same whether there were feed heating or not. Thus, for a given $\dot{m}_G$, the value of $\dot{m}_S$ would be the same in both cases. However, the bleeding of steam from the turbine for feed heating would reduce the mass flow rates in later stages of the turbine, so reducing the rate of work output $\dot{W}_S$ for a given value of $\dot{m}_S$ (and also of $\dot{m}_G$, since $\dot{m}_S/\dot{m}_G$ is the same in both cases). Hence, for a given value of $\dot{m}_G$, while $\dot{W}_G$ would be the same when feed heating as when there were no feed heating, $\dot{W}_S$ would be lower if there were feed heating.

Thus feed heating would reduce the total rate of work output, $\dot{W}_G + \dot{W}_S$, and therefore the overall efficiency of the plant. This fact, when the pinch point governs, has not always been appreciated, as is demonstrated in Problem 9.8, the answers to which will be found to be instructive. For convenience of working, the gas-turbine plant in that problem has been taken as a true *cyclic* (closed-circuit) plant, with *heat* supply in the heater replacing the *energy* supply in the combustion chamber of the open-circuit plant depicted in Fig. 9.14. That also enables an interesting application to be made of the expressions derived in the next section.

9.15. Combined plant incorporating cyclic gas-turbine plant

In §6.16, a reminder was given that, in practice, most gas-turbine plant are of the internal-combustion, open-circuit type. At the same time, it was noted that the theoretical studies of closed-circuit (cyclic) plant give qualitative insight into some aspects of the performance of open-circuit plant. For that reason, it is instructive to make a brief study of a combined plant in which both the gas-turbine and the steam-turbine plant are cyclic, as in Fig. 9.16 and also in Problem 9.8. This is then truly a *combined cycle* or *binary* plant. One must therefore warn here that it is a very common, but misleading practice, to describe combined gas/steam plant incorporating *non-cyclic*, open-circuit gas-turbine plant as "combined-*cycle*" plant.

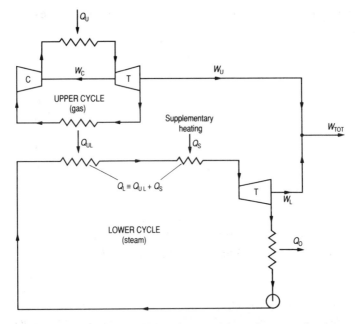

Fig. 9.16. Combined-cycle (binary) plant with supplementary heating.

In the following brief study, we derive a simple but informative relation, first devised by the author, between the overall thermal efficiency of the plant and the respective thermal efficiencies of the upper (gas-turbine) cycle and the lower (steam) cycle. With reference to Fig. 9.16, there is *supplementary heating* of magnitude Q_S, in addition to the heat quantity Q_{UL} transferred to the lower cycle. The purpose of the former is to enable steam at a higher temperature to be supplied to the steam turbine in the lower cycle, thus improving the thermal efficiency η_L of that cycle.

The thermal efficiency of the combined cycle, and the individual thermal efficiencies of the two cycles are defined respectively as follows:

$$\eta_U \equiv \frac{W_U}{Q_U}, \quad \eta_L \equiv \frac{W_L}{Q_L}, \quad \eta_{\text{COMB.}} \equiv \frac{W_U + W_L}{Q_U + Q_S},$$

where $Q_L \equiv Q_{UL} + Q_S$. Hence:

$$\eta_{\text{COMB.}} = \frac{\eta_U Q_U + \eta_L Q_L}{Q_U + Q_S}. \tag{9.34}$$

It is convenient and informative to define two quantities, μ_U and μ_L, which may appropriately be called *weighting factors*, thus:

$$\mu_U \equiv \frac{Q_U}{Q_U + Q_S} \quad \text{and} \quad \mu_L \equiv \frac{Q_L}{Q_U + Q_S}.$$

Substituting for these in eqn. (9.34), we obtain the following expression for $\eta_{COMB.}$:

$$\boxed{\eta_{COMB.} = \mu_U Q_U + \mu_L Q_L} \qquad (9.35)$$

The significance and utility of the weighting factors, μ_U and μ_L, is most readily appreciated from inspection of the answers to Problem 9.8. There they are immediately seen to provide the respective weighted contributions of each cycle to the overall thermal efficiency, $\eta_{COMB.}$, of the complete combined plant. For example, in case (b) in Problem 9.8, although the lower (steam) cycle has the high value of $\eta_L = 44.6\%$, its contribution to $\eta_{COMB.}$ is severely restricted by the very low value of $\mu_L = 0.19$.

9.16. Simple recuperative plant with facility for supplementary firing—cogeneration application

It was noted in §4.7 that, for a metalurgically permissible temperature of about 700 °C at entry to the gas turbine, as much as 300% or more of excess air would have to be supplied to the combustion chamber when burning a typical fuel (cf. Problem 4.7). There will then be much unconsumed oxygen in the exhaust products from the gas turbine. Advantage is taken of that fact in the plant depicted diagrammatically in Fig. 9.17, a detailed study of which forms the subject of Problem 9.9.

This is the type of *cogeneration plant* mentioned briefly in §§7.19 and 9.7. Steam is generated in the HRSG, not for the purposes of power production, but as process steam. At times of low demand for process steam, the waste thermal energy in the exhaust products from the gas

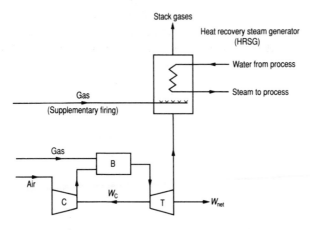

FIG. 9.17. Gas/steam cogeneration plant for power and process steam.

turbine is sufficient to meet the demand for steam, while peak demands for process steam are met by burning supplementary fuel with some of the unconsumed oxygen in those exhaust products as they pass through the HRSG.

The data presented in Problem 9.9 correspond approximately to the design conditions for the DOMO plant supplying process steam to a dairy factory in Beilen, in the Netherlands.[H,V] From the answers and solution to Problem 9.9, it will be seen that, at exhaust from the gas turbine, the unconsumed oxygen comprises about 16% of the gaseous products of combustion. After burning more fuel in the HRSG at peak demand for process steam, there is still about 10% unconsumed oxygen in the stack gases leaving the HRSG.

9.17. High-efficiency combined plant with supplementary firing

In the simple type of recuperative combined plant discussed in §9.14, having a single-pressure steam cycle, the steam pressure and temperature are limited by the temperature level of the exhaust gases from the gas turbine. That results in a rather low thermal efficiency of the steam cycle. There are two alternative means of improving on this situation, namely either by the use of *supplementary firing* or by the use of a *dual-pressure* (or even triple-pressure) steam cycle. We shall now consider the first of these two options, to which reference has already been made in §§9.6 and 9.7.

A combined gas/steam plant, in which the supplementary fuel supplied to the combustion chamber of the steam boiler is just sufficient to burn up *all* of the excess oxygen in the exhaust products from the gas turbine, has been called by Wood[72] a *high-efficiency combined cycle* (better called a combined *plant*), as depicted in Fig. 9.18.

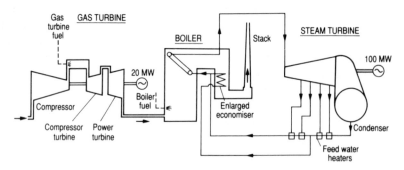

Fɪɢ. 9.18. High-efficiency combined plant (basic concept)—Wood.[72] (Reprinted by permission of the Council of the Institution of Mechanical Engineers from *Modern Steam Plant Practice*, I. Mech. E., C76/71, Apr. 1971.)

The calculations in Problem 9.7 show that if, instead of exhausting to the furnace of the HRSG, the combustion products from the gas turbine in Problem 9.6 were supplied to an *adiabatic* combustion chamber supplied with just sufficient supplementary C_8H_{18} to burn up the excess oxygen, the final products temperature would be about 2300 °C. If those final products were to be used for the transfer of heat to generate steam in a cyclic steam plant, that temperature would be much too high for the heat transfer to be allowed to take place in a simple tubular heat exchanger. The supplementary fuel and the exhaust products from the gas turbine are therefore fed into the furnace of a conventional water-tube boiler, as depicted in Fig. 9.18. Not shown in the simplified diagrammatic arrangement of the boiler are the *water-wall* tubes which surround the furnace, and in which much of the evaporation occurs. These *riser tubes* are fed by large *downcomers* from the steam drum (also not shown in the diagram) in a process of *natural circulation* (see Appendix D). Those features are shown clearly in Fig. 9.22.

Two further important features of the plant of Fig. 9.18 are worthy of special comment. Firstly, since there is a very large difference between the flame temperature in the furnace and the water/steam mixture circulating inside the water-wall tubes, the possibility of a pinch point does not arise. These is consequently nothing to inhibit the provision of bled-steam feed heating, as there was in the simple recuperative plant discussed in §9.14.

The second feature that is worthy of special note is the "enlarged economiser", and the fact that it is fed with feed water drawn off at an intermediate point in the feed train. In this respect, the plant differs from a conventional steam power plant in which the flue gases from the boiler are cooled before discharge to the stack by being passed through a surface *air preheater*. There they transfer heat to the cool incoming air which is fed into the boiler by the *forced-draught fans*, so improving the boiler efficiency. (That procedure is illustrated in the more complex plant depicted in Fig. 9.22.) Since the furnace in Fig. 9.18 is not fed with cool incoming air, but with the oxygen-bearing exhaust products from the gas turbine at a fairly high temperature (cf. Problem 4.7), the "enlarged economiser" has to take the place of an air preheater. Furthermore, the cooling water is drawn off from an intermediate point in the feed train where the water temperature is not cool enough to result in condensation of the H_2O in the flue gases, so avoiding corrosion on the exterior surfaces of the economiser tubes. About one-third of the flow is drawn off from this intermediate point and bypasses the later feed heaters.

Because of the supplementary firing in the boiler, the condition of the steam supplied to the turbine can be on a par with that in a conventional high-efficiency steam plant. Thus the combined gas/steam plant will have an overall efficiency greater than that of a high-efficiency steam plant alone. Some typical figures have already been quoted in §9.7.

We next turn to the second alternative means of improving the performance of combined gas/steam plant beyond that of the single-pressure type of plant discussed in §9.14; namely, the provision of a dual-pressure, or even a triple-pressure, steam cycle.

9.18. Dual-pressure and triple-pressure steam plant, without supplementary firing

Reference has already been made in §9.7 to the dual-pressure plant at Donge in the Netherlands, variously reported[75,76] as having an overall efficiency of 44.4%, with W_G/W_{TOT} = 0.64, and 46.1% with W_G/W_{TOT} = 0.62. Again, the only extraction of bled steam for feed-heating is to a direct-contact (DC) deaerator heater, from which the feed water is pumped to the low-pressure economiser in the steam-raising tower (boiler).[76] As with the dual-pressure steam cycle in the gas-cooled nuclear-power plant at Calder Hall, and depicted in Fig. 8.1, superheated steam from the low-pressure drum is supplied to a steam chest located part way down the turbine. In view of this similarity, the reader should have no difficulty in making a sketch of the circuit diagram for the Donge plant. Instead of presenting that diagram here, it will therefore be more instructive to present the flow diagram for a triple-pressure plant,[76] as depicted in Fig. 9.19.

This plant is of unusual interest because the triple-pressure system in this particular application serves a special purpose, namely the provision of a supply at medium pressure for the injection of *steam* into the gas-turbine combustion chamber (GTCC) of the gas-turbine plant. The reason for such an unusual procedure is discussed in the next section. Inspection of Fig. 9.19 reveals the following key points:

(1) The LP evaporator serves solely to supply the steam (dry saturated) to the deaerator heater. This is different from the Donge plant, in which the deaerator heater took steam bled from the turbine.
(2) The MP evaporator serves solely to provide injection of dry saturated steam into the gas-turbine combustion chamber (GTCC).
(3) The HP economiser, evaporator and superheater serve solely to supply superheated steam to the turbine.
(4) Neither the LP nor HP evaporators supply steam to a later stage in the turbine, as was provided by the LP steam supply in the dual-pressure cycles at Donge and Calder Hall (Fig. 8.1).

We can now pass on to consider briefly the purpose of steam injection in *steam-injection gas-turbine* (STIG) plant.

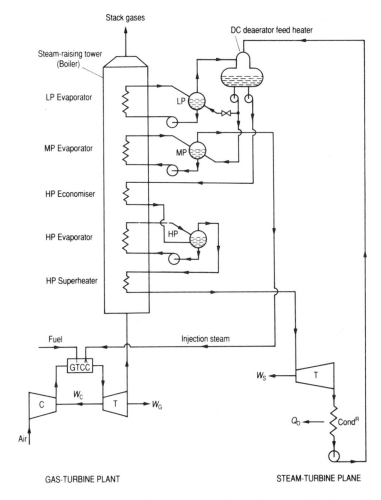

FIG. 9.19. Triple-pressure combined gas/steam plant, with steam injection to the gas-turbine combustion chamber. — After Wunsch.[76]

9.19. Steam-injection gas-turbine (STIG) plant

Fig. 9.20 outlines, in its simplest terms, a STIG unit in which a CGT gas-turbine plant exhausts into a heat recovery steam generator (HRSG), as in the simple recuperative plant of Fig. 9.14. However, there is an additional feature in the plant of Fig. 9.20, in that some of the steam generated in the HRSG is injected into (or upstream of) the GTCC. The mass flow rate of injected steam is typically of the order of 15% of the mass flow rate of air supplied to the GTCC.[W]

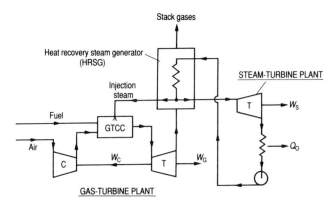

Fɪɢ. 9.20. Steam-injection, gas/steam combined plant (STIG).

In a STIG plant, steam injection has two beneficial influences:

(1) As a result of the cooling effect of the steam in the primary-flame zone of the combustor, it results in a reduction in the emission of noxious oxides of nitrogen, NO_x, from the plant.

(2) It provides an increase in both power output and overall efficiency. For a given temperature at inlet to the gas turbine, extra fuel has to be supplied in order to heat the injected steam to that temperature, but the additional power arising from the expansion of the injected steam as it passes through the gas turbine more than offsets the otherwise adverse effect on the overall efficiency of the plant of the increase in fuel supply.

As a result of increased attention to environmental considerations, it became evident that greater efforts had to be made to minimise the emission of noxious gases such as NO_x and SO_2 from plant burning fossil fuels. It has thus been the need to reduce NO_x emissions that has stimulated interest in the development of STIG plant, though currently with fairly restricted practical application. Detailed analysis of such plant is beyond the scope of the present volume, so that other sources must be consulted for further information.[X,Y]

As our final study of combined gas/steam plant, we turn to consideration of another method of reducing NO_x emissions which has become of increasing interest, namely the use of *fluidised-bed combustion* (FBC).

9.20. Fluidised-bed combustion (FBC)

A means of reducing the emission of NO_x and SO_2 in coal-fired plant is provided by *fluidised-bed combustion*, FBC. In this, a bed of fine particles is kept in a state of constant agitation by the action of jets of combustion air blown through the bed as coal is fed in. The combustion chamber is at about atmospheric pressure in *atmospheric fluidised-bed combustion* (AFBC), or is supplied with air under pressure in *pressurised fluidised-bed combustion* (PFBC)[(Z)] that must not be confused with *pulverised fuel* (PF).

In both types of FBC, the bed is "cooled" by the transfer of heat from the burning coal to a tubular heat exchanger immersed in the bed, and through which a coolant fluid is passed. In the "boiling" bed, heat transfer is highly efficient. The coolant may be either H_2O or air, depending on the type of plant. Fig. 9.21 gives a simplified schematic representation of an FBC combustor. The means of continuous ash removal are not shown. With the relatively low combustion temperature of about 850 °C, the emission of oxides of nitrogen, NO_x, is lower than from a conventional water-tube boiler. The emission levels of SO_2 are also lower, in consequence of the addition of *sorbent* material, such as limestone or dolomite, to the bed. That absorbs the sulphur compounds formed during the combustion of sulphur-bearing coal.

It was noted in §1.2 of Chapter 1 that the use of coal as the fuel for open-circuit gas-turbine plant can cause a fouling problem on the turbine blades. One of the stimuli for the development of fluidised-bed combustion has been the fact that the combustion products from such plant are less obnoxious than in conventional firing, in which sodium salts distilled from the ash tend to condense on the turbine blades and cause corrosion. With the lower combustion temperature in a cooled fluidised bed, vaporisation of sodium salts is less, while a cyclone separator between the combustor and the turbine further helps to reduce carry-over into the turbine. An interesting review of small-scale applications of FBC will be found in Ref. AA.

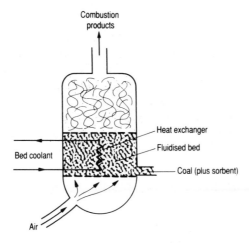

Fig. 9.21. Fluidised-bed combustor.

An alternative means of accommodating the gas turbine to products resulting from the burning of coal is to supply the turbine from a *coal-gasification plant*,[BB] but this is beyond the scope of the present volume. We therefore turn instead to a more detailed study of FBC combined gas/steam plant.

9.21. PFBC combined plant, with H₂O as the coolant fluid

When *pressurised* fluidised-bed combustion (PFBC) is utilised in a combined gas/steam plant, the air supplied to the combustor comes from the compressor of a CBT gas-turbine plant, in which the gas turbine drives the compressor as well as providing electrical power.[CC] The reader should have no difficulty in picturing the complete layout of such a plant. A more interesting and informative picture is provided by the plant which is illustrated fully in the next section. That uses *air* as the coolant fluid in an *atmospheric* fluidised-bed combustion (AFBC).

9.22. Conceptual AFBC plant, with air as the coolant fluid

Fig. 9.22 provides a fitting tailpiece to our studies of combined gas/steam plant, since it has many features of special interest. This conceptual plant may be looked upon as a composite of the ideas that have been studied in previous sections. It also provides an excellent object of study for the student and engineer alike, since it shares with more conventional boiler plant features of interest that have not, up to this point, received the mention which they deserve. Attention is drawn to the following points:

BOILER PLANT

(1) The centrepiece is a *fluidised-bed boiler* in which air, not H₂O, is the "coolant" fluid flowing through the coils of the heat exchanger which is immersed in the bed.

(2) As is the practice in conventional water-tube boilers, *atmospheric pressure* is maintained in the boiler by the combined action of the *forced-draught* (FD) and *induced-draught* (ID) fans, producing what is called *balanced draught*.

(3) Again as in conventional boilers, the air for combustion is heated in an *air preheater* on its way from the *forced-draught* fan to the furnace. Heat is transferred to the incoming air from the flue gases, which pass through the air preheater as they are drawn out of the furnace by the *induced-draught* fan and discharged to the stack.

(4) Before discharge to the stack, the flue gases pass in succession through a *cyclone separator* and an *electrostatic precipitator*. These reduce the emission of particulate matter into the environment, the

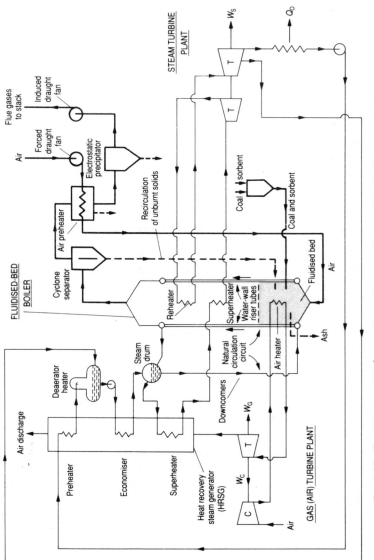

FIG. 9.22. Combined gas(air)/steam plant with atmospheric fluidised-bed combustion (AFBC).

fluidised-bed combustion having already taken care of reduction in NO_x and SO_2 emissions.

(5) In classical FBC, the fluidising velocity of the air is about 2 m/s, and the bed has a well-defined "upper surface" that "bubbles"; that is the case in what is called *bubbling fluidised-bed combustion* (BFBC). If the velocity is increased to about 8 m/s, unburnt solids are carried upwards and out from the top of the combustor. These solids are collected in the cyclone separator and are reinjected at the bottom of the combustor, as in Fig. 9.22, giving what is termed *circulating fluidised-bed combustion* (CFBC). This requires a taller combustor than BFBC, but fuel burn-out and sulphur-trapping is improved.

(6) As in conventional water-tube boilers, the furnace is lined with *water-wall* riser tubes, in which the generation of steam takes place. The resulting steam/water mixture is discharged to the *steam drum*, the unevaporated water returning through *downcomer* pipes to the feet of the water-wall by a process of *natural circulation* around the circuit. (See Appendix D.)

GAS-TURBINE PLANT

(7) The gas-turbine part of the plant is unusual in two respects. Its working fluid is *air*. Moreover, energy input between the compressor and the turbine is not by the burning of fuel, but by the transfer of heat as the air flows through the tubes of the heat exchanger immersed in the fluidised bed of the boiler. Since the temperature of the air at exhaust from the gas turbine is still high, it is cooled in the HRSG by transferring heat to the steam circuit before discharge to atmosphere. That discharge is of a perfectly clean fluid — air.

STEAM CIRCUIT

(8) After being heated in the water *preheater* placed in the tail end of the HRSG, the feed water from the condenser of the steam plant is deaerated in the *deaerator heater*, before passing through the *economiser* on its way to the *steam drum*. In the light of earlier discussions, the reader should consider why there is only a single feed heater, the deaerator.

(9) From the steam drum, the dry saturated steam passes successively through two *superheaters*, the first within the high-temperature end of the HRSG and the second within the boiler furnace, in which there is also a steam *reheater*. This enables steam to be delivered at conditions which are competitive with modern steam plant of the more conventional type. The overall efficiency of the combined gas

(air)/steam plant would therefore be expected to be something in excess of 40%.

Whilst the plant just described is only conceptual in origin, a considerable variety of FBC plants of various sizes have been installed or are under construction. In 1989, the 80 MW pilot PFBC rig at Grimethorpe,[Z] funded by British Coal, was still the largest operating unit in the world. For an up-to-date presentation of world-wide progress at that time, the reader may consult the Proceedings of the 10th International Conference on Fluidised Bed Combustion — *FBC: Technology for Today*, San Francisco, 1989.

9.23. Conclusion

One of the driving forces for the renewal of interest in combined gas/steam plant, in fluidised-bed combustion of coal and in coal gasification has been the increased attention to the problems of environmental pollution. As demonstrated in Problem 9.10, for a given energy release the burning of natural gas produces only about half the quantity of carbon dioxide produced by the burning of coal, and CO_2 is one of the main contributors to the so-called *greenhouse effect*, leading to global warming. However, in the United Kingdom, it appears that this fact may have provided a convenient cover for more subtle political decisions, leading to appreciable unease and some anger.

Not so many years ago, it was accepted wisdom that natural gas, as a premium fuel of limited reserves, was far too valuable as a chemical feedstock to allow it simply to be burnt in large industrial and power-station plants. Indeed, the European Commission introduced a formal ban on such use, an action countered by a request from the British Government that the ban be lifted because more gas was available than had previously been thought! Until the late Eighties, such a ban was indeed in force in the UK itself, but that ban was withdrawn, by political direction, about two years before the electricity generating industry was scheduled for privatisation by a Government with an insatiable appetite for selling off national assets into private hands, on the pretext of introducing competition. In such an event, the only way in which smaller-scale newcomers to the industry could hope to compete with the two large generating companies into which it was proposed to break up the Central Electricity Generating Board was the building of smaller combined gas/steam plant, which is markedly less capital-intensive and quicker to install than large conventional steam plant. Somewhat ironically these two larger suppliers inevitably followed suit in order themselves to compete with the new smaller suppliers, and even foresaw the possible future demise of the very large steam plant that had been the life-blood of the turbine manufacturing industry for many years. There followed a minor rush of applications from international consortia with a British component, and others, for permission to construct combined gas/steam power plant burning natural gas. Applications were duly granted by the Government even before the date at which the generating industry was due to be sold off to the private sector. In the light of the earlier comment on the value of natural gas as a chemical feedstock, the granting of permission for the building of a very large plant by a US/UK consortium for supplying power to one of the country's largest chemical companies may seem a particular irony. The reader must be left to discover in due course whether such policies will prove to have been in the national and wider interest.

It should finally be noted from the preceeding sections that the industrial gas turbine is not projected to displace the steam turbine, but to supplement it in combined gas/steam plant. Thus, as with nuclear power, mankind will still continue to live in the Age of Steam so far as the large-scale generation of electrical power is concerned. Again the student should be warned that, as with nuclear power, technological considerations do not always win against commercial pressures and political dogma and ideology.

9.24. Footnote

Very shortly after the manuscript for this 4th Edition was delivered to the Publisher, an important conference on *Power Generation and the Environment* was held in November 1990 at the Headquarters of the Institution of Mechanical Engineers in London. The conference was sponsored jointly by the Power Industries Division of the Institution, the Institution of Electrical Engineers and Verein Deutscher Ingenieure. Its Proceedings constitute an excellent collection of 28 papers, many of which deal with the types of plant discussed in Chapters 7, 8 and 9 herein. Thus the material in those three chapters, and particularly in this chapter, will prove invaluable in aiding understanding of the papers presented at the conference. The Proceedings are worthy of study as a whole, and particularly the following two papers relating to the future of coal-fired power plants:

(a) Schemenau, W. and van den Berg, C., The future of coal-fired power plants. *Conf. Proc.* C 410/053, 1–12, I. Mech. E., London, 1990.

(b) Dawes, S. G. *et al.*, Options for advanced power generation from coal, *Conf. Proc.* C 410/042, 123–134, I. Mech. E., London, 1990.

There is considerable variety in possible circuit layouts of combined gas/steam plants, and these do not necessarily follow exactly those illustrated in this chapter. The two listed papers depict and discuss some of them, including some in which the gas for combustion would be supplied by coal-gasification in an Integrated Gasification Combined "Cycle" (IGCC) plant.

Problems

9.1. For the ideal super-regenerative steam cycle illustrated in Figs. 9.1 and 9.2, show that

$$\frac{M_B}{M_G} = \frac{\Delta s_3}{\Delta s_2} \quad \text{and} \quad \frac{M_A}{M_G} = \frac{\Delta s_3}{\Delta s_1}.$$

In an ideal super-regenerative steam cycle, evaporation takes place at $10 \, \text{MN/m}^2$, the maximum cycle temperature is $550 \, ^\circ\text{C}$, the exit pressure from the reheated turbine is $1.0 \, \text{MN/m}^2$ and the condenser pressure is $4.0 \, \text{kN/m}^2$. Determine:

(a) M_B/M_G and M_A/M_G,
(b) the thermal efficiency of the cycle,
(c) the Carnot efficiency for the same extreme temperature limits.

Noting that the difference between the Carnot efficiency and the cycle efficiency arises from two causes, namely (1) the fact that the heat to the cycle is not all added at the top temperature, and (2) the irreversibility of the process occurring in heat exchanger X of Figs. 9.1 and 9.2, evaluate:

(d) the *entropy creation due to irreversibility* (§A.6 of Appendix A) in each of these processes, expressed per unit mass of steam passing through X,
(e) the fraction of the total loss in cycle efficiency below η_{Carnot} that is attributable to each of these causes.

Answer: (a) 0.406, 0.229; (b) 61.2%; (c) 63.3%; (d) 0.022 kJ/kg K, 0.063 kJ/kg K; (e) 0.26, 0.74.

9.2. At the design load of a Field super-regenerative steam cycle such as that illustrated in Figs. 9.3 and 9.4, the steam flow rate M_B to the condensing turbine is such that the quantity of feed water injected into the spray desuperheater is just sufficient to ensure that the condition of the steam at the compressor outlet is dry saturated. It may be assumed that there is complete mixing of the injected water and the steam before entry to the compressor. The pressures and temperatures at the points specified are:

Position	Pressure (MN/m²)	Temperature (°C)
2	0.7	100
4	7.0	(dry saturated)
6	7.0	550
8	2.5	550
10	0.7	300
11	0.004	(wet)

The enthalpy rise of the feed water is the same in both direct-contact heaters, and β (as defined in §7.9) may be taken as equal to 2270 kJ/kg in each. The isentropic efficiency of each turbine, and of the compressor, is 85%. Pressure drops in the piping may be neglected.

Determine the ratios M_B/M_G and M_B/M_A. Calculate the thermal efficiencies η_G, η_S and η_F as defined in §9.4, and check that they satisfy eqn. (9.6).

Answer: 0.281; 1.135; 19.8%; 26.4%; 40.9%.

9.3. Derive eqn. (9.15) for the *exhaust-heated* combined gas–steam plant described in §9.6, the relevant calorific value being the combined heating value of the fuel burnt in the two combustion chambers.

9.4. In a mercury–steam binary vapour cycle the mercury leaves the mercury boiler dry-saturated at $900 \, \text{kN/m}^2$, and is condensed in the mercury–steam condenser–boiler at $17 \, \text{kN/m}^2$. Dry saturated steam leaves the condenser–boiler at $4 \, \text{MN/m}^2$, and is heated to $425 \, °\text{C}$ in a superheater placed in the flue-gas passes of the mercury boiler.

The steam is supplied to a regenerative steam cycle in which the condenser pressure is $4 \, \text{kN/m}^2$. The feed water from the condenser is heated to $170 \, °\text{C}$ in three bled-steam feed-heating stages, and is then further heated to the steam saturation temperature by being passed through an economiser placed in the exit flue gas stream from the mercury boiler. For the purposes of calculation it may be assumed that three direct-contact feed heaters are used, that the enthalpy rise of the feed water is the same in each, and also that β (as defined in §7.9) is equal to $2215 \, \text{kJ/kg}$ in each heater. The isentropic efficiency of the mercury turbine is 75% and of the steam turbine 82%.

The properties of saturated mercury are given below:

Pressure	Enthalpy/(kJ/kg)		Entropy/(kJ/kg K)	
kN/m^2	h_f	h_g	s_f	s_g
900	68.4	358.1	0.1415	0.5121
17	36.9	331.1	0.0940	0.6362

Calculate:

(a) the ratio of the mass flow rates of steam through the condenser-boiler and steam condenser respectively;
(b) the ratio of the mass flow rates of mercury and steam through the condenser-boiler;
(c) the thermal efficiency η_{CY} of the binary cycle;
(d) the overall efficiency η_o and heat rate of the plant (in Btu/kW h), given that the efficiency of the mercury boiler is 85%;
(e) the thermal efficiencies η_M and η_S defined in §9.11;
(f) the value of q defined by eqn. (9.27); check that η_{CY} satisfies eqn. (9.26);
(g) the value of x defined by eqn. (9.31); check that η_o satisfies eqn. (9.29).

Answer: (a) 1.295; (b) 6.84; (c) 45.1%; (d) 38.3%, 8910 Btu/kW h; (e) 22.0%, 34.7%; (f) 0.723; (g) 0.135.

9.5. Derive equation (9.25).

Additional problems

9.6. The open-circuit (internal combustion) gas-turbine plant of Problem 4.7 exhausts to a heat recovery steam generator (HRSG), in which steam is generated in a single-pressure steam cycle, thus forming a simple recuperative combined plant of the type depicted in Fig. 9.14. The HRSG supplies superheated steam to the turbine at $2 \, \text{MN/m}^2$, $325 \, °\text{C}$ and the turbine exhausts to the condenser at $7 \, \text{kN/m}^2$. There is no bled-steam feed heating.

With the conditions in the gas-turbine plant the same as those in Problem 4.7, the temperature difference at the pinch point in Fig. 9.15 is 20 K. Taking the specific heat capacity of the combustion products passing through the HRSG to be constant, and equal to $1.075 \, \text{kJ/kg K}$, calculate the following quantities:

(a) The mass of combustion products per unit mass of steam generated.
(b) The temperature of the combustion products leaving the HRSG.

It can readily be shown (Ref. A in Preface, Problem 4.7) that, per kmol of C_8H_{18} burnt, the kilomoles of combustion products are: 47.5 excess O_2, 225.5 N_2^*, 8 CO_2, 9 H_2O. Hence determine, per kilogram of C_8H_{18} burnt:

(c) The mass m_G of combustion products.
(d) The mass m_S of steam generated.
(e) The heat quantity Q_S transferred in the HRSG.

Evaluate the following quantities, as defined in §9.7:

(f) $x \equiv W_G/\text{LCV}$. $\qquad\qquad\qquad\qquad\qquad\qquad\qquad\qquad\qquad$ (9.8)
(g) Q_S/LCV.

(h) $\eta_B = x + \dfrac{Q_S}{\text{LCV}}$. $\qquad\qquad\qquad\qquad\qquad\qquad\qquad\qquad$ (9.10)

(i) $\eta_S \equiv W_S/Q_S$, $\qquad\qquad\qquad\qquad\qquad\qquad\qquad\qquad\qquad$ (9.11)

given that the isentropic efficiency of the steam turbine is 82%.

(j) $\eta_o' \equiv \eta_B \eta_S$. $\qquad\qquad\qquad\qquad\qquad\qquad\qquad\qquad\qquad\qquad$ (9.14)

(k) $\eta_o \equiv (W_G + W_S)/\text{LCV} = \eta_o' + x(1 - \eta_S)$. $\qquad\qquad\qquad\quad$ (9.15)

Answer: (a) 8.274;   (b) 148.6 °C;   (c) 73.5 kg;   (d) 8.89 kg;   (e) 25.95 MJ;   (f) 0.193;
(g) 0.584;   (h) 77.7%;   (i) 26.7%;   (j) 20.7%;   (k) 34.0%.

9.7. Show that if, instead of exhausting to the HRSG, the combustion products leaving the gas turbine in Problem 9.6 were supplied to an *adiabatic* combustion chamber supplied with just sufficient additional C_8H_{18} to burn up all the excess oxygen, the final products temperature would be about 2300 K. The effects of dissociation are to be neglected.

The molar enthalpies of the respective products at 2300 K are given below; molar enthalpies at 750 K and 298 K (25 °C) are listed in the table of Problem 4.7.

Product	N_2^*	CO_2	H_2O
Molar enthalpy, MJ/kmol, at 2300 K	75.70	119.28	98.27

9.8. Fig 9.23 depicts a hypothetical combined-cycle (binary) plant in which a cyclic (closed-circuit) gas-turbine plant with air as its working fluid, is superposed on a cyclic steam plant. The notation is similar to that in Fig. 9.16.

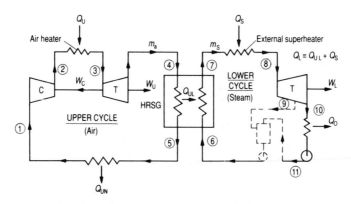

FIG. 9.23. Combined-cycle (binary) plant.

In a published theoretical study[DD] of such a plant, an analysis was made of the calculated overall efficiency when the feed water was heated alternatively:

(a) by steam bled from the turbine;
(b) in a solar heater over the same range of temperature.

No study was made of the overall efficiency in the absence of any feed heating.

Given the data below, and basing the calculations on unit mass of air circulating around the upper cycle, determine the thermal efficiency, $\eta_{COMB.}$, of the combined plant when (a) feed heating by bled steam; (b) feed heating over the same temperature range by solar heating, regarding the heat supplied by the solar heater as "free"; (c) there is no feed heating. Parasitic pressure drops in the ducting and heat exchangers are to be assumed to be negligible.

UPPER CYCLE

The compression ratio $r_p = 17$, and the air is to be treated as a perfect gas with $c_p = 1.01 \text{ kJ/kg K}$.

$t_1 = 15\,°C$; $t_3 = 1000\,°C$; $\eta_C = 85\%$; $\eta_T = 87\%$.

LOWER CYCLE

$p_8 = 6.8\text{ MN/m}^2$; $t_7 = 325\,°C$; $t_8 = 425\,°C$; $p_{10} = 7\text{ kN/m}^2$; $\eta_T = 87\%$.

HEAT RECOVERY STEAM GENERATOR (HRSG)

$t_7 = 325\,°C$.
Temperature difference at pinch point (cf. Fig. 9.15) = 22 K.

FEED HEATING

(a) *By bled steam*

$p_9 = 2\text{ MN/m}^2$; $h_9 = 2960\text{ kJ/kg}$; $t_6 = 212.4\,°C$ (sat. temp. at 2 MN/m^2).

(b) *By solar heater*

$t_6 = 212.4\,°C$.

In each case, evaluate $\eta_{COMB.}$ from equation (9.35), namely:

$$\eta_{COMB.} = \mu_U \eta_U + \mu_L \eta_L,$$

having first evaluated η_U, η_L, μ_U and μ_L, where μ_U and μ_L are the *weighting factors* defined in §9.16, namely:

$$\mu_U \equiv \frac{Q_U}{Q_U + Q_S} \quad \text{and} \quad \mu_L \equiv \frac{Q_L}{Q_U + Q_S}.$$

Check the calculated values of $\eta_{COMB.}$ by direct calculation from the respective work and heat quantities. What conclusions may be drawn from a comparison of the respective values of $\eta_{COMB.}$?

In each case, determine the temperature of the air leaving the HRSG.

Answer:

Case	(a)	(b)	(c)
η_U	34.1%	34.1%	34.1%
μ_U	0.9764	0.9764	0.9764
η_L	35.8%	44.6%	33.8%
μ_L	0.1901	0.1901	0.2512
$\eta_{COMB.}$	40.1%	41.8%	41.8%
t_5	289.4 °C	289.4 °C	252.9 °C

9.9. Figure 9.24 depicts a cogeneration plant in which an open-circuit gas-turbine plant exhausts to a heat recovery steam generator (HRSG), which supplies steam to a dairy factory in the Netherlands. The HRSG is provided with gas burners for supplementary firing when the demand for process steam is high.

The data given below correspond approximately to the design conditions for the DOMO plant in Beilen.[H,V]

GAS TURBINE PLANT
Pressure ratio $r_p = 7$, $\eta_C = 85\%$
$t_1 = 25\,°C$, $t_3 = 850\,°C$, $t_4 = 490\,°C$
Air supply rate = 20.45 kg/s
Fuel: Natural gas (CH_4) at 25 °C

HRSG
$t_4 = 490\,°C$, $t_5 = 138\,°C$, $t_6 = 90\,°C$
$p_7 = 1.3\,MN/m^2$, dry saturated steam
Fuel: Natural gas (CH_4) at 25 °C

Neglecting all pressure drops in the ducting and in the heat exchanger, calculate:

(a) The value of t_2, treating the air as a perfect gas.
(b) The molar ratio of air to fuel supplied to the gas-turbine combustion chamber, and the percentage of excess air supplied.
(c) The percentage of oxygen in the products leaving the gas-turbine plant.
(d) Per kmol of air supplied, the following work quantities:

   (1) Compressor work input, W_C.
   (2) Turbine work output, W_T.
   (3) The net work output, W_G, from the gas-turbine plant.

(e) For the given rate of air supply, the net power output from the gas-turbine plant, and the rate of fuel supply to it.
(f) The overall efficiency, η_o, of the gas-turbine plant, based on the lower calorific value.

WITHOUT SUPPLEMENTARY FIRING

(g) The rate of heat transfer in the HRSG, in MW.
(h) The rate of steam delivery from the HRSG, in tonnes per hour.
(i) The total efficiency η_{TOT} (energy utilisation factor, EUF) of the complete plant in the absence of supplementary firing.

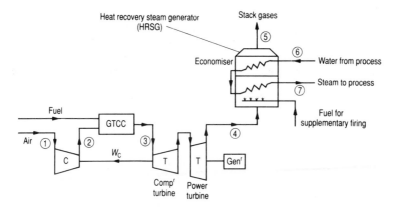

FIG. 9.24. Gas/steam cogeneration plant for power and process steam.

WITH SUPPLEMENTARY FIRING

With supplementary firing in the HRSG to increase to 35 t/h the delivery rate of dry saturated steam at the same pressure, determine:

(j) The rate of fuel supply to the HRSG, and the percentage of oxygen in the products leaving the HRSG.

(k) The total efficiency η_{TOT} (energy utilisation factor, EUF) of the complete plant.

The molar enthalpies of the products of combustion at the respective temperatures are given in the following table (from Ref. 1, Table 3):

Product	CO_2	H_2O	N_2	O_2
Temperature		Molar enthalpy		
°C		MJ/kmol		
25	9.37	9.90	8.67	8.66
138	13.85	13.74	11.97	12.02
490	30.32	26.50	22.57	23.27
850	49.57	41.09	34.22	35.70

Atmospheric nitrogen, N_2^* (Ref. 1, Table 2), may be taken to have the same molar enthalpy as nitrogen, N_2.

Answer: (a) 285.8 °C; (b) 41.77, 339%; (c) 15.8%; (d) 7.64 MJ, 12.38 MJ, 4.74 MJ; (e) 3.34 MW, 0.270 kg/s; (f) 24.75%; (g) 7.90 MW; (h) 11.81 t/h; (i) 83.2%; (j) 0.132 kg/s, 10.2%; (k) 91.9%.

9.10. For a given energy release (in terms of lower calorific value), calculate the ratio of the mass of CO_2 produced in the complete combustion of natural gas (treated as methane, CH_4) to the mass of CO_2 produced in the complete combustion of a coal in which the mass fraction of carbon is 0.8 and having a lower calorific value of 26.7 MJ/kg.

Answer: 0.50.

CHAPTER 10

Advanced refrigerating and gas-liquefaction plant

10.1. Introduction

In Chapter 5 discussion was confined to simple vapour-compression cycles such as are encountered in commercial plant providing a modest degree of refrigeration. This chapter commences with a study of the *absorption* refrigeration cycle, an alternative to and, in certain respects, a variant of the vapour-compression cycle. It finds some commercial application.

Vapour-compression refrigeration plant normally operate over a range of temperature which extends below that of the atmosphere by only a relatively modest amount. Much lower temperatures can be obtained, however, by operating two or more such cyclic plants in cascade, each using a different refrigerant, and this method of cascade refrigeration is next considered.

None of the foregoing types of plant are capable of operating in the *cryogenic* range of temperatures. This range may be arbitrarily defined as lying between absolute zero and the liquefaction temperature (i.e. boiling-point) of methane at atmospheric pressure, namely from zero to about 110 K (or about 200 R). The study of plant capable of providing refrigeration and gas liquefaction in this range will take up a substantial proportion of this chapter.

10.2. Cyclic absorption refrigeration plant

An absortion refrigeration cycle constitutes a variant of the vapour-compression cycle, in that it utilises a refrigerant (usually ammonia or lithium bromide) which is readily absorbed in water and which exists in the vapour phase at some points in the cycle and in the liquid phase at other points. Such a plant is shown schematically in Fig. 10.1, in which it is first helpful to note that the plant within control surface Y merely replaces the

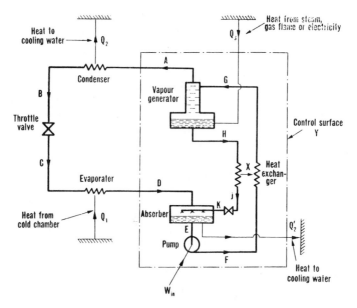

Fɪɢ. 10.1. Ammonia absorption refrigerating plant.

vapour compressor, the remainder of the plant being identical to that in the simple vapour-compression cycle illustrated in Fig. 5.3(a).

In the ammonia absorption cycle, instead of the ammonia being compressed in the gaseous state after exit from the *evaporator*, it is first taken into solution in water in an *absorber*. In consequence of the much smaller specific volume of this *aqua solution*, the work of compression ($\int v\,dp$ in reversible, steady-flow compression) is only a small fraction of the work that would have been required to compress the same amount of ammonia had it been in the vapour phase. Since the absorption reaction is exothermic, heat must be transferred continuously from the absorber to a cooling water supply in order to maintain steady conditions of operation.

The plant within control surface Y depends for its operation on the fact that the solubility of ammonia vapour in water decreases with rise in temperature of the solution. The pump therefore withdraws the strong ammonia solution from the absorber and delivers it to a *vapour generator* in which, as the result of a supply of heat from an external source (e.g. a steam-heating coil or a gas flame) some of the ammonia is driven out of solution in consequence of the rise in temperature. As this is an endothermic reaction, the supply of heat also serves to maintain steady conditions. The mixture of ammonia vapour and water vapour so driven off at high pressure passes then through the remaining items of a conventional plant, namely the condenser, throttle-valve and evaporator.

The weakened ammonia solution from the generator is returned through a throttle-valve to the absorber, passing on its way through heat exchanger X, which is included for reasons of economy, since the heat transferred in it to the high-pressure solution on its way to the generator reduces the amount of heat Q_3 required to be supplied to the generator. The reader may best follow the processes occurring within the plant by working through Problem 10.1. The solution of this problem requires only the application of the continuity (mass conservation) and steady-flow energy equations, together with a knowledge of the properties of aqua-ammonia solutions at different concentrations, temperatures and pressures. A more detailed treatment of absorption refrigeration cycles will be found in a specialist text.[93]

Absorption refrigeration plant is not of high performance and usually only finds commercial application in situations in which, for example, there is available for supplying Q_3 an adequate supply of heating steam which would otherwise go to waste. There is, however, one situation in which absorption refrigeration finds widespread application, namely in the Electrolux domestic refrigerator. As the result of the adoption of an ingenious system of circulation of the fluids by thermal siphon and bubble lift, this operates without any input of mechanical work at all. It is widely used for domestic purposes since there the power consumption (through electricity or gas) is sufficiently small not to be of first importance. A description of the mode of operation will be found in a specialist text.[94]

10.3. Performance measure for absorption refrigerators

If the same performance measure were used for absorption refrigerators as for vapour-compression plant, the coefficient of performance of the plant illustrated in Fig. 10.1 would be expressed as

$$CP \equiv \frac{Q_1}{W_{in}}. \tag{10.1}$$

However, this would be a misleading measure of the plant's performance, for although the work input has been reduced much below that of the comparable vapour-compression plant, a further source of energy input has been called upon in the form of Q_3, the heat supplied from an external source. Moreover, in the Electrolux refrigerator, W_{in} is zero. Thus a more rational measure of performance is obtained by writing

$$CP \equiv \frac{Q_1}{Q_3 + W_{in}}. \tag{10.2}$$

Since W_{in} is very small compared with Q_3, this may be written, with little error, as

$$\text{CP} = \frac{Q_1}{Q_3}. \tag{10.3}$$

For the Electrolux refrigerator, W_{in} is zero and eqn. (10.3) involves no approximation.

10.4. Performance criterion for absorption refrigerators

It will be seen that absorption refrigerators are cyclic plant in which the work input is either very small or zero, and which, in effect, exchange heat with three reservoirs of thermal energy at different temperatures; the cold chamber, or brine circulating through it, serves as a low-temperature source, a supply of cooling water serves as a sink at intermediate temperature, and a steam-heating coil or gas flame serves as a high-temperature source. If T_1, T_2 and T_3 are the absolute temperatures corresponding respectively to the conditions of operation in the evaporator, in the condenser (and absorber) and in the generator, an expression for the ideal coefficient of performance of a reversible cyclic plant exchanging heat reversibly with reservoirs at these temperatures, while absorbing zero work, can be obtained in the following manner.

From the First Law of Thermodynamics, for a system (e.g. unit mass of fluid) taken round a cycle,

$$\oint (dQ - dW) = 0,$$

and since W is zero and Q_2 is of opposite sign to Q_1 and Q_3,

$$Q_1 - Q_2 + Q_3 = 0. \tag{10.4}$$

As a corollary of the Second Law,

$$\oint \left(\frac{dQ}{T}\right)_{REV} = 0,$$

whence

$$\frac{Q_1}{T_1} - \frac{Q_2}{T_2} + \frac{Q_3}{T_3} = 0. \tag{10.5}$$

Eliminating Q_2 from eqns. (10.4) and (10.5), the coefficient of performance is given by

$$\text{CP} = \frac{Q_1}{Q_3} = \frac{[1 - (T_2/T_3)]}{[(T_2/T_1) - 1]}. \tag{10.6}$$

Since T_1 and T_2 are related respectively to the required temperature of the cold chamber and to the temperature of the cooling water supply, only the vaue of T_3 is to any extent open to choice. Equation (10.6) shows that the higher the value of T_3 the greater will be the ideal coefficient of performance, the limiting value being that of a reversed-Carnot refrigerating plant operating between T_1 and T_2, namely

$$\text{Limiting CP} = \frac{T_1}{T_2 - T_1}, \tag{10.7}$$

However, T_3 is determined in practice by the properties of the aqua-ammonia solution, and is, moreover, limited by the temperature of the heating medium, which may be a supply of steam at near-atmospheric pressure. The corresponding ideal CP is thus much smaller than this limiting CP [Problem 10.1(f)]. The actual CP is, in turn, considerably smaller than the corresponding ideal CP [Problem 10.1(e) and (f)], so accounting for the limited use of absorption refrigeration plant. It will be seen from Problem 10.1(a) that a not inappreciable fraction of Q_3 is required merely to heat up the weak solution returned to the absorber, which constitutes a high proportion of the feed to the generator; the coefficient of performance would consequently be improved by the installation of heat exchanger X in Fig. 10.1, since part of this energy would then be transferred to the incoming feed to raise its temperature and so reduce pressure Q_3. However, the coefficient of performance would still be less than unity.

10.5. Multiple vapour-compression cycles operating in cascade

In a simple vapour-compression cycle of the kind described in Chapter 5, the lowest temperature in the cycle occurs in the evaporator. For a given refrigerant, the lower the required refrigeration temperature the lower will be the required saturation pressure of the refrigerant in the evaporator; at the same time, the greater will be the specific volume of the vapour entering the compressor and so the greater the physical size of the plant. The situation can be relieved to some extent by the choice of a refrigerant with the most suitable properties, but there is clearly a limit to the lowest temperature that is both practically and economically acceptable with a simple cyclic plant using a single refrigerant. However, much lower temperatures can be obtained by operating two or more such plant *in cascade*, to give multiple vapour-compression cycles. Two such cycles operating in cascade, as in Fig. 10.2, would consititute a binary plant comparable to the binary power plant cycles described in Chapter 9.

It is not difficult to show (Problem 10.2) that the overall coefficient of performance C_o of this binary cycle is given by

$$\left(1 + \frac{1}{C_o}\right) = \left(1 + \frac{1}{C_1}\right)\left(1 + \frac{1}{C_2}\right). \tag{10.8}$$

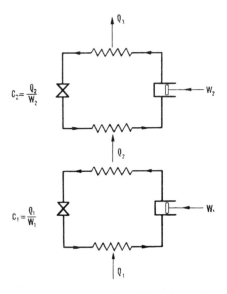

Fig. 10.2. Binary cascade refrigeration cycle.

This expression is analogous to eqns. (9.22) and (9.25) for the binary power plant cycles discussed in Chapter 9, but it will be noted that, whereas η_{CY} of the binary power plant is greater than the efficiency of either of the component cycles, C_0 is less than either C_1 or C_2: this is, of course, in accord with the fact, noted in §5.4, that the coefficient of performance of a refrigeration cycle decreases with increase in the difference between the reservoir temperatures.

   Cascade refrigerating plant is used in the liquefaction of natural gas, which consists principally of hydrocarbons of the paraffin series, of which methane has the lowest boiling point at atmospheric pressure. Refrigeration down to that temperature can be provided by a ternary cycle using propane C_3H_8, ethane C_2H_6 and methane CH_4, whose boiling points at standard atmospheric pressure are respectively 231.1 K, 184.5 K and 111.7 K. A simplified flow diagram for such a plant is shown in Fig. 10.3. The reader should explain for himself why progressive cooling and condensation of the gas in the manner shown is thermodynamically advantageous. In order to gain further thermodynamic advantage, the methane circuit would, in practice, be slightly different from that shown in the figure. The compressed methane vapour would first be cooled by heat exchange with the propane in the propane evaporator before being condensed by heat exchange with the ethane in the ethane evaporator, so reducing the degree of irreversibility involved in the cooling and condensation of the methane. Also, because of the high temperature after

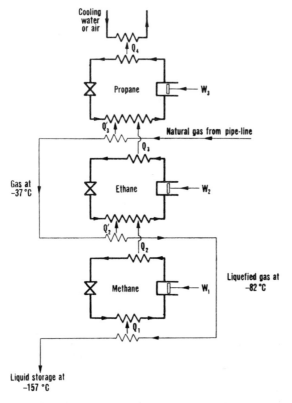

Fig. 10.3. Ternary cascade refrigeration plant for the liquefaction of natural gas.

compression, the gas leaving each compressor would pass first through a water-cooled after-cooler. In large-scale plant of this type, the compressors themselves would be rotary turbo-machines instead of the reciprocating type shown diagrammatically in the figure.

In a plant for liquefying natural gas from the Sahara,[95] ethylene, C_2H_4, with a boiling point of 169.4 K at standard atmospheric pressure, replaces ethane as the second-stage refrigerant, and the liquefied gas is finally cooled by the flashing process described in §5.8 and illustrated further in the next section of this chapter; in this way, what little nitrogen there is in the gas can be separated out, since the temperature is not low enough to cause it to liquefy.

It will be clear that the principle advantages of cascading are the reasonably small pressure range of the refrigerant in any one cycle and the ability to choose refrigerants that have the most suitable properties within each of the comparatively narrow temperature ranges. Figure 10.4 illustrates a still more sophisticated plant[96] for the liquefaction of natural

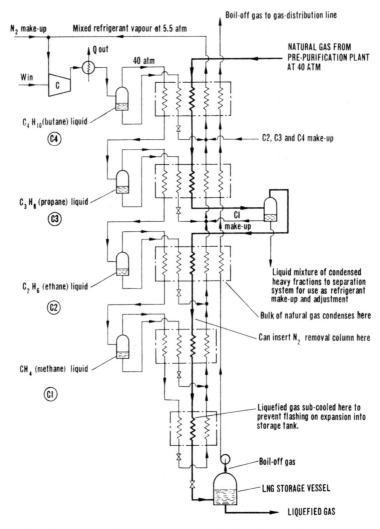

Fɪɢ. 10.4. Mixed-refrigerant cascade cycle for liquefaction of natural gas.

gas. This is described as a *mixed-refrigerant cascade plant*. The mixed refrigerant is composed of four hydrocarbons of progressively lower boiling-point temperature; these pass round a complex cycle which, in effect, comprises four vapour-compression refrigeration processes in cascade, each with its own condensing, throttle and evaporating stages but without separation of the refrigerants from each other, the mixed refrigerant vapour being compressed in a single compressor. In the same way as direct-contact feed heaters are thermodynamically superior to surface heaters in a complex regenerative steam cycle, this direct mixing of the refrigerants is thermodynamically superior to confinement in the separate closed circuits of Fig. 10.3, in spite of the irreversibility of the mixing process.

10.6. Multiple cascade plant for the production of solid carbon dioxide (dry ice)

A three-stage process which bears some resemblance to the cascade plant of Fig. 10.3 is employed in the production of solid carbon dioxide, or "dry ice". Since carbon dioxide provides the working fluid throughout, inter-stage flash tanks can be used in place of surface heat exchangers, in the manner shown in Fig. 10.5. In this respect the plant also bears a certain resemblance to the mixed-refrigerant cascade cycle of Fig. 10.4, though

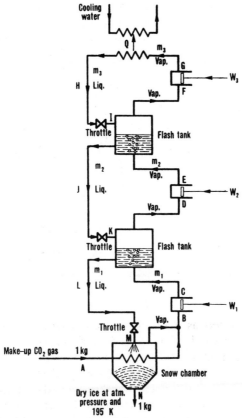

FIG. 10.5. Three-stage cascade plant for the production of solid carbon dioxide (dry ice).

gaining still further thermodynamic advantage in the direct-contact condensation process occurring in each flash tank.

The entire process can be followed on the temperature–entropy diagram of Fig. 10.6. It will be seen from this diagram that the triple-point pressure of CO_2 is above normal atmospheric pressure, so that CO_2 never occurs in the liquid phase at atmospheric pressure; hence the name "dry ice" for solid CO_2, for at atmospheric pressure it sublimates directly into CO_2 vapour without passing through the liquid phase.

The topmost pressure in the CO_2 circuit must be such that the saturation temperature at this pressure will be above the temperature of the cooling water used to condense the CO_2 vapour in the condenser between G and H; this necessitates a pressure of some 70 atm. The specific reversible work of compression in steady flow is equal to $\int v \, dp$, so that, in consequence of the reduction in specific volume of the vapour in the intercooling process, a reduction in the required work input results from multi-stage compression with intercooling between stages. This cooling of the vapour after compression in a stage takes place by direct contact between the superheated vapour and liquid at the same pressure, as the vapour is bubbled through the liquid in an inter-stage flash tank. This method solves the problem of intercooling at sub-atmospheric temperatures.

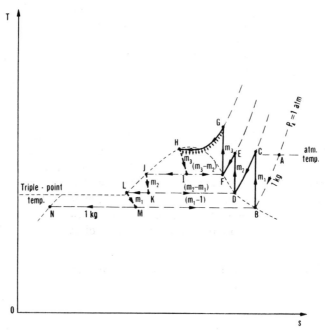

Fig. 10.6. Temperature–entropy diagram for the dry-ice process.

The liquid entering each of the throttles at H and J is flashed into a mixture of liquid and vapour at I and K respectively, in the manner described in §5.8. However, when the liquid from L passes through the first-stage throttle into the "snow" chamber, it flashes into a solid-vapour mixture at state M, since the pressure in the chamber is below the triple-point pressure. The solid CO_2 at state N is collected from the snow chamber and the vapour at state B is drawn into the suction of the first-stage compressor, together with make-up vapour precooled by passing it through a coil in the snow chamber.

Application of the Steady-flow Energy Conservation Equation successively to the snow chamber and the two flash tanks yields the following equations:

$$m_1(h_B - h_L) = (h_A - h_N), \tag{10.9}$$

$$m_2(h_D - h_J) = m_1(h_C - h_L), \tag{10.10}$$

$$m_3(h_F - h_H) = m_2(h_E - h_J). \tag{10.11}$$

Using the flow quantities thus calculated, the total work input to the compressors per unit mass of dry ice formed may be calculated and the rational efficiency thence determined (Problem 10.3).

10.7. The rational (exergetic) efficiency of the dry-ice process

The dry-ice plant is seen to operate as an open-circuit, steady-flow work-absorbing device which exchanges heat with only a single reservoir, the cooling water, which we may call the environment at absolute temperature T_0. The device takes in gaseous CO_2 at atmospheric pressure and temperature (state A) and delivers solid CO_2 at the saturation temperature corresponding to atmospheric pressure (state N). This is the kind of device discussed in §§A.9 and A.10 of Appendix A, from which it will be seen that, per unit mass of dry-ice produced, the minimum work input for such a device operating between these two given states is

$$(W_{in})_{REV} = (b_N - b_A), \tag{10.12}$$

where b is the *steady-flow availability function*, $(h - T_0 s)$.

Since the actual work input, $(W_{in})_{ACTUAL}$, will be greater than $(W_{in})_{REV}$, the *rational (exergetic) efficiency* of the process will be given by

$$\eta_R \equiv \frac{(W_{in})_{REV}}{(W_{in})_{ACTUAL}} = \frac{b_N - b_A}{(W_{in})_{ACTUAL}}, \tag{10.13}$$

and the criterion of excellence against which to judge the performance of the actual plant will be a value of η_R equal to 100%.

REFRIGERATION AND GAS LIQUEFACTION AT CRYOGENIC TEMPERATURES

The processes discussed in the remainder of this chapter are those found in refrigeration and gas liquefaction in the cryogenic range of temperature, which we defined arbitrarily in §10.1 as from zero to about 110 K (about 200 R), the boiling-point of methane at atmospheric pressure. In this range lie the boiling points at atmospheric pressure of the so-called "permanent" gases. Some typical figures are:

Oxygen	90.2 K
Air	78.8 K
Nitrogen	77.3 K
Hydrogen	20.4 K
Helium	4.2 K

10.8. Liquefaction of gases by the throttle-expansion Linde process

The Linde process, which produces liquefied gas at atmospheric pressure, provides a simple, though not highly efficient, means of liquefying gases having very low boiling-point temperatures at atmospheric pressure. In essentials, though an open-circuit process, it bears a certain resemblance to the vapour-compression cycle, and the reader will benefit from a close study of the similarities and differences, as revealed by the diagrams for the two types of plant set side-by-side in Fig. 10.7. The most important difference between the two lies in provision of the heat exchanger X in the Linde plant, whereby, through internal heat exchange, the low-temperature, low-pressure fluid from the flash-tank at 6 is used to cool the "high-temperature", high-pressure fluid from 2′ before it enters the throttle at 3. The regenerative heat exchanger X in Fig. 10.7(c) may conveniently be regarded as separating the high-temperature part of the plant to the right from the lower-temperature part of the plant to the left.

Unlike the vapour-compression cycle of Fig. 10.7(a), there is no external heat absorption at the lowest temperature in the Linde liquefaction process (though there would be such heat absorption at the lowest temperature if the plant were used to provide refrigeration by evaporation of the two-phase mixture from 4 to 6, instead of being used for gas liquefaction; in the latter case, saturated liquid and saturated vapour are withdrawn separately from the flash-tank at 5 and 6 respectively).

Figure 10.7(d) shows the gas compressed reversibly and adiabatically in a single stage of compression and then cooled to atmospheric temperature in a precooler before entering heat exchanger X. If the precooler uses a supply of cooling water, the temperature at 2′ will necessarily be somewhat

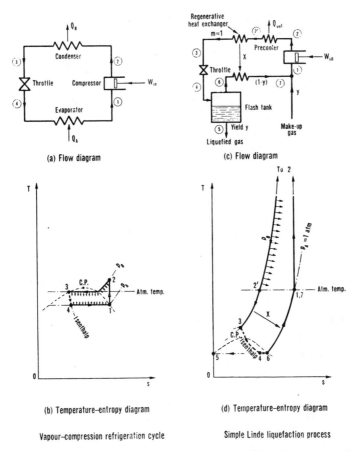

FIG. 10.7. Comparison of the vapour-compression refrigeration cycle and the simple Linde liquefaction process.

greater than the atmospheric temperature T_1, while it can be made less than T_1 if, instead of cooling water, a low-temperature fluid from an auxiliary refrigeration cycle is used as the cooling agent. The kind of situation in which a need for the latter will arise is discussed later in §§10.9 and 10.15.

It will be shown shortly that, for maximum yield of liquefied gas, the gas needs to be compressed in the compressor to a high super-critical pressure, which may be of the order of one or two hundred atmospheres. Centrifugal compressors are unsuited to such high compression ratios, so that reciprocating compressors are used. Figure 10.7(d) shows that adiabatic compression in a single stage would result in an excessively high delivery temperature. This is avoided in practice by water-jacketing the compressor and compressing in two or more stages, with intercooling between stages.

Figure 10.8 shows two-stage compression with intercooling. For purposes of illustration, the compression in each stage is shown in Fig. 10.8(b) as being adiabatic and reversible, with the gas leaving both the intercooler and precooler at a temperature equal to T_1. The ideal mode of operation for minimum work input would, of course, be reversible, isothermal compression at atmospheric temperature T_1 along the path 1–3–5.

Compared with Fig. 10.7(c), Fig. 10.8(a) emphasises better the importance of the regenerative heat exchanger X in serving to separate the high-temperature part of the plant above X from the low-temperature part below it. Such a regenerative heat exchanger features in all refrigeration and liquefaction plant operating at really low (i.e. cryogenic) temperatures.

10.9 Operational requirements in the Linde process

The temperature–entropy diagram of Fig. 10.8(b) relates to conditions obtaining when the plant has reached a steady state of operation, with continuous supply of make-up gas and continual withdrawal of liquified

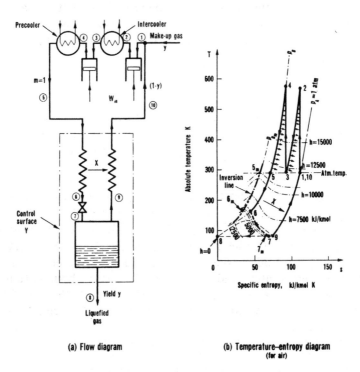

(a) Flow diagram

(b) Temperature–entropy diagram
(for air)

FIG. 10.8. Simple Linde liquefaction process—condition for maximum yield of liquefied gas.

gas. Before studying this steady state of operation it is necessary to consider that particular property of the gas which enables the plant to settle down to this steady state of gas liquefaction.

After starting the compressor to take in gas at atmospheric pressure and temperature, the liquefaction process will only be self-initiating if the fluid experiences a fall in temperature as a result of the pressure drop occurring across the throttle; this requires that the state of the fluid entering the throttle must at all times be such that its *isenthalpic Joule–Thompson coefficient* $[\mu_h \equiv (\partial T/\partial p)_h]$ is positive,† for a drop in temperature will then accompany a drop in pressure. Under those circumstances, the temperature level of the system will fall progressively until a steady state of operation, with liquid withdrawal, is reached.

It can readily be shown (Problem 10.5) that, for a fluid substance, μ_h is given by

$$-c_p\mu_h = \mu_T = v - T\left(\frac{\partial v}{\partial T}\right)_p,$$ (10.14)

where μ_T is the *isothermal Joule–Thompson coefficient* $[\equiv (\partial h/\partial p)_T]$. For a perfect gas, which has the equation of state $pv = RT$, it is easy to show from this expression that μ_h is zero (Problem 10.5), so that such a gas could not be liquefied by the Linde process. For real gases, the sign of μ_h depends on the relative magnitudes of the two terms on the right-hand side of eqn. (10.14), and these magnitudes change with change in state of the gas. Van der Waals' equation

$$[p + (a/v^2)](v - b) = RT,$$

is known to reproduce approximately the behaviour of real gases. Using this equation of state, expression may be deduced, in terms of v, for the values of p and T for which μ_h is zero (Problem 10.5). When plotted on the $p-T$ plane, these values give the *inversion line*, in crossing which μ_h changes sign. Figure 10.9, plotted in reduced coordinates, shows the inversion line for a van der Waals gas, and also for a real gas, nitrogen.

It is seen that the stipulation that μ_h must be positive for ensured initiation of the Linde liquefaction process requires that the state of the gas at throttle inlet must be such that it lies in the area below the inversion line in Fig. 10.9. For both oxygen and nitrogen (and therefore for air) at atmospheric temperature and any pressure less than about 400 atm, μ_h is positive, so that the above requirement is satisfied. On the other hand, as can be seen from the table in Fig. 10.9, the critical temperatures for hydrogen and helium are so low that the state points for these gases at atmospheric temperature lie outside the inversion curve at all pressures, so

---

†In point of fact, for the temperature to drop, it is the *overall* value, $(\Delta T/\Delta p)_h$, that must be positive. If μ_h is positive at throttle inlet, this will always be so.

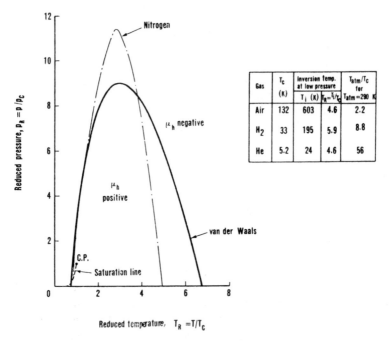

Fig. 10.9. Inversion line for a van der Waals gas and for nitrogen. p_c = critical pressure, T_c = critical temperature.

that μ_h is negative at atmospheric temperature and any pressure. Hydrogen and helium must consequently be cooled in the precooler of Fig. 10.8(a) to a temperature below the inversion point if they are to be liquefied by the Linde process. The required precooling to the very low inversion temperature may, as mentioned in §10.8, be provided by a low-temperature fluid from an auxiliary refrigerating plant. Liquid nitrogen can be used for precooling hydrogen, while helium can be precooled with liquid hydrogen. An alternative method of precooling is described later in §10.18.

10.10. Conditions for maximum liquid yield in the Linde process under steady operation

If interest centres on obtaining the maximum *liquid yield*, y, per unit mass of gas supplied to the throttle, it will be found that it is necessary to design the plant to operate at a particular delivery pressure, We shall now study this condition. For this purpose we shall consider a gas requiring no precooling to sub-atmospheric temperature, so that we shall assume that the high-pressure fluid leaves the precooler at a temperature equal to the atmospheric temperature T_1. We shall also assume that heat exchanger X

is ideal to the extent of permitting zero temperature difference between the two fluids at the warm end, so that the low-pressure fluid will leave the heat exchanger also at temperature T_1. We shall see shortly that the state of the fluid before the throttle at 6 then follows.

Assuming that kinetic energy changes are negligible, the Steady-flow Energy Conservation Equation for the control volume bounded by control surface Y in Fig. 10.8(a) gives

$$h_5 + q_{in} = yh_8 + (1 - y)h_1,$$

where q_{in} is the stray "heat leak" into the control volume from the environment per unit mass of gas drawn into the plant. Thus

$$y = \frac{(h_1 - h_5) - q_{in}}{h_1 - h_8}, \tag{10.15}$$

Since h_1 is the specific enthalpy of the gas at atmospheric pressure and temperature and h_8 is the specific enthalpy of saturated liquid at atmospheric pressure, these are both fixed. It is thus evident from eqn. (10.15) that the greatest yield will be obtained when q_{in} is zero and when the compressor delivery pressure p_B is such that h_5 is a minimum. A glance at the lines of constant enthalpy on Fig. 10.8(b) shows that, as p_B is increased and point 5 moves along the isotherm through 1, h_5 will be a minimum **when 5 lies on the inversion line** at 5_m. This may be confirmed by noting that on the inversion line $\mu_T \equiv (\partial h/\partial p)_T = 0$. Thus the greatest yield will be obtained when the fluid is compressed to p_{Bm} in Fig. 10.8(b).

With h_{5m} inserted in eqn. (10.15), and taking q_{in} to be zero, y_{max} is thus determined. Thence

$$h_{6m} = h_{7m} = y_{max}h_8 + (1 - y_{max})h_9. \tag{10.16}$$

This equation gives the state of the fluid at throttle inlet for which the throttle orifice should be designed to enable the plant to operate steadily at the prescribed conditions. The reader may confirm that, with q_{in} zero, eqns. (10.15) and (10.16) are compatible with the energy conservation equation for the flow through heat exchanger X, namely:

$$h_{5m} - h_{6m} = (1 - y_{max})(h_1 - h_9).$$

10.11. Conditions for minimum work input per unit liquid yield (maximum rational efficiency)

If interest centres on designing the plant for minimum work input per unit mass of gas liquefied (that is, for the maximum value of the *rational efficiency* defined in the next section), it will be found that the optimum value of p_B is less than that for maximum yield per unit mass of gas supplied to the throttle. This may be deduced from a study of Fig. 10.8(b).

We shall again consider a gas that requires no precooling to sub-atmospheric temperatures and so will again take the temperature at 5 to be the same as that at 1. It will be seen from a study of Fig. 10.8(b) that the rate of decrease of h_5 with increase of pressure will fall off as point 5_m is approached, and so the rate at which the yield increases with increase in p_B will fall off. Thus it will be apparent that, as point 5_m is approached, the rate of increase in yield y with increase in pressure will at some point become less than the corresponding rate of increase in work input W, so that the point of minimum work input per unit yield w will occur before 5_m is reached.

$$\left(\text{Note that } w \equiv \frac{W}{y}, \text{ so that } \frac{dw}{w} = \frac{dW}{W} - \frac{dy}{y}. \right)$$

These conclusions may be verified by reference to Problem 10.7, which relates to a quasi-ideal plant, in that compression is assumed to be isothermal and reversible, and heat into the plant is ignored. Departures from both of these simplifying assumptions would have to be taken into account in assessing the optimum value of p_B for an actual plant. It is also important in practice to remember that, as in the case of power plant (cf. Appendix C), it would not be economic to push up the operating conditions to those corresponding to the peak of the efficiency curve (Problem 10.7). Thus the **economic** compressor delivery pressure is less than the thermodynamic optimum value, being in practice about 200 atm.

10.12. The rational (exergetic) efficiency of the Linde process

It should be noted that, as in the "dry ice" plant discussed in §10.6, a plant operating on the Linde process is not cyclic; it would only become so if it were run, not as a liquefaction plant but as a refrigerating plant, with the supply of make-up gas shut off and with heat absorbed from a refrigeration chamber to evaporate the liquid–vapour mixture from state 7 to state 9 in Fig. 10.8(b). As a liquefaction process, the plant operates as an *open-circuit*, *steady-flow*, *work-absorbing device* capable of exchanging heat with a single reservoir, namely the cooling water, which we shall characterise as the *environment* at absolute temperature T_0. From §§A.9 and A.10 in Appendix A, it will thus be seen that the minimum work required to liquefy unit mass of the substance from gas in state 1 at atmospheric pressure and temperature to saturated liquid in state 8 at atmospheric pressure will be

$$(w_{\text{in}})_{\text{REV}} = (b_8 - b_1), \tag{10.17}$$

where $b \equiv (h - T_0 s)$, the *steady-flow availability function*. This would be the work input when all processes, including heat exchange with the environment, were reversible (Problem 10.6).

$(w_{in})_{REV}$ provides a rational criterion against which to judge the work input required to produce unit mass of liquid in the actual plant, so that we may define the *rational (exergetic) efficiency* as

$$\eta_R \equiv \frac{(w_{in})_{REV}}{(w_{in})_{ACTUAL}}. \tag{10.18}$$

From Problem 10.8, it will be seen that, even when the gas is compressed reversibly and isothermically at atmospheric temperature, so that heat transfer with the environment is reversible, the efficiency of the Linde process is low. As can be seen from a study of the answers to Problem 10.8, this poor performance is due to the irreversibility of the throttle process and of the heat transfer in heat exchanger X. In an actual plant the efficiency would be lower still on account of further irreversibilities since, for example, the compression process would be neither reversible nor isothermal, the heat exchange process would be less perfect and there would be some stray heat leak into the plant from the environment. The serious effect of the latter is next studied.

10.13. The effect of heat leak into the plant

Because a cyclic plant capable of giving refrigeration in cryogenic temperatures operates over such a wide temperature range, its coefficient of performance is low (cf. §5.4). It follows that the extra refrigeration required to compensate for any heat leak from the environment into the refrigerating chamber will result in an appreciable increase in the required work input. The same may be said of the effect of heat leak into those parts of the open-circuit Linde plant which are at sub-atmospheric temperatures, particularly heat leak into the flash tank. It is also evident from eqn. (10.15) that any heat leak into the plant will adversely affect the liquid yield; alternatively, if the yield is to be maintained, the pressure will have to be raised in order to lower h_5, though this possibility is restricted by the limitation discussed in §10.10. The effect of heat leak on the required work input may be evaluated as follows.

Suppose that, per unit mass of gas liquefied, there is a heat leak of magnitude δq from the environment at absolute temperature T_0 to the fluid in the flash tank at absolute temperature T_f. The increase in the required work input as a result of the irreversibility of this heat transfer with the environment may be evaluated from the entropy creation (as defined in §A.6 of Appendix A) resulting from this particular irreversibility; it is equal to the *extra work input due to irreversibility*, evaluated by the method presented in §A.7 of Appendix A.

Per unit mass of gas liquefied, the entropy creation due to the irreversibility of this heat transfer is given by

$$\delta S_C = \frac{\delta q}{T_f} - \frac{\delta q}{T_0}.$$

Thus, from eqn. (A.14), the increase in the required work input will be given by

$$\delta w_{in} = T_0\left[\frac{\delta q}{T_f} - \frac{\delta q}{T_0}\right] = n\,\delta q, \tag{10.19}$$

where

$$n = \left(\frac{T_0}{T_f} - 1\right). \tag{10.20}$$

The quantity n is seen to be equal to the reciprocal of the coefficient of performance of a Carnot refrigerating cycle operating between T_f and T_0. The reader would profit by explaining the reason for this.

For $T_0 = 288$ K and $T_f = 77.3$ K, the boiling-point of nitrogen at atmospheric pressure, $n = 2.7$, while if $T_f = 4.2$ K, the boiling-point of helium at atmospheric pressure, $n = 67.6$. It must be noted that these figures relate to the **mimimum** extra work input resulting from heat leak into the plant, since implicit in the above derivation is the assumption that other processes in the plant are reversible. On account of the inefficiency of the actual compression process the extra work input would, in practice, be greater than the figures quoted. It is evident that it is very important to ensure near-perfect insulation of plant operating at really low cryogenic temperatures.

10.14. Modifications to the Linde process to give plant of higher performance

The Linde process of gas liquefaction is simple but relatively inefficient, since it requires a high work input unit yield. Plant of improved performance can be designed by elaborating on the simple Linde process in a number of ways. The following three developments will be considered in turn:

(1) *The simple Linde process with increased precooling by auxiliary refrigeration*, giving an increased yield with a small increase in work input.
(2) *The dual-pressure Linde process*, in which a reduction in work input is obtained by not allowing all the compressed fluid to expand down to liquefaction pressure; this results in a *dual-pressure* plant. Precooling by auxiliary refrigeration as in (1) may also be included.

(3) *The Claude and Heylandt processes*, with increased precooling by incorporation of an expansion engine, giving both an increased yield and a lower net work input.

10.15. The simple Linde process with increased precooling by auxiliary refrigeration

When there is negligible heat leak into the plant, and in the general case when T_{10} does not equal T_1, eqn. (10.15) for the yield in the simple Linde process becomes

$$y = \frac{h_{10} - h_5}{h_{10} - h_8},$$ (10.21)

in which the numerical subscripts relate to Fig. 10.8.

A study of the shape of the isenthalps on the temperature–entropy diagram of Fig. 10.8(b), which is drawn for air,[97] shows that if, keeping p_A and p_B the same, we progressively reduce T_5, and assume that the temperature T_{10} of the low-pressure stream at exit from heat exchanger X at all times equals T_5, then $(h_{10} - h_5)$ will be increased, so increasing the yield y; at the same time $(h_{10} - h_8)$ will fall, so further increasing y. Furthermore, the reduction in temperature and therefore in specific volume at compressor inlet, in consequence of the mixing of the lower temperature stream from 10 with the make-up gas from the atmosphere, will bring some reduction in work input. T_5 may be reduced below atmospheric temperature by making the precooler coil of the Linde plant the evaporator of an auxiliary ammonia vapour-compression refrigerating plant. In this way, T_5 may be reduced to a typical value of about −45 °C. The work input to the ammonia cycle has, of course, to be debited, but this cycle has a high coefficient of performance so that, as can be seen from the typical figures quoted in Table 10.1, the effect of this is relatively small and the overall gain due to the increased precooling is appreciable.

TABLE 10.1. *Comparison of the performance of Linde air liquefaction processes (after Ruhemann*[98]*)*

	Ideal reversible process	Simple Linde	Simple Linde with NH$_3$ cycle	Dual-pressure Linde	Dual-pressure Linde with NH$_3$ cycle
Yield	1	0.1	0.2	0.08	0.17
$\eta_R \equiv \dfrac{w_{REV}}{w_{ACTUAL}}$	1	0.07	0.135	0.12	0.21

w_{REV} = work input per unit yield for ideal reversible liquefaction process at atmospheric pressure, with the environment at atmospheric temperature [eqn. (10.17) and Problem 10.6].

10.16. The dual-pressure Linde process

Figure 10.10(a) shows a flow diagram of the dual-pressure (cascade) Linde process, which gives an improvement in the work input per unit yield in consequence of a reduction in work input per unit mass of gas delivered by the compressor; as can be seen from Table 10.1, however, this is achieved at the expense of some reduction in the yield y.

The plant incorporates two Joule–Thompson expansion stages, flash-tank A providing the liquefied gas and operating in the normal way at atmospheric pressure, while receiver B is at an intermediate pressure. When liquefying air, the pressure in B is usually between 40 and 50 atm when the compressor delivery pressure is about 200 atm. Since the critical pressure for air is 37.25 atm, the pressure in B under normal operating conditions is thus supercritical, so that there is then no separation into liquid and gaseous phases within receiver B; thus in Fig. 10.10(b), which does not show the compression processes, points 4, 5 and 6 are coincident, the compressor being designed to withdraw from receiver B a fraction $(1 - y_B)$ of the fluid entering it. If, under certain operating conditions, p_B falls below the critical pressure, then receiver B serves as a second

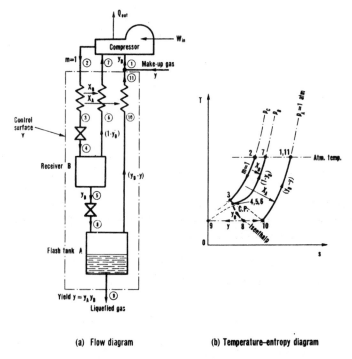

(a) Flow diagram   (b) Temperature–entropy diagram

F<small>IG</small>. 10.10. Linde dual-pressure (cascade) liquefaction process.

flash-tank; the reader is advised to sketch the process on the temperature–entropy diagram in these circumstances and to consider what modification would result to the equations given later in this section if the plant were designed to operate with a subcritical pressure in *B* (Problem 10.10).

Although the compressor delivery pressure in the dual-pressure Linde process is not any higher than that in the simple Linde process, the dual-pressure process is frequently, and rather misleadingly, referred to as the *high-pressure Linde process* because of the addition of the second Joule–Thompson expansion stage discharging into receiver *B* at a pressure in excess of atmospheric.

A study of Fig. 10.10 reveals that the reduction in work input, to which reference has already been made, arises from the fact that only a fraction y_B of the fluid delivered by the compressor is compressed over the full pressure range from 1 to 200 atm, the remainder being compressed over only a fiftieth of this pressure ratio when the intermediate pressure is 50 atm (note that the work input is a function of the pressure ratio of compression). This reduction in work input per unit mass is obtained at the expense of a certain reduction in yield, because only a fraction y_B of the fluid from the compressor reaches the flash-tank *A*. There, a fraction y_A of y_B leaves as liquid, so that the yield y of the plant is given by

$$y = y_A y_B.$$

However, Table 10.1 shows that the yield in the dual-pressure process is not greatly below that in the simple Linde process, so that, as a result of the considerable reduction in work input per unit mass of gas delivered by the compressor, the work input per unit yield in the dual-pressure process is appreciably less, and the rational efficiency therefore appreciably greater, than in the simple Linde process. Table 10.1 also shows that the addition of an auxiliary ammonia refrigerating cycle to precool the fluid entering heat exchanger *X* so increases the yield that the resulting further reduction in the work input per unit yield leads to a considerable further increase in the rational efficiency.

Referring to Fig. 10.10, and noticing that $h_3 = h_4 = h_5 = h_6 = h_8$ when p_B is supercritical, application of the Steady-flow Energy Equation to the flow through control surface *Y* gives

$$h_2 = (1 - y_B)h_7 + (y_B - y)h_{11} + yh_9. \qquad (10.22)$$

Similarly, for flow through flash-tank *A*, noting that $h_8 = h_3$, we have

$$y_A = \frac{y}{y_B} = \frac{h_{10} - h_3}{h_{10} - h_9}. \qquad (10.23)$$

Given a value for y_B and the operating pressures p_A, p_B and p_C, and assuming heat exchangers *X* to be ideal to the extent that the temperature differences at the high-temperature end are negligible, so that $T_1 = T_2 = T_7 = T_{11}$, eqns. (10.22) and (10.23) may be solved for h_3 and y (Problem 10.9). Given sufficient information about the compression processes, the work input to the compressor may be calculated. In this way, a series of optimisation calculations may be carried out with varying values of y_B, p_B and p_C. In practice, dual-pressure Linde plants for air liquefaction usually work at the pressures previously quoted and a value of $y_B \sim 0.2$, giving the value of 0.08 for the yield y quoted in Table 10.1.

10.17. The Claude and Heylandt liquefaction processes, combining work expansion and throttle expansion

The Claude and Heylandt processes are basically the same, being different from each other in only one particular, which is described later. Like the simple Linde process, they operate as single-pressure plant, but give a work input per unit yield less than that of the simple Linde process in consequence both of an increase in yield and of a reduction in work input per unit mass of gas handled by the compressor.

A flow diagram for the Claude process is shown in Fig. 10.11(a), from which it will be seen that the high-pressure flow from the compressor divides at 3, only a fraction x proceeding to the throttle valve. The remainder bypasses the flash chamber and is taken instead to an expansion engine. The gas suffers an appreciable temperature drop in passing through the engine in consequence of the work performed therein. This cold fluid from the engine passes back to the compressor through the cold side of heat exchanger X_B, thereby serving to provide a means of precooling the high-pressure fluid before it enters control volume Y_2 at 4. Control volume Y_2 is seen to correspond to the control volume Y of a simple Linde plant, and so this precooling serves exactly the same purpose of increasing the yield as was served by the ammonia precooling in §10.15. The resulting high value of y_2 ($= y_1/x$) helps to offset the fact that only a fraction x of the compressor flow enters the flash chamber, while the work output from the expansion engine serves to reduce the net work input to the plant. In consequence the plant achieves a performance at least as good as that of the dual-pressure Linde process with ammonia pre-cooling, without requiring the provision of any auxiliary refrigerating cycle.

The selection of optimum operating conditions cannot be treated analytically without numerical calculation, but a little thought will enable certain qualitative conclusions to be drawn. For example, a higher value of x will permit of a higher yield y_1 but will, on the other hand, result in a smaller work output from the expansion engine. In consequence of these two opposing tendencies, there will, for any selected compressor delivery pressure, clearly be an optimum value of x for which the net work input per unit yield will be a minimum and the rational efficiency therefore a

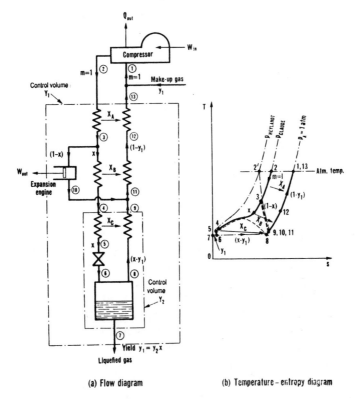

(a) Flow diagram   (b) Temperature - entropy diagram

FIG. 10.11. Claude and Heylandt liquefaction processes. (*Note*: There is no heat exchanger X_A in the Heylandt plant)

maximum. It is also to be expected that the optimum value of x will be greater the higher the compressor delivery pressure, since then the beneficial effect of the greater yield will help to offset the adverse effect of the greater work input per unit mass of gas passing through the compressor. These conclusions are borne out by the curves of Fig. 10.12.

We next consider the selection of an appropriate value for T_3. A higher value of T_3 will result in a greater work output from the expansion engine ($W_{REV} = -\int v \, dp$, and v is greater at higher temperature) but will, on the other hand, result in a higher temperature of the fluid entering control volume Y_2 and therefore in a smaller yield. In consequence of these two opposing tendencies there will thus, for a given compressor delivery pressure p_B, be an optimum value of T_3. Alternatively, as can be seen from Fig. 10.11(b), the higher the value of p_B the higher will have to be the value of T_3 if the temperature T_4 at entry to control volume Y_2 is to be about the same for all values of p_B.

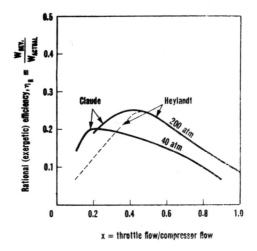

x = throttle flow/compressor flow

FIG. 10.12. Rational (exergetic) efficiency of the Claude and Heylandt air liquefaction processes. (From the data of Lenz, *Handb. d. exp. Phys.* **9**. 1929.)

We are now in a position to discuss the only essential difference between the Claude and Heylandt processes. For the Claude air-liquefaction process, which was the earlier to be used, a compressor delivery pressure of about 40 atm was selected. For this value of p_B the optimum value of T_3 is found to be about $-80\,°C$. In the later Heylandt process the selected value of p_B was about 200 atm. As would be expected from the foregoing discussion, at this higher value of p_B the optimum value of T_3 is higher, in fact about 20 °C. This is about equal to atmospheric temperature and so enables heat exchanger X_A to be dispensed with. This is the only respect in which the Claude and Heylandt circuits differ. Approximate figures for the respective optimum operating conditions are given in Table 10.2.

It is seen that, in spite of the higher value of p_B in the Heylandt process, the higher yield associated with the higher optimum value of x results in the work input per unit yield in the Heylandt process being less and the

TABLE 10.2. *Comparison of the performance of Claude and Heylandt air liquefaction processes (after Ruhemann[98])*

Process	$\dfrac{p_B}{\text{atm}}$	Optimum t_3 (°C)	Optimum x	$\eta_R \equiv \dfrac{w_{\text{REV}}}{w_{\text{ACTUAL}}}$
Claude	40	−80	0.2	0.21
Heylandt	200	15	0.45	0.25

w_{REV} = work input per unit yield for ideal reversible liquefaction process at atmospheric pressure, with the environment at atmospheric temperature [eqn. (10.17) and Problem 10.6].

rational efficiency therefore greater, than in the Claude process. A comparison with the figures in Table 10.1 shows that the efficiency is also greater than that of any of the Linde processes. The Heylandt process finds application in the commercial production of liquid oxygen.

The reader will obtain a better understanding of the two processes by working through Problems 10.11 and 10.12, in which some simplifying assumptions have been made for ease of calculation. These problems may readily be solved by successive application of the Steady-flow Energy Equation to control volume Y_1 and then to individual items of the plant.

10.18. More complex plant combining work and throttle expansion

It has already been noted in §10.9 that before hydrogen and helium can be liquefied at atmospheric pressure by the Joule–Thompson throttle-expansion process, the gas must be precooled to a temperature below the inversion point. For helium at atmospheric pressure this temperature is about 24 K. In a laboratory-scale plant, the required precooling may be achieved by passing the gaseous helium through baths of liquid nitrogen and hydrogen before it enters the regenerative heat exchanger in which the gas flowing to the throttle is cooled by the return flow of unliquefied gas from the flash chamber. On a larger scale, it is more convenient to dispense with the liquid nitrogen and hydrogen and to precool the gas to be liquefied by passing it through a series of counter-flow heat exchangers in which it transfers heat to coolant gas which flows in a separate circuit and which has had its temperature lowered by passage through work-producing turbines. The formation of liquid in the turbine circuit is avoided, the only liquefaction occurring in the flash-chamber of the throttle-expansion circuit.

Figure 10.13 gives a flow diagram for such a plant which can be used either as a helium liquefier or to provide cryogenic refrigeration.[99] The plant incorporates two turbines operating in series, but many variants are possible. With turbines operating in parallel it is possible to arrange for apportionment of the respective flow rates in such a way as to ensure optimum operating conditions in the counter-flow heat exchangers. Detailed consideration of the alternative possibilities is beyond the scope of the present volume and the reader is directed elsewhere.[100,101]

10.19. Gas refrigerating machines for small-scale refrigeration and gas liquefaction at cryogenic temperatures

In §5.5, reference was made to refrigeration cycles using as the working substance a fluid which remains gaseous throughout the cycle. It was there

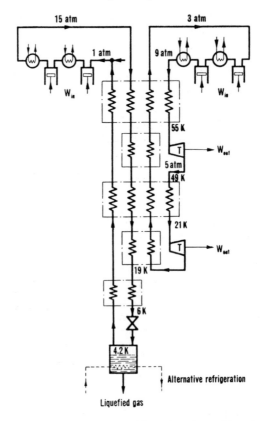

Fig. 10.13. Helium liquefaction and refrigerating plant with series operation of expansion turbines. (Courtesy Sulzer Bros. Ltd., Winterthur, Switzerland.)

stated that while these were not suited to large-scale commercial refrigeration, they had found application on a smaller scale. Two such types of *refrigerating machine* are the Philips Cryogenerator developed by Kohler and Jonkers[102] in the Netherlands and the machines developed by Gifford and his associates in the USA.

The Philips machine is illustrated in Fig. 10.14. It operates on a principle similar to that of the air refrigeration system developed by the Scottish engineer, Alexander Kirk, in the mid nineteenth century. Two articulated pistons, the main piston and above it a *displacer piston*, operate out of phase with each other in a single cylinder. The space between the main piston and the displacer serves to compress the gas, which is later expanded in the space between the top of the displacer and the cylinder head. As the relative motion of the two pistons causes the gas to be passed to and fro between these two spaces, it flows through the matrix of a regenerative

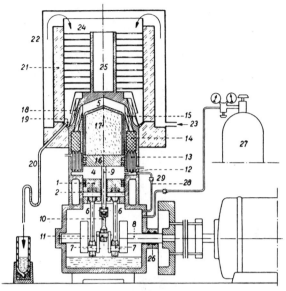

FIG. 10.14. *Philips Gas Refrigerating Machine for the liquefaction of air.* (Courtesy *Philips Technical Review*[103]) 1, Main piston. 2, Cylinder. 3, Displacer piston. 4, Compression space. 5, Expansion space. 6, Main connecting rods. 7, Main cranks. 8, Crankshaft. 9, Displacer rod. 10, Displacer connecting rod. 11, Displacer crank. 12, Ports. 13, Cooler. 14, Regenerator. 15, Freezer. 16, Displacer piston. 17, Insulating cap. 18, Condenser for air liquefaction. 19, Annular channel for the liquefied air. 20, Swan-neck tapping pipe. 21, Insulating screening cover. 22, Mantle. 23, Air inlet. 24, Ice separator plates. 25, Central tube connecting the plates to the freezer (15). 26, Gas-tight shaft seal. 27, Gas cyclinder supplying the refrigerant. 28, Refrigerant interconnecting pipe. 29, Non-return valve.

heat exchanger housed external to the cylinder, so being alternately cooled and heated by alternately transferring heat to and from the matrix.

In being transferred from the compression space to the regenerator, the warm gas passes through a surface heat exchanger, the high-temperature recuperator or "cooler" which, being supplied with cooling water, serves to counter the temperature rise during compression. Substantial further cooling then takes place when the gas passes through the regenerator in the forward direction into the expansion space where, as a result of subsequent expansion brought about by the downward motion of the lower main piston, the temperature drops further. In consequence of the upward motion of the upper displacer piston, the cooled gas is then transferred back through the regenerator, passing on the way through a further surface heat exchanger, the low-temperature recuperator or "freezer", in which heat is transferred to it from the low-temperature refrigeration source. Substantial further heating of the gas then takes place as it passes back through the regenerator matrix to the compression space. The regenerator, like the regerative heat exchanger of the Linde process described in §10.8, is thus seen to serve the important function of separating those parts of the plant which operate around atmospheric from those operating at cryogenic temperatures.

If the working gas in the machine is hydrogen or helium, the machine may be used as an air liquefier by passing the air to be liquefied over the external surface of the low-temperature recuperator or "freezer"; this air thus takes the place of the low-temperature refrigeration source. Figure 10.14 shows the machine adapted for such a purpose.

It will be seen that the working gas passes round a closed thermodynamic cycle, and it is consequently appropriate to consider what type of ideal reversible cycle would provide the best criterion against which to judge the performance of the actual machine. In this context, reference is usually made[†] to either the ideal reversed Stirling cycle or the ideal reversed Ericsson cycle. In these cycles, for which the reader may himself draw p–v and T–s diagrams (Problem 10.13), **isothermal** compression and expansion alternate with regenerative cooling and heating, the latter taking place at constant volume in the reversed Stirling cycle and at constant pressure in the reversed Ericsson cycle. Since both cycles are completely reversible and all external heat transfers take place at constant temperature, they both have a coefficient of performance equal to that of a reversed Carnot cycle, namely

$$CP = \frac{T_A}{T_B - T_A},$$ (10.24)

where T_A and T_B are the absolute temperatures of heat reception and rejection respectively.

In the actual machine, neither the gas volume nor the pressure remain constant during the external heat transfers. For more detailed consideration of the actual processes occurring within the Philips machine the reader should consequently consult the quoted references.[102,103]

The Gifford machines also incorporate a regenerator but have only a displacer piston and use a separate reciprocating compressor, with mechanically operated inlet and exhaust valves between the compressor and the displacer cylinder. The regenerator is placed between these valves and the displacer cylinder, in which expansion occurs. The principle of operation differs appreciably from that of the Philips machine and is described in the quoted reference.[104]

Whilst the Philips machine described above is best suited to refrigeration between 190 K and liquid-air temperature (70 K), multi-stage Gifford

---

†This traditional practice in relation to the Stirling cycle engine has been criticised by Organ.[EE]

machines[105] with inter-stage regenerators can operate down to liquid-helium temperature (4 K). A later development of the Philips machine,[106] incorporating two stages of compression and expansion by the provision of a smaller extension piston on top of the displacer piston, permits temperatures down to 12 K to be achieved. At the same time the machine has also been developed to a size suitable for industrial refrigeration.

Problems

10.1. In an ammonia absorption refrigerating plant (Fig. 10.1), the absolute pressures in the condenser and evaporator are respectively $1.17 \, \text{MN/m}^2$ and $0.24 \, \text{MN/m}^2$. Stray heat transfers with the environment and pressure drops in all items of the plant other than across the throttle valves may be neglected. The plant is not fitted with the heat exchanger X shown in the figure.

(a) A saturated solution of ammonia and H_2O liquid leaves the absorber at E at 32 °C. A saturated mixture of ammonia and H_2O vapour leaves the generator at A, and a saturated solution of ammonia and H_2O liquid leaves at H, both streams being at 105 °C. Calculate, per kg of mixture leaving the generator at A, (i) the quantity of aqua-ammonia solution entering the generator, (ii) the quantity of heat Q_3 supplied to the generator, (iii) the fraction of Q_3 that is used to raise the temperature of the weak solution returned to the absorber.

(b) An aqua-ammonia solution leaves the condenser at 32 °C. Calculate the quantity of heat Q_2 transferred to the cooling water passing through the condenser.

(c) The mixture leaves the evaporator at −7 °C. Calculate the liquid fraction at D, and the quanity of heat Q_1 supplied to the evaporator from the cold chamber.

(d) Calculate, per kg of mixture entering the absorber, the quantity of heat Q_2' transferred to the cooling water passing through the absorber.

(e) Calculate the coefficient of performance of the plant, neglecting the work input to the pump.

(f) Calculate the coefficient of performance of the corresponding ideal plant in which T_1, T_2 and T_3 are the absolute temperatures corresponding respectively to the temperatures of the aqua ammonia leaving the evaporator, condenser (and absorber) and generator. Calculate also the coefficient of performance of a reversed-Carnot refrigerator operating between the same values of T_1 and T_2.

The equilibrium properties of aqua ammonia are given in the following table. The enthalpy of a liquid mixture of given concentration and temperature may be assumed to be independent of pressure.

Pressure (MN/m^2):	0.24		0.24	1.17	
Temperature (°C):	−7		32	105	
	Liquid	Vapour	Liquid	Liquid	Vapour
Concentration x (kg NH_3/kg mixture)	0.755	0.999	0.402	0.318	0.925
Specific enthalpy h (kJ/kg)	−200	1270	−110	280	1570

The specific enthalpy of an aqua-ammonia solution of concentration 0.925 at 32 °C is 100 kJ/kg.

Answer: (a) 7.23 kg; 4109 kJ; 0.591. (b) 1470 kJ. (c) 0.303; 725 kJ. (d) 3364 kJ. (e) 0.176. (f) 1.32; 6.82.

10.2. Derive eqn. (10.8).

10.3. In the "dry-ice" process illustrated in Figs. 10.5 and 10.6, gaseous CO_2 is drawn in at A at a pressure of 1 bar and a temperature of 20 °C, and solid carbon dioxide is withdrawn from the snow chamber, in which the solid and vapour are in equilibrium at a pressure of 1 bar. The pressures at exit from each of the compressor stages are respectively 6, 20 and 70 bar. In each case, dry saturated vapour enters the compressor stage and saturated liquid enters the throttle. Stray heat transfers with the environment may be neglected.

Calculate, per kg of solid CO_2 produced, (a) the values of m_1, m_2 and m_3; (b) the total work input; (c) the theoretical work input for a completely reversible process operating between the same initial and final states of the CO_2 in the presence of an environment at 20 °C; (d) the rational efficiency of the actual process; (e) the coefficient of performance of the plant if, instead of producing dry ice, it is used as a cyclic refrigerating plant with the supply of make-up gas from A shut off, and the solid–vapour mixture at M is turned into saturated vapour at B as a result of the transfer of heat from a refrigerating chamber to the snow chamber.

Check that the coefficient of performance of the cyclic plant calculated in (e) above is also given by the expression

$$\left(1 + \frac{1}{C_0}\right) = \left(1 + \frac{1}{C_1}\right)\left(1 + \frac{1}{C_2}\right)\left(1 + \frac{1}{C_3}\right),$$

where C_1, C_2 and C_3 are the respective coefficients of performance of the three vapour-compression refrigerating cycles BCDLMB, DEFJKD and FGHIF, so that its performance is equivalent to that of a ternary vapour-compression cycle. Explain why this result holds.

The relevant properties of CO_2 are as follows:

$h_A = 803.5$, $h_C = 804.0$, $h_E = 782.6$, $h_G = 789.1$ kJ/kg; $s_A = 4.841$ kJ/kg K.

Pressure	Saturated solid		Sat. liquid	Sat. vapour
	h	s	h	h
bar				
	kJ/kg	kJ/kg K	kJ/kg	kJ/kg
1	151.5	1.568	—	723.0
6	—	—	386.5	729.6
20	—	—	452.0	735.0
70	—	—	592.5	—

From *Thermophysical Properties of Carbon Dioxide*, Vukalovich, M. P. and Altunin, V. V.[3]

Answer: (a) 1.938 kg; 2.914 kg; 6.761 kg. (b) 677.2 kJ/kg. (c) 307.5 kJ/kg. (d) 45.4%. (e) 0.963; $C_1 = 4.15$; $C_2 = 5.24$; $C_3 = 2.63$.

10.4. In a gas-liquefaction plant, gas at the environment temperature T_0 and 1 atm pressure is first compressed to N atm and cooled in an after-cooler to T_0. The gas then enters a refrigerating plant in which it is wholly liquefied, leaving as saturated liquid at 1 atm. The compressor has an *isothermal* efficiency of η_C. The rational efficiency of the refrigerating process, excluding the compressor and aftercooler, is η_R.

The gas may be treated as a perfect gas at all pressures when at temperature T_0, and also at all temperatures when gaseous at 1 atm pressure. At 1 atm pressure, the boiling point is T_0/n and the specific enthalpy of evaporation is L.

Derive an expression for the total work input to the plant per unit mass of gas liquefied.

Answer: $\dfrac{c_p T_0}{\eta_R}\left(\ln N + \dfrac{1}{n} - 1\right) + \dfrac{L}{\eta_R}(n-1) - \left(\dfrac{1}{\eta_R} - \dfrac{1}{\eta_C}\right) RT_0 \ln N.$

10.5. Prove eqn. (10.14), and show that μ_h is zero for a perfect gas. Noting that, at the critical point, $p = p_c$, $T = T_c$, $v = v_c$ and

$$\left(\frac{\partial p}{\partial v}\right)_T = \left(\frac{\partial^2 p}{\partial v^2}\right)_T = 0,$$

express van der Waals' equation of state (§10.9) in terms of the reduced coordinates, $p_R \equiv p/p_c$, $T_R \equiv T/T_c$ and $v_R \equiv v/v_c$. Show that points on the inversion line (for which the isenthalpic Joule–Thompson coefficient μ_h is zero) are given by the relations

$$T_R = \frac{3(3v_R - 1)^2}{4v_R^2} \quad \text{and} \quad p_R = \frac{9(2v_R - 1)}{v_R^2}.$$

Sketch the inversion line on the pressure–volume and pressure–temperature planes.

Answer: $p_R = \dfrac{8T_R}{3v_R - 1} - \dfrac{3}{v_R^2}.$

Problems 10.6 to 10.12 relate to processes for the production of saturated liquid air at a pressure of 1 bar from air at 1 bar and 290 K, which is also the temperature of the environment.

10.6. Calculate the ideal work input for a completely reversible steady-flow process operating between the end states specified above, with the environment at the same temperature. Express the result both in MJ/kg and kW h/litre of liquid air produced.

Answer: 0.701 MJ/kg, 0.171 kW h/litre.

10.7. In a simple Linde process, the high-pressure air entering the regenerative heat exchanger and the low-pressure air leaving it are at 290 K. For a range of compressor delivery pressure from 250 to 500 bar, draw a graph showing the yield y plotted against the delivery pressure, and so determine the optimum pressure for maximum yield. At this pressure calculate the temperature of the air entering the throttle.

For the same range of compressor delivery pressure plot the work input per unit yield when the compression process is isothermal and reversible. Thence estimate the optimum pressure for maximum efficiency (i.e. minimum work input per unit yield), and the value of this work input, expressed both in MJ/kg and kW h/litre. Calculate the corresponding rational efficiency of the process.

Answer: About 400 bar, 167 K; about 335 bar, 4.36 MJ/kg, 1.06 kW h/litre; 16.1%.

10.8. In a simple Linde process, the high-pressure air entering the regenerative heat exchanger and the low-pressure air leaving it are at 290 K. The pressure at compressor delivery is 200 bar and the compression process is iothermal and reversible.

Calculate (a) the yield; (b) the temperature of the air entering the throttle; (c) the work input per unit yield, expressed both in MJ/kg and kW h/litre; (d) the rational efficiency of the process.

Evaluate the extra work input due to irreversibility arising from (i) the heat transfer process in the regenerative heat exchanger, (ii) the throttling process, expressing these quantities as a percentage of the actual work input.

If, instead of being used for air liquefaction, the plant were used as a cyclic refrigerating plant, what would be its coefficient of performance? Express this as a percentage of the coefficient of performance of a reversed-Carnot refrigerating plant operating between the same extremes of temperature. Take the saturation temperature of air at 1 bar to be 81.7 K, the dew point of the vapour at that pressure.

Answer: (a) 0.0890. (b) 170 K. (c) 4.92 MJ/kg; 1.20 kW h/litre. (d) 14.2%. (i) 20.1%. (ii) 65.7%. 0.0849, 21.6%.

10.9. In a dual-pressure Linde process (Fig. 10.10), the intermediate pressure p_B is supercritical. The compression process is isothermal and reversible, and the operating conditions are as follows: $p_B = 50$ bar, $p_C = 200$ bar, $T_{11} = T_7 = T_2 = T_1$ and $y_B = 0.2$.

Calculate (a) the yield y; (b) the values of T_3 and T_4; (c) the wetness y_A of the mixture leaving the low-pressure throttle; (d) the work input per unit yield, expressed both in MJ/kg and kW h/litre; (e) the rational efficiency of the process.

Evaluate the extra work input due to irreversibility arising from (i) the heat transfer process in the regenerative heat exchangers, (ii) the HP throttling process, (iii) the LP throttling process, expressing these quantities as a percentage of the actual work input.

> *Answer*: (a) 0.0685. (b) 146.7 K; 135.1 K. (c) 0.342. (d) 2.62 MJ/kg; 0.638 kW h/litre.
> (e) 26.8%. (i) 27.6%. (ii) 29.6%. (iii) 16.0%.

10.10. In a dual-pressure Linde process (Fig. 10.10), the intermediate pressure is subcritical. The compression process is isothermal and reversible, and the operating conditions are as follows: $p_B = 30$ bar, $p_C = 200$ bar, $T_{11} = T_7 = T_2 = T_1$.

Calculate (a) the wetness y_A of the mixture leaving the LP throttle; (b) the value of y_B for which the plant must be designed in order to satisfy the above conditions; (c) the yield y; (d) the value of T_3; (e) the work input per unit yield, expressed both in MJ/kg and kW h/litre; (f) the rational efficiency of the process.

Evaluate the extra work input due to irreversibility arising from (i) the heat transfer process in the regenerative heat exchangers, (ii) the HP throttling process, (iii) the LP throttling process, expressing these quantities as a percentage of the actual work input.

> *Answer*: (a) 0.431. (b) 0.178. (c) 0.0766. (d) 168.7 K. (e) 2.70 MJ/kg; 0.657 kW h/litre.
> (f) 26.0%. (i) 24.4%. (ii) 40.1%. (iii) 9.5%.

10.11. In a Claude process (Fig. 10.11), the pressure at compressor delivery is 40 bar, the temperature of the air entering the expansion engine is $-80\,°C$, $T_{13} = T_2 = T_1$, $T_9 = T_{10}$ and $x = 0.2$. The compression process is isothermal and reversible, and the isentropic efficiency of the engine process is 75%.

Calculate (a) the value of T_{10}; (b) the yield y_1; (c) the value of T_{12}; (d) the wetness of the mixture leaving the throttle; (e) the specific enthalpy of the fluid at points 5 and 4 respectively; (f) the net work input per unit yield, expressed both in MJ/kg and kW h/litre; (g) the rational efficiency of the process.

> *Answer*: (a) 90.0 K. (b) 0.184. (c) 156.8 K. (d) 0.920. (e) 16.4 kJ/kg; 17.1 kJ/kg.
> (f) 1.29 MJ/kg; 0.314 kW h/litre. (g) 54.3%.

10.12. In a Heylandt process (Fig. 10.11), the pressure at compressor delivery is 200 bar, $T_{13} = T_{2'} = T_1$, $T_9 = T_{10}$ and $x = 0.45$. The compression process is isothermal and reversible, and the isentropic efficiency of the expansion engine is 75%.

Calculate (a) the value of T_{10}; (b) the yield y_1; (c) the wetness of the mixture leaving the throttle; (d) the values of T_5 and T_4 respectively; (e) the net work input per unit yield, expressed both in MJ/kg and kW h/litre; (f) the rational efficiency of the process.

Evaluate the extra work input due to irreversibility arising from (i) the heat transfer process in heat exchangers X_B and X_C respectively, (ii) the throttling process, (iii) the engine process, expressing these quantities as a percentage of the actual work input.

> *Answer*: (a) 104.0 K. (b) 0.289. (c) 0.642. (d) 120.5 K; 124.0 K. (e) 1.226 MJ/kg;
> 0.299 kW h/litre. (f) 57.2%. (i) 1.75%; 0.75%. (ii) 13.7%. (iii) 26.6%.

10.13. Draw pressure–volume and temperature–entropy diagrams for the reversed Stirling cycle and the reversed Ericcson cycle (§10.19).

APPENDIX A

Thermodynamic availability and irreversibility

A.1. Introduction

This appendix concerns *non-cyclic* work-producing and work-absorbing devices which may exchange heat with the *environment*, treated as an essentially infinitely large source (or sink) of thermal energy which can be freely drawn upon (or burdened) without charge or debit and without changing its thermodynamic state. Availability studies relate always to situations in which it is desired to know the maximum possible work output (or minimum possible work input) when a system or fluid is taken by some *non-cyclic* process from a given *specified* state 1 to a second *specified* state 2 in the presence of an environment at *specified* temperature T_0 and *specified* pressure p_0. These studies enable appropriate *performance measures and criteria* to be set up with which respectively to measure and judge the performance of such non-cyclic work-producing and work-absorbing plants and devices. Electrical storage batteries, fuel cells and other internal-combustion power plants are examples of these work-producing devices, while examples of such work-absorbing devices are provided by the Linde and other open-circuit gas-liquefaction processes. Although none of these are cyclic in operation, some are studied in this book and the concept of *available energy* finds applications in the chapters on internal-combustion plant, advanced and nuclear steam plant and advanced refrigerating plant.

The study will need some revision of thermodynamic principles and some precise definition of thermodynamic terms. Of particular importance are the concepts of *reversibility and irreversibility*, since *irreversible processes always result in lost opportunities for producing work* (or in augmenting the required work input).

A.2. Aspects of reversibility

Irreversible process. A process undergone by a system is irreversible if all its effects on both the system and its environment *cannot be effaced* without calling in aid a Perpetual Motion Machine of the Second Kind (a PMM 2),[†] in contravention of the Second Law of Thermodynamics. All real-life processes are in some measure irreversible.

Reversible process. In postulating that a process is reversible, we imply that all effects of the process on the system *and* its environment *can* be effaced by some effacing process (an *effacer*) which is known to be possible of execution.

Internal reversibility implies that all processes *within* the system are reversible (i.e. that there is an absence of friction, of diffusive heat transfer and of diffusive mixing within the system, so that the system passes through a succession of *equilibrium* states in a *quasi-static* process).

External reversibility implies that all heat exchanges between the system and its environment are made reversibly. This requires either that the temperature of any part of the system that exchanges heat with the environment is equal to (i.e. infinitesimally different from) that of the environment [Fig. A.1(a)] or that such heat exchange takes place via auxiliary, reversible cyclic devices [Fig. A.1(b)].

Full reversibility implies the existence of both internal and external reversibility, in the sense defined above.

If any part of the system that was at bulk temperature T exchanged heat directly (e.g. via a thermally resistive boundary layer) with an environment

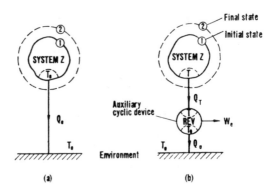

Fig. A.1. External reversibility.

[†]A PMM 2 is a continuously operating cyclic device which produces positive net work while exchanging heat with only a single thermal reservoir. The Second Law states that it is impossible to construct such a device.

at temperature T_0, the opportunity would be lost for producing the *external work*, W_e, that is obtained from the auxiliary cyclic device in Fig. A.1(b).

A.3. Different forms of work output

In Fig. A.2, System Z undergoes an *internally reversible* process between specified states 1 and 2, and *full reversibility* of the process is ensured by the *external reversibility* provided by such auxiliary cyclic devices as are needed to ensure reversibility of all heat exchanges with the environment. In consequence, we define the Extended System Z^+, enclosing within it System Z and these auxiliary cyclic devices. It is the *gross* work output, $[(W_g)_{REV}]_1^2$, from this Extended System that is of primary interest in availability studies, but we need also to define explicitly the other forms of work output involved in the overall process.

Displacement work output, W_d, is the work performed by System Z in expansion against the environment. To achieve internal reversibility, the process must be quasi-static; that is, the expansion must take place infinitely slowly. When this is the case $(W_d)_{REV} = p_0(V_2 - V_1)$.

Internal work output, W_i, is the total work output (mechanical and/or electrical) produced *directly* by System Z, *including W_d*.

External work output, W_e, is the work produced by such auxiliary cyclic devices as are needed to ensure external reversibility. This is wholly in the form of useful *shaft work*.

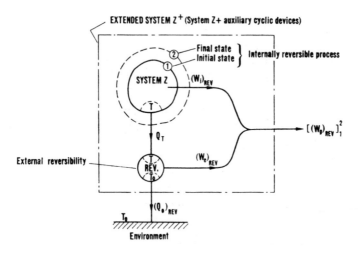

FIG. A.2. Different forms of work output.

Gross work output, $W_g = W_i + W_e$.

Total **shaft** *work output,* $W_x = (W_i - W_d) + W_e = W_g - W_d$.

Note that W_g and W_x are respectively the gross work output and the total shaft work output produced by the Extended System Z^+. *It is this system, rather than the initial system Z, that is the focus of attention in availability studies.*

Note also that, in *steady-flow processes*, $W_d = 0$, so that $W_x = W_g$.

A.4. Important theorems in availability

The following three theorems are of particular relevance to our studies, though there are others which are equally important.[†]

Theorem 1. *Gross work output between specified end states*
For a system that can exchange heat with a single thermal reservoir (e.g. the environment at temperature T_0), the gross work output is the same for all *fully reversible* processes *between the same specified end states 1 and 2.* (This quantity will be termed the *reversible gross work* and will be given the symbol $[(W_g)_{REV}]_1^2$.) During any *irreversible* process between these same specified end states in the presence of the specified environment at temperature T_0, the gross work output is always less than $[(W_g)_{REV}]_1^2$.

Theorem 2. *Loss of gross work output and entropy creation*
For a system undergoing an irreversible process between specified end states in the presence of an environment at temperature T_0, the loss of gross work output due to irreversibility is equal to $T_0 \Delta S_C$, where ΔS_C is the *entropy creation* due to irreversibility within the system.

Theorem 3. *Entropy conservation in a fully reversible process*
In a fully reversible process, there is zero *entropy creation* (i.e. entropy is conserved).

A definition of the term *entropy creation due to irreversibility* will be given shortly.

A.5. Proof of Theorem 1

The concept of the availability of energy for work production arises from this theorem, which follows as a corollary of the Second Law.

Figure A.3(a) represents two alternative processes undergone by a

---

[†]An exhaustive treatment of the theorems of thermodynamic availability appears in a second book by the author[5,D,E] and also in the authors's critical review paper on the subject.[6]

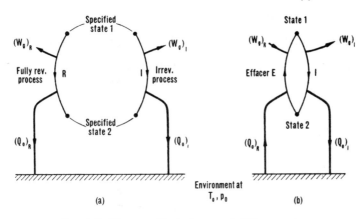

FIG. A.3. Diagrams illustrating proof of Theorem 1.

system between specified end states 1 and 2, R being *fully reversible* and I irreversible, the *gross* work outputs being respectively $(W_g)_R$ and $(W_g)_I$. From the definition of a reversible process given in §A.2, since R is postulated to be reversible it must be possible to devise a process which could act as an *effacer* E of all the effects of process R on both the system and its environment; that is, in the effacing process E, the system would absorb $(W_g)_R$ and $(Q_0)_R$ from the environment when taken back to state 1 from state 2. Processes E and I together could then take this system round a cyclic process, as depicted in Fig. A.3(b).

From the First Law energy conservation equation for the system,

$$(Q_0 + W_g)_R = (Q_0 + W_g)_I$$
$$\therefore (Q_0)_I - (Q_0)_R = (W_g)_R - (W_g)_I. \qquad (A.1)$$

Now, if $(W_g)_I > (W_g)_R$, then $(Q_0)_R > (Q_0)_I$, in which case the cyclic process constitutes a PMM 2, so contravening the Second Law. Hence:

$$(W_g)_I \leq (W_g)_R \qquad (A.2)$$

However, if $(W_g)_I$ were to equal $(W_g)_R$ it would be possible to prove that process I was reversible, since process E could then also act as the effacer of process I. Hence:

$$(W_g)_I < (W_g)_R. \qquad (A.3)$$

Were I and R replaced in Fig. A.3(a) by alternative fully reversible processes R_A and R_B respectively, expression (A.2) would become $(W_g)_{R_A} \leq (W_g)_{R_B}$, whereas if they were replaced respectively by R_B and R_A it would become $(W_g)_{R_B} \leq (W_g)_{R_A}$; since both these statements are true, to satisfy them both we must have

$$(W_g)_{R_A} = (W_g)_{R_B}. \tag{A.4}$$

However, R_A and R_B are *any* two fully reversible processes between the specified end states, so that we may write

$$(W_g)_{R_A} = (W_g)_{R_B} \equiv [(W_g)_{\text{REV}}]_1^2. \tag{A.5}$$

Expressions (A.3) and (A.5) together establish the truth of Theorem 1.

In general, $[(W_g)_{\text{REV}}]_1^2$ will clearly be a function of the thermodynamic properties of the system in states 1 and 2, *and of the environment temperature T_0*. Expressions for this gross reversible work in non-flow and steady-flow processes will be established after presenting the proofs of Theorems 2 and 3, for which an understanding of the terms *entropy creation due to irreversibility* and *thermal entropy flux* will first be needed.

A.6. Entropy creation and thermal entropy flux

The concept of entropy creation due to irreversibility is most easily introduced by reference first to a *thermally isolated* system. We know, as a corollary of the Second Law, that

$$(\Delta S_{\text{isol.}})_{\text{REV}} = 0 \tag{A.6}$$

and

$$(\Delta S_{\text{isol.}})_{\text{IRREV}} > 0. \tag{A.7}$$

Thus, when an isolated system undergoes an irreversible process, the system always suffers an increase in entropy, despite the absence of heat transfer to it during the process. We call this increase in entropy the *entropy creation* due to irreversibility within the isolated system.

To extend this concept to a system which does not remain thermally isolated, but experiences heat transfers across its boundary, consider Fig. A.4.

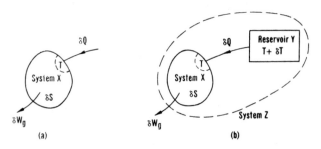

Fig. A.4. Heat transfer to a system during an infinitesimal irreversible process.

In Fig. A.4(a), system X suffers an entropy change δS while undergoing an infinitesimal irreversible process. In the course of this process it receives, *across a boundary at which the temperature is* T, an amount of heat δQ.

In order not to introduce effects due to extraneous irreversibilities, we suppose δQ to be taken in from a thermal reservoir Y which is at a constant temperature infinitesimally in excess of T and is subject only to reversible processes. This situation is depicted in Fig. A.4(b).

The combined system Z, comprising X and Y together, constitutes a thermally isolated system in which the entropy creation due to irreversibility is given by

$$\delta S_C = \delta S_Z = \delta S_X + \delta S_Y = \delta S - \frac{\delta Q}{T} \tag{A.8}$$

Now reservoir Y was only introduced for analytical convenience, and we recall that the only irreversibility that occurs is within the original system X. Thus, if we treat the quantity $\delta Q/T$ as the *thermal entropy flux* brought into system X in consequence of the transfer to it of the heat quantity δQ flowing in across a boundary at which the temperature is T, then the right-hand side of equation (A.8), as well as being the entropy creation in isolated system Z, also represents *the entropy creation due to irreversibility in the original system X.*

In general, system X may be subject to any number of heat transfers across points in its boundary which are at a variety of temperatures. Each will contribute to the total thermal entropy flux, $\Delta S_Q = \Sigma \delta Q/T$, and a general expression for the entropy creation due to irreversibility during a finite change of state of the system will be

---

Entropy creation, $\Delta S_C \equiv \Delta S - \Delta S_Q$, $\qquad\qquad$ (A.9)

where ΔS = entropy change of the system

and ΔS_Q = *thermal entropy flux* = $\sum \delta Q/T$ $\qquad$ (A.10)

The summation in equation (A.10) must be taken over all heat quantities at all points on the boundary of the system at which heat transfers occur, the temperature T being, in each case, *the particular temperature of that part of the boundary at which the heat transfer occurs*.

We may thus define the entropy creation ΔS_C due to irreversibility within a system as *that part* of the entropy change of the system *which cannot be accounted for by the thermal entropy fluxes associated with the heat transfers experienced by the system*.

A.7. Proof of Theorem 2

To establish the relation between entropy creation and the loss of gross work output due to irreversibility, consider two alternative processes undergone by a system between specified end states 1 and 2 in the presence of an environment at temperature T_0, process I being *irreversible* and process R being *fully reversible*, as depicted in Fig. A.5. During these processes the system suffers an entropy change ΔS.

The energy conservation equation for the system gives

$$(W_g)_R - (Q_0)_R = (W_g)_I - (Q_0)_I. \tag{A.11}$$

For process I, noting the irreversibility of heat transfer occurs *within the system*, so that the system *boundary* is at T_0,

$$\text{thermal entropy flux, } \Delta S_Q = \frac{(Q_0)_I}{T_0}. \tag{A.12}$$

For process R, with the system at some varying temperature T different from T_0, the situation is depicted in more detail in Fig. A.6, in which T is arbitrarily taken as being less than T_0. For the process to be fully reversible, the heat exchange between the system and environment must take place via a reversible auxiliary cyclic device, as shown in the figure. The process in question is consequently that taking place within control surface X. We then have:

(a) For the system,

$$\Delta S = \int \frac{(dQ_T)_R}{T}.$$

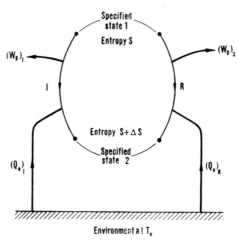

Fig. A.5. Alternative irreversible and fully reversible processes between specified end states.

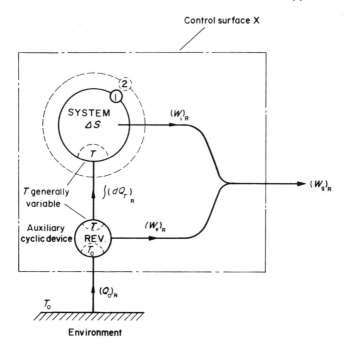

Fig. A.6. Detailed representation of fully reversible process R.

(b) For the fluid passing round the reversible auxiliary cyclic device,

$$\frac{(Q_0)_R}{T_0} - \int \frac{(dQ_T)_R}{T} = 0$$

in an integral number of completed cycles.

Whence, for process R,

$$\Delta S = \frac{(Q_0)_R}{T_0}. \tag{A.13}$$

Thus, from equations (A.11), (A.12) and (A.13), the loss of gross work output due to irreversibility is given by

$$(W_g)_R - (W_g)_I = (Q_0)_R - (Q_0)_I$$
$$= T_0 \Delta S - T_0 \Delta S_Q$$
$$= T_0(\Delta S - \Delta S_Q).$$

$$\therefore \quad \boxed{(W_g)_R - (W_g)_I = T_0 \Delta S_C} \tag{A.14}$$

where ΔS_C is the entropy creation due to irreversibility defined in §A.6 by equations (A.9) and (A.10).

Equation (A.14) establishes the truth of Theorem 2. It is easy to see that this result is also applicable to a control volume in steady-flow processes, the corresponding extensive quantities then being expressed either per unit time or per unit quantity of fluid passing through the control volume. It is, of course, equally applicable to work-absorbing plant, in which case $T_0 \Delta S_C$ is equal to the *extra gross work input* due to irreversibility. This finds application in the studies of gas liquefaction plant in Chapter 10.

The direct connection between lost work and entropy creation is a consequence of the fact that work production is only achieved with maximum effect when the process is *fully ordered* (i.e. when the system passes through a succession of equilibrium states; namely a *reversible* process). The loss of work in an irreversible process thus results from a failure to maintain a fully ordered transition from one energy form to another. Since increase in entropy between two specified equilibrium states is associated with an increase in *disorder* of the system, it is not surprising to find that the lost work (or *dissipation*, as in certain circumstances it is sometimes called) is related directly to the *entropy creation* during the process.

Theorem 2 is very important in that it provides a simple way of assessing the relative seriousness of the individual irreversibilities in each sub-process of a multi-process plant; it is so used in the studies of gas liquefaction plant in Chapter 10 (§10.13 and Problems 10.8–10.12).

Theorem 2 is also important in leading on to Theorem 3, with the aid of which we shall find it simple to write down expressions for reversible gross work output.

A.8. Proof of Theorem 3

When a process is *fully reversible* in the sense defined in §A.2, there is no loss of gross work output due to irreversibility. It is then evident from equation (A.14) that

$$\boxed{(\Delta S_C)_{\mathrm{REV}} = 0} \tag{A.15}$$

Equation (A.15) establishes the truth of Theorem 3. For convenience, and by analogy with the energy conservation equation, it may be described as the *entropy conservation equation* for the system (or control volume). It is only applicable in the event of full reversibility.

A.9. Derivation of expressions for reversible gross work output

Theorem 3 provides a simple means of deriving expressions for the reversible gross work output, $[(W_g)_{REV}]_1^2$, for a process between specified end states in the presence of a specified environment. When these are specified the problem involves only two unknowns, the gross work output and the heat quantity $(Q_0)_{REV}$ exchanged with the environment. The *entropy conservation equation* and the *energy conservation equation* together provide the means of solving for these two unknowns. Table A.1 gathers together the results of applying these two equations to both non-flow and steady-flow processes in which changes in kinetic and potential energies are taken to be negligible.

A.10. Expressions for reversible shaft work output— available energy

Of greater practical interest than $[(W_g)_{REV}]_1^2$ is the reversible *shaft* work output $[W_x)_{REV}]_1^2$, which is less than $[(W_g)_{REV}]_1^2$ by the amount of *displacement work*, $(W_d)_{REV}$, performed in expansion against the pressure of the environment. The latter is zero in steady-flow processes, in which we are dealing with a fixed control volume, while in reversible non-flow processes it is equal to $p_0(V_2 - V_1)$. Since we are dealing with processes between specified end states in the presence of a specified environment, the values of p_0, V_1 and V_2 are all fixed; hence $[(W_x)_{REV}]_1^2$, like $[(W_g)_{REV}]_1^2$, is the same for all *fully reversible* processes between the specified end states. The quantity $[(W_x)_{REV}]_1^2$ is called the *available energy*[†] in the given situation. Expressions for the available energy also appear in Table A.1.

Particular attention is drawn to the expression $(B_1 - B_2)$, for the *steady-flow available energy*, where $B \equiv H - T_0 S$, the *steady-flow availability function*. This expression features in Chapters 2, 7 and 8, while $(B_2 - B_1)$ features in Chapter 10, which deals with work-*absorbing* processes.

A.11. The available energy in chemical processes

Of particular interest is the special case of a chemical process in which both the reactants initially and the products finally are at the temperature T_0 and pressure p_0 of the environment. Expressions for the available energy in these circumstances are given in the last line of Table A.1. It will be seen that, in the case of steady flow, the available energy in these

[†]It is necessary to warn that some earlier writers have applied this term to a different physical quantity.

TABLE A.1. *Expressions for reversible work.*

	NON-FLOW	STEADY FLOW
DEFINITIONS OF CERTAIN 'FUNCTIONS' $A^* \equiv U - T_0 S$ $A \equiv U + p_0 V - T_0 S$ (Non-flow avail^y function) $B \equiv H - T_0 S$ (Steady-flow avail^y function) $\left. \begin{array}{l} F \equiv U - TS \\ \text{(Helmholtz function)} \\ G \equiv H - TS \\ \text{(Gibbs function)} \end{array} \right\}$ Properties	 FOR EXTENDED SYSTEM Z^+	 FOR EXTENDED CONTROL VOLUME C^+
ENERGY CONSERVATION EQN.	$[(W_s)_{\mathrm{REV}}]_1^2 + (Q_0)_{\mathrm{REV}} = U_1 - U_2$ (When KE and PE changes negligible)	$[(W_s)_{\mathrm{REV}}]_1^2 + (Q_0)_{\mathrm{REV}} = H_1 - H_2$
ENTROPY CONSERVATION EQN.	$\dfrac{(Q_0)_{\mathrm{REV}}}{T_0} = S_1 - S_2$	

From these two equations, we can write down the following results:

Rev. work outputs		A₁* − A₂*	B₁ − B₂
	GROSS ≡ $[(W_g)_{REV}]_1^2$	$A_1^* - A_2^*$	$B_1 - B_2$
	Disp. work on envir⁴ ≡ $(W_d)_{REV}$	$p_0(V_2 - V_1)$	—
	SHAFT ≡ $[(W_x)_{REV}]_1^2$	$A_1 - A_2$ (Non-flow *AVAILABLE ENERGY*)	$B_1 - B_2$ (Steady-flow *AVAILABLE ENERGY*)
	Special case when $\left.\begin{array}{l}T_1=T_2=T_0\\ p_1=p_2=p_0\end{array}\right\}$ e.g. $\left\{\begin{array}{l}\text{Reactants in state 1}\\ \text{Products in state 2}\end{array}\right\}$ both at T_0, p_0	$[(W_g)_{REV}]_1^2 = (F_1 - F_2)_{T_0,\,p_0} \equiv -\Delta F_0$ $[(W_x)_{REV}]_1^2 = (G_1 - G_2)_{T_0,\,p0} \equiv -\Delta G_0$	$[(W_x)_{REV}]_1^2 = (G_1 - G_2)_{T_0,\,p_0} \equiv -\Delta G_0$

circumstances is equal to $(G_1 - G_2)_{T_0, p_0}$, written for short as $-\Delta G_0$, where $G \equiv H - TS$, the *Gibbs function*. This result features in Chapter 4.

A.12. Rational efficiency

It will be seen from Theorem 1, and the argument in §A.10, that the available energy, $[(W_x)_{\text{REV}}]_1^2$, is the *theoretical maximum* shaft-work output that is obtainable as a result of the specified change of state in the presence of the specified environment. It will therefore be a rational procedure to define the efficiency of *non-cyclic* work-producing and work-absorbing devices in the following way, and to call this efficiency the *rational efficiency*,[†] η_R, of the device:

$$\text{(a) } \textit{Work-producing device, } \eta_R \equiv \frac{[W_{\text{actual}}]_1^2}{[(W_x)_{\text{REV}}]_1^2}, \tag{A.16}$$

where, in steady-flow, $[(W_x)_{\text{REV}}]_1^2 = (B_1 - B_2)$. (A.17)

$$\text{(b) } \textit{Work-absorbing device, } \eta_R \equiv \frac{[(W_x)_{\text{REV}}]_1^2}{[W_{\text{actual}}]_1^2}, \tag{A.18}$$

where, in steady-flow, $[(W_x)_{\text{REV}}]_1^2 = (B_2 - B_1)$. (A.19)

In the foregoing, W denotes work *output* and W denotes work *input*. In *non-flow* processes, as will be seen from Table A.1, the *non-flow availability function A* replaces the *steady-flow availability function B*, where $A \equiv U + p_0 V - T_0 S$ and $B \equiv H - T_0 S$.

A.13. Exergy and the dead state

In recent years, the concepts presented in the foregoing sections have been used to bring new terms into the vocabulary of thermodynamic availability. These relate to the evaluation of the available energy when, from an intial state 1, the system (or fluid, in steady flow) is brought to a condition described as the *dead state* 0, in which it is in thermal and mechanical equilibrium with the environment, so that it is then at T_0 and p_0. The available energy for this change of state is called the *exergy*[‡] in

[†]The *rational efficiency* of a *non-cyclic* process, as here defined, has also been called *thermodynamic efficiency* and also *exergetic efficiency*; it must not be confused with the *thermal efficiency* of a *cyclic* heat power plant.

[‡]In relation to a system subject to a non-flow process, earlier writers have called this quantity variously the "available energy of the body and medium" (Gibbs), the "énergie utilisable" (Gouy) and the "availability" (Keenan). The latter word is reserved by the present author for a *concept*, not a physical quantity. Reference 6 includes a historical review of the development of the subject of thermodynamic availability.

state 1, a term first coined by Rant in 1953 and now widely adopted; it is expressive of the potentiality of a system for work extraction. In the present notation, exergy is thus defined by the expression:

$$\text{in state 1, } \textit{exergy} \equiv [(W_x)_{\text{REV}}]_1^0. \tag{A.20}$$

The expression for exergy will be different for non-flow and steady-flow processes. From Table A.1, it will be evident that, when changes in kinetic and potential energies are negligible,

$$\textit{non-flow exergy} \text{ in state 1}, \mathcal{E}_{n_1} = (A_1 - A_0), \tag{A.21}$$

$$\textit{steady-flow exergy} \text{ in state 1}, \mathcal{E}_{f_1} = (B_1 - B_0). \tag{A.22}$$

Some writers have used the term *exergy* only in relation to the case of steady flow.

The magnitude of the steady-flow exergy $\mathcal{E}_{f_1}$ in state 1 can conveniently be represented on an enthalpy-entropy diagram for the fluid by means of the simple construction illustrated in Fig. A.7.

On such a diagram, the slope of an isobar at any point is given by $(\partial H/\partial S)_p = T$, so that the line shown as a tangent to the isobar p_0 at the *dead state* 0 has a slope equal to T_0. Thus it is seen that the magnitude of $\mathcal{E}_{f_1}$ can be represented by the distance shown in the figure.

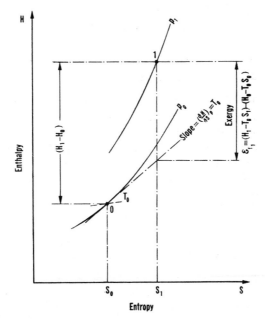

FIG. A.7. Depiction of steady-flow exergy on the enthalpy–entropy diagram.

A.14. Anergy

Following his introduction of the new term *exergy* into the vocabulary of thermodynamic availability in 1953, Rant brought in a further new term, *anergy*, in 1962. However, whereas *exergy* is an excellent word for its purpose, the new term *anergy* has not found the same wide acceptance, not only because the concept with which it deals is of smaller interest to engineers but probably also because *anergy* is phonetically too close to *energy* to be readily differentiated from it in the spoken tongue. A brief reference to this new term is, however, not out of place in the present context, although the term is not used in the body of the book.

In the case of steady-flow with which we have just been dealing, whereas for *exergy* we have

$$exergy \ in \ state \ 1 = B_1 - B_0 = (H_1 - T_0 S_1) - (H_0 - T_0 S_0),$$

the definition of *anergy*[†] is such that

$$anergy \ in \ state \ 1 = (H_1 - exergy) = H_0 + T_0(S_1 - S_0). \quad \text{(A.23)}$$

Thus, in this context, anergy is that part of the initial enthalpy that cannot be transformed into useful shaft work. Rant, in fact, gave a wider connotation to it than this and the term has been more fully discussed in a paper by Baehr.[107]

A.15. Available energy, exergy and lost work due to irreversibility in an adiabatic steady-flow process

To obtain a better appreciation of some of the terms introduced in this appendix, the construction given in Fig. A.7 is used again in Fig. A.8, which relates to an irreversible, adiabatic, steady-flow process undergone by a fluid between specified states 1 and 2 in the presence of an environment at T_0 and p_0, when overall changes in kinetic and potential energies are negligible.

It is seen that this construction makes it possible to illustrate the relative magnitudes of quite a number of the quantities involved, bearing in mind the following relations:

$$Steady\text{-}flow \ available \ energy \equiv [(W_x)_{REV}]_1^2 = (B_1 - B_2) = (\mathcal{E}_{f_1} - \mathcal{E}_{f_2})$$
$$\text{(A.24)}$$

$$Lost \ work \ due \ to \ irreversibility \equiv [(W_x)_{REV}]_1^2 - [(W_x)_{IRREV}]_1^2$$
$$= T_0(S_2 - S_1). \quad \text{(A.25)}$$

[†]In Appendix A of the First Edition of this book, the specific anergy was incorrectly defined, being erroneously given as $T_0(S_1 - S_0)$; this error arose from the fact that the author had received verbally an incorrect definition of the newly coined term at the time that the book went to press.

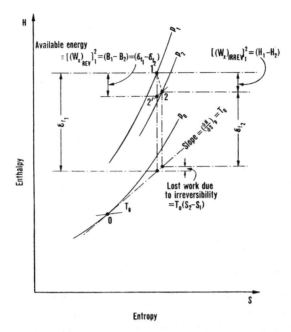

FIG. A.8. Available energy, exergy and lost work due to irreversibility in an irreversible, adiabatic steady-flow process.

It will be recalled that, in steady-flow processes, there is no difference between gross work, W_g, and total shaft work, W_x.

APPENDIX B

The advance in operating conditions in steam power stations

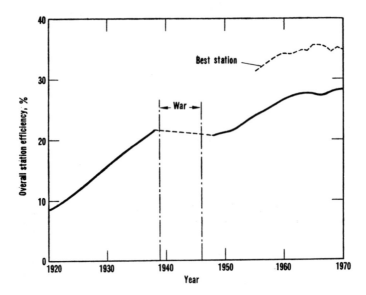

FIG. B.1. Average annual station efficiency of all power stations in the United Kingdom.

TABLE B.1. *Advance in terminal conditions and machine outputs in fossil-fuel power stations*

Approximate of installation	Units	United Kingdom								Supercritical pressure machines				
										UK	USA			Japan
		1907	1919	1938	1950	1958	1959	1966	1973	1965	1959	1972	1975	1990
Output	MW	5	20	30	60	120	200	500	660	375	450	800	1300†	700
Gauge pressure	lbf/in² (MN/m²)	190 (1.3)	200 (1.4)	600 (4.1)	900 (6.2)	1500 (10.3)	2350 (16.2)	2300 (15.9)	2300 (15.9)	3500 (24.1)	3500 (24.1)	3500 (24.1)	3500 (24.1)	4500 (31.0)
Initial temp.	°F (°C)	500 (260)	600 (316)	850 (454)	900 (482)	1000 (538)	1050 (566)	1050 (566)	(1049) 565	1100 (593)	1050 (566)	1000 (538)	1000 (538)	1050 (566)
1st reheat temp.	°F (°C)	—	—	—	—	1000 (538)	1000 (538)	1050 (566)	(1049) 565	1050 (566)	1050 (566)	1025 (552)	1000 (538)	1050 (566)
2nd reheat temp.	°F (°C)	—	—	—	—	—	—	—	—	—	1050 (566)	1050 (566)	1000 (538)	1050 (566)
Final feed temp.	°F (°C)	—	175 (79)	340 (171)	385 (196)	435 (224)	460 (238)	485 (252)	(485) 252	505 (263)	547 (286)	548 (287)	554 (290)	
No. of feed heaters		—	2	3	4	6	6	7	8	8	9	7	8	8
Condenser: Vacuum (Pressure)	inHg (kN/m²)	26 (13.5)		28.5–29.0 (5.1–3.4)		28.9 (3.7)	28.9 (3.7)	28.7 (4.4)	28.4 (5.4)	28.7 (4.4)	28.5 (5.1)	28.5 (5.1)	28.5 (5.1)	
Overall efficiency	%		~17	27.6	30.5	35.6	37.5	39.8	39.5		~40			41.7

This table is far from being comprehensive, but shows some of the major steps over the years. The table indicates the approximate dates at which installation of machines at the conditions listed became commercially attractive; single pioneering machines of advanced design in many cases antedated those shown in the table.

†Cross-compound two-shaft machine at 3600 rev/min. Double-flow H.P. turbine and two double-flow L.P. turbines on one shaft, double-flow reheat turbine and two double-flow L.P. turbines on the other shaft.[108]

APPENDIX C

Some economic considerations

C.1. The determination of the economic operating conditions for steam power plant

Practical optimisation calculations for power plant always involve consideration of economic as well as thermodynamic factors. It is rarely an economic proposition to design a plant for the conditions which thermodynamic calculations indicate are those which would give the highest plant efficiency. As the peak of the efficiency curve is approached, the further gain due to an additional advance in operating conditions progressively falls off, and the cost of achieving this further gain will not be economic if the resulting fuel saving is offset by the effect of the resulting increase in capital cost of the plant. The **economic** (as distinct from the thermodynamic optimum) operating conditions will be those at which the one just counterbalances the other; their determination is illustrated by the following calculation showing how the economic steam pressure in a non-reheat plant may be estimated, but the method is of general application. The following factors additional to the thermodynamic parameters enter the calculation:

S = cost of fuel, £ per tonne (10^3 kg).

V = calorific value of fuel, MJ/kg.

L = plant load factor, %.

R = % rate of capital charges (interest, sinking fund, etc.).

E = capital cost of plant, £ per MW of output.

It will be assumed that the plant operates at full design load all the time that it is running, but that it is in operation for only L% of the year. Thus the calculation is only applicable to a base-load station.

Following Baumann,[13] the first four of the foregoing parameters may be grouped to form an *operation factor m* given by

$$m = \frac{SL}{VR}.$$ (C.1)

Brief consideration will show that the greater the value of m, the higher will be the economic pressure. (It may be noted that the cost of fuel to a bulk user would be linked to the calorific value of the fuel; S and V would then not be independent variables, and the fuel might be charged at so many pence per thousand MJ.)

It is desired to determine how far beyond an arbitrarily chosen *datum pressure* it will be economic to go in advancing the steam pressure for which the plant is to be designed. The following data from thermodynamic calculations will be required:

η_D = % station overall efficiency at the datum steam conditions ($5\,\text{MN/m}^2$, 425 °C in the given example),

δ = % reduction in station "heat rate" (as plotted in Fig. C.1) as the result of an advance beyond the datum steam conditions, where the station "heat rate" is the reciprocal of the station overall efficiency.

Then, **per MJ generated**,

Reduction in fuel consumption

$$= \frac{\delta}{\eta_D V} \text{ kg/MJ}.$$

Monetary saving

$$= \frac{S\delta}{\eta_D V} \times 10^{-3} \text{ £/MJ}. \qquad (a)$$

Also, **per MW of installed capacity**,

MJ generated per annum $= 8760 \times 3600 \times \dfrac{L}{100} = 315.4 \times 10^3 \ L.$ \qquad (b)

Monetary saving per annum $=$ (a) $\times$ (b).

The capital value of the saving, $K = \dfrac{(a) \times (b)}{R/100}$,

giving $K = 31\,540\dfrac{m\delta}{\eta_D}$ £/MW. \qquad (C.2)

It will be an economic proposition to increase the design pressure further so long as the capital value of the saving for a given increase in pressure exceeds the additional capital outlay involved on account of the rise in pressure. Hence, if E is the capital cost of the plant, expressed in £ per MW of output, then, at the *economic pressure*,

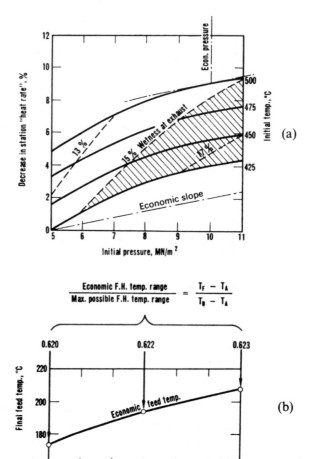

FIG. C.1. The decrease in station "heat rate" for advance in initial steam conditions beyond 5 MN/m², 425 °C. For 50 MW non-reheat plant with 5-stage feed heating and 3.4 kN/m² condenser pressure. [After Baumann, *Proc. I. Mech. E.* **155**, 125 (1946).]

$$\frac{dE}{dp} = \frac{dK}{dp} = 31\,540\,\frac{m}{\eta_D}\frac{d\delta}{dp}.$$

But $d\delta/dp$ is the slope of a graph of δ against p, such as is given in Fig. C.1(a), which is based on data given by Baumann.[13] Hence the economic pressure will be that at which a line at the *economic slope* is a tangent to the curve, where

Economic slope: $\quad \dfrac{d\delta}{dp} = \dfrac{\eta_D}{31\,540\,m}\dfrac{dE}{dp}.$ \hfill (C.3)

In a period of high inflation, numerically quoted values of S and dE/dp would soon be out-of-date. However, as inflation progresses, the values of these two quantities might be expected to keep roughly in step with each other. Consequently, the ratio of these quantities is given in the following example, instead of their respective values.

EXAMPLE. Determine the economic pressure for a non-reheat plant to which Fig. C.1 is applicable, and which is to operate at an initial temperature of 500 °C, given the following data:

$$V = 25 \text{ MJ/kg}, \quad L = 50\%, \quad R = 8\%.$$

$$\eta_D = 30\% \text{ at } 5 \text{ MN/m}^2, 425 \text{ °C}.$$

$\dfrac{dE/dp}{S} = 110$, where dE/dp is expressed in £ per MW of installed capacity per MN/m^2 increase in pressure at 500 °C, and S is expressed in £ per tonne.

From eqn. (C.3) we have, in the above units:

$$\frac{d\delta}{dp} = \frac{\eta_D VR}{31\,540\,L} \frac{dE/dp}{S}$$

$$= \frac{30 \times 25 \times 8}{31\,540 \times 50} \times 110$$

$$= 0.42\% \text{ per MN/m}^2.$$

Hence, from Fig. C.1(a), the economic pressure for such a plant would be about 10 MN/m^2. The values of the economic feed temperature plotted in Fig. C.1(b) were estimated by similar means. It is of interest to note that the economic temperature range of feed heating is a practically constant fraction of the maximum possible range, independently of the steam pressure and temperature; this is a result which stems from the approximate constancy of x_n for a given number of heaters which was established in §7.12. Other features of Fig. C.1 are discussed in Chapter 7.

In practice, more sophisticated accounting methods would be used than are here indicated, but the foregoing calculation serves to draw attention to the importance of distinguishing between theoretical **optimum** steam conditions for maximum efficiency and actual **economic** design conditions.

C.2. Loan redemption—discounted cash flow (DCF) analysis

The foregoing discussion related to a base-load station for which it was assumed that the load factor L would remain essentially constant throughout the life of the plant. However, because plant becomes increasingly obsolete with advancing age, priority in operation is given to more recently constructed plants of higher efficiency. The assumption of constancy of load factor is consequently unrealistic when making calculations relating to loan-redemption payments over the life of a plant. Because economic assessments are just as important as thermodynamic analyses, we turn briefly to a consideration of the factors involved in loan redemption.

Having built a plant at a certain capital cost of £C, and having borrowed a large sum of money in order to do so, one would then be faced with the need to repay the loan, with appropriate interest at a rate of $r\%$ over a period spanning the expected life of the plant. Such repayment would be made in annual instalments with money provided by assessing, on each unit of electricity sold, a constant capital charge of £c towards redemption of the loan. In calculating the charge to be so assessed, one would apply the technique of *Discounted Cash Flow* (DCF). This takes account of the fact that £x_n paid at the end of year n towards loan redemption would not cancel £x_n worth of the original loan, but only the smaller quantity £$x_n/(1 + r/100)^n$. In this context, the following names are given to the respective quantities listed:

$$£x_n = \textit{Cash flow for year } n.$$

$$£x_n/(1 + r/100)^n = \textit{Discounted cash flow for year } n.$$

$$1/(1 + r/100)^n = \textit{Discount factor for year } n \text{ (also described as the}$$
$$\textit{present value of £1 which is due for repayment in } n$$
$$\text{years time).}$$

The quantity £x_n would be the product of the capital charge per unit of electricity sold, namely £c, and the number of units that would be produced in year n with the plant operating at the particular load factor L_n predicted for that year. The value of c would be assessed in the manner next described.

It is clear from the foregoing discussion that, in order to redeem, over a period of N years (the expected life of the plant), a total debt equal to the capital cost £C of the plant, the following expression will have to be satisfied:

$$C = \sum_{n=1}^{n=N} [x_n/(1 + r/100)^n]. \qquad (C.4)$$

Since x_n is a function both of the capital charge c (chosen to be constant) and the load factor L_n, which will vary with n, the determination of the value of c that will satisfy equation (C.4) will need to be made by calculation of the individual terms in the summation of equation (C.4), each term corresponding to a year in the expected life of the plant. It is evident from the form of equation (C.4) that the contributions to the above summation from years in the latter part of the working life of the plant will be relatively small. This is a fact in favour of DCF analysis, because it minimises the financial effect of predictions which relate to years which are furthest into the future and which are therefore least reliable.

Various values of the interest rate r are used in DCF calculations, according to the purpose of the analysis and whether allowance is to be made for the possibility of inflation. In this context, it is important to note that the *apparent* fall in the value of the £1 with time that results from the application of the discount factor is only the result of applying compound interest and is not connected with inflation.

The assessment of economic steam conditions in §C.1 was carried out in terms of pounds of fixed value. In the absence of inflation, the relatively low rate of interest quoted would be acceptable to a reasonable investor. However, in an inflationary period, an investor would expect an appreciably larger return on his capital in order to protect the purchasing power of the money lent. It would then be appropriate to use a noticeably higher rate of interest in the DCF calculations when assessing the amount that purchasers of electricity should be charged during the life of the plant. On top of this, the charge would be adjusted on a year-by-year basis to compensate for rising fuel costs resulting from inflation.

APPENDIX D

Boiler circulation theory

D.1. Introduction

The process of natural circulation in a boiler circuit results from the differential effects of gravity on the denser fluid in the unheated leg of the circuit (the *downcomer*) and the less dense fluid in the heated leg (the *riser*) of a U-tube circuit. That process occurs in the circuits of (unpumped) domestic central-heating plant and, to a much more pronounced extent, in the tubes of a boiler or steam-raising tower, as has already been noted in §§8.15 and 9.22.

This appendix is an updated version of the material contained in a paper written by the present author[FF] at a time when there were conflicting theories on the process of boiler circulation. In papers by Lewis & Robertson (1940), Silver (1946) and Davis (1947), those theories were called respectively *hydraulic*, *thermodynamic* and *expansion* theories, and they all gave different answers. The author succeeded in showing that, when errors in those theories were eliminated, they could be united into a single theory, which the author called a *hydrodynamic* theory of circulation in order to distinguish it from the others. That is also a more appropriate title, since the basic equation derives from mechanical (rather than thermodynamic) considerations, although it can alternatively be derived thermodynamically, as is shown below.

The hydrodynamic theory and the three preceding theories all made the simplifying assumption that the steam/water mixture in the riser moves as a homogeneous fluid, without bubble slip. The effect of the latter was subsequently investigated in an extensive experimental study made by the author and his colleagues on a large, high-pressure rig in the University Engineering Department in Cambridge[GG]. That project was sponsored by the Water Tube Boilermakers' Association and the analysis was later carried further by Thom.[HH]

D.2. Frictional flow of a homogeneous fluid along a heated pipe

It is necessary first to derive the fundamental equation for frictional flow before applying it to the special case of circulation, without bubble slip, in a simple boiler U-tube circuit.

It is instructive to derive the equation first from *thermodynamic* considerations (i.e. from the *energy* equation) and then from *mechanical* considerations (i.e. from the *momentum* equation).

D.2.1. Thermodynamic derivation

The Steady-Flow **Energy** Equation for flow along an element δz of a vertical heated pipe is:

$$\delta Q = \delta h + \delta\left(\frac{V^2}{2}\right) + g\,\delta z = \delta u + p\,\delta v + v\,\delta p + V\,\delta V + g\,\delta z. \quad (D.1)$$

We also have the first $T\,\delta s$ equation for the fluid:

$$T\,\delta s = \delta u + p\,\delta v. \quad (D.2)$$

Frictional flow is irreversible and, from equation (A.8) in Appendix A, the *entropy creation*, δs_c, due to irreversibility is given by:

$$\delta s_c = \delta s - \frac{\delta Q}{T},$$

so that:

$$T\,\delta s = \delta Q + T\,\delta s_c. \quad (D.3)$$

This equation tells us that the increase in entropy δs of the fluid is greater, by an amount δs_c, than that produced by the addition of heat δQ. This excess is occasioned by the dissipation of mechanical energy by friction, so that, if we write

$$T\,\delta s_c \equiv \delta W_f, \quad (D.4)$$

we may describe δW_f as that part of the *available mechanical energy* which is dissipated by friction. Equations (D.2) and (D.3) then give us:

$$\delta u + p\,\delta v = \delta Q + \delta W_f. \quad (D.5)$$

Substituting equation (D.5) in equation (D.1), we arrive at the required basic equation for frictional flow of a homogeneous fluid along a pipe, namely:

$$\boxed{\delta W_f + v\,\delta p + V\,\delta V + g\,\delta z = 0} \quad (D.6)$$

All of these quantities have the dimensions of energy for unit mass. If we divide throughout by g, we have the following alternative form of the basic equation:

$$\frac{\delta W_f}{g} + \frac{v\,\delta p}{g} + \frac{V\,\delta V}{g} + \delta z = 0 \qquad (D.7)$$

All of these quantities now have simply the dimension L, and the hydraulic engineer would describe the first as the *loss of total-head due to friction*, and the others as respectively the changes in *pressure head, kinetic head* and *hydrostatic head*.

It will be noticed that δQ does not appear in equations (D.6) and (D.7), thus indicating that the flow conditions are controlled by *mechanical* considerations, not thermodynamic. However, v (and therefore V) at any point in a heated pipe will be dependent on the heat input.

In order to obtain an expression for dW_f, we have recourse to mechanics, through the Steady-Flow **Momentum** Equation. This leads us to the following alternative derivation of equation (D.6), which is indeed simpler and more direct.

D.2.2. Mechanical derivation

For steady flow through a specified control surface, the *Steady-Flow Momentum Equation* states that:

> *In a given direction, the net applied force exerted on a control volume within a specified control surface is equal to the net rate of efflux of momentum from the control surface in that direction.*

We now apply this theorem to upward flow through a vertical heated pipe of cross-sectional area A, by defining the chain-dotted control surface bounding the elementary control volume depicted in Fig. D.1. The mass of fluid within this control surface if δM. The mass flow rate of fluid through the control surface is $\dot{m}$, the velocity of the fluid at entry is V and the frictional force exerted on the control surface by the pipe wall is δF. We then have:

Net upward applied force on the control volume depicted in Fig. D.1:

$$= [pA - (p + \delta p)A] - g\,\delta M - \delta F$$
$$= -A\,\delta p - g\,\delta M - \delta F. \qquad (D.8)$$

Net rate of efflux of momentum from the control surface:

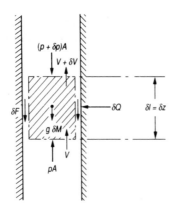

F<small>IG</small>. D.1.

$$= \dot{m}[(V + \delta V) - V] = \dot{m}\,\delta V. \tag{D.9}$$

Equating these two quantities gives:

$$- A\,\delta p - g\,\delta M - \delta F = \dot{m}\,\delta V. \tag{D.10}$$

We also have:

$$\delta M = \rho A\,\delta l = \frac{A\,\delta l}{v} \tag{D.11}$$

and

$$\dot{m} = \rho A V = \frac{A V}{v}. \tag{D.12}$$

Hence, from equations (D.10), (D.11) and (D.12), and writing $\delta l = \delta z$, we obtain the following equation:

$$\frac{v}{A}\frac{\delta F}{g} + \frac{v\,\delta p}{g} + \frac{V\,\delta V}{g} + \delta z = 0. \tag{D.13}$$

It is immediately evident that this equation, which was derived from mechanical considerations (momentum), is exactly analogous to equation (D.7), which was derived from thermodynamic considerations (energy).

In relation to equation (D.7), we noted that the first term, $\delta W_f/g$, represented the hydraulic engineer's *loss of total-head due to friction*, to which we can give the symbol δH_f. We then see, from a comparison of equations (D.7) and (D.13), that δH_f is related to δF through the expression:

$$\delta H_f \equiv \frac{\delta W_f}{g} = \frac{v}{A}\frac{\delta F}{g}. \tag{D.14}$$

D.2.3. Expressions for losses of total-head due to friction

(a) *Pipe-wall friction*

δF is conventionally expressed in terms of a dimensionless friction coefficient c_f, defined as follows:

$$c_f = \frac{\tau_0}{\frac{1}{2}\rho V^2},\qquad(\text{D.15})$$

where τ_0 is the shear stress in the fluid at the pipe wall. Hence, for a section of pipe of length δl and diameter d, it is easy to show that δH_f is given by the following expression, familiar to the hydraulic engineer:

$$\delta H_f \equiv \frac{\delta W_f}{g} - \frac{4 c_f \delta l}{gd}\frac{V^2}{2}.\qquad(\text{D.16})$$

(b) *Other losses of total-head due to friction*

Losses of total-head occur not only due to friction at the walls of a pipe, but also at entry and exit, in pipe bends and in fittings such as valves. Noting that δH_f in equation (D.16) is proportional to $V^2/2$, the specific kinetic energy (i.e. per unit mass), it is common practice to relate other losses of total-head similarly to $V^2/2$, through an experimentally determined frictional *loss coefficient*, K. For these, δH_f is then given by the expression:

$$\delta H_f \equiv \frac{\delta W_f}{g} = K\frac{V^2}{2}.\qquad(\text{D.17})$$

We are now in a position to deduce the hydrodynamic theory of circulation in a simple boiler circuit, and then to demonstrate its application.

D.3. The hydrodynamic theory of circulation

Natural circulation occurs in a simple U-circuit when the densities of the fluid in the two legs of the U are different. In a domestic central-heating circuit, the fluid is single-phase in both legs, but less dense in the warmer leg. In the simple boiler circuit depicted, in Fig. D.2, the difference in density arises principally from the generation of steam in the heated leg.

In order not to obscure the main theme with complicating side issues, and to show as simply as possible the equivalence of the *hydrodynamic* and the *thermodynamic* (or *expansion*) theories of circulation, the following approximating assumptions are made:

(1) The steam/water mixture in the riser moves as a homogeneous fluid, without bubble slip, and there is a uniform increase in specific volume with distance from D to E.

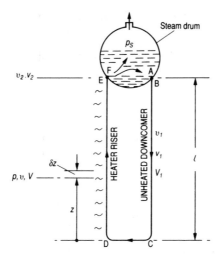

Fig. D.2. Simple U-circuit. [After Haywood[FF]]

(2) Frictional effects from C to D, and in the bends at C and D, are negligible.

(3) $p_A = p_F = p_S$, and the kinetic energy of the fluid is assumed to be negligible at A and F.

The hydrodynamic theory is concerned with the variation of *pressure* (not "head"), the condition to be fulfilled being that, around the circuit:

$$\int_A^F dp = 0.$$ (D.18)

From equation (D.6) we shall have:

$$-\int dp = g\int \frac{dz}{v} + \int \frac{V\,dV}{v} + \int \frac{dW_f}{v}$$ (D.19)

Thus, from equations (D.18) and (D.19), we may write down the *governing equation* for the calculation of the rate of circulation according to the *hydrodynamic* theory, namely:

$$g\int_C^B \frac{dz}{v} - g\int_D^E \frac{dz}{v} = \int_A^F \frac{dW_f}{v} + \int_A^F \frac{V\,dV}{v}$$ (D.20)

All the terms in equation (D.20) represent *pressure*, so that the hydrodynamic theory of circulation was given formal expression by the present author[FF] in the following words:

The rate of circulation adjusts itself until a state of equilibrium has been reached such that the difference between the "hydrostatic" pressures at the feet of the downcomer and riser legs is equal to the sum of the total drops in pressure due to friction and acceleration in the circuit.

"Hydrostatic" pressure is the pressure which would exist at the point in question if there were a stationary column of fluid above the point, whose density varied in the same way as it does under the actual dynamic conditions of circulation.

Integration of the individual terms of equation (D.20) is a simple matter. Those integrations are therefore left for the reader to perform in Problem D.1, in which an expression for the velocity of circulation in the downcomer is obtained.

D.4. The thermodynamic (or expansion) theory of circulation

There was no real difference between the *thermodynamic* theory of Silver (1945) and the *expansion* theory of Davis (1947) for the case when the fluid was assumed to flow as a homogeneous mixture. Basically, the same answer was arrived at by slightly different reasoning, but the work of both was subject to certain errors. Those are now only a matter of history. The corrected theory presented here results from an essentially *thermodynamic* approach to the problem, while at the same time it is very expressive of the physical nature of the process to describe it as an *expansion* theory.

The theory is dependent upon an understanding of the significance of the expansion-work term, $p\, \delta v$, in a frictional, steady-flow process. This may be understood by substituting in equation (D.6) the relation

$$v\, \delta p = \delta(pv) - p\, \delta v \qquad (D.21)$$

giving:

$$\boxed{p\, \delta v = \delta W_f + \delta(pv) + \delta\!\left(\frac{V^2}{2}\right) + g\, \delta z} \qquad (D.22)$$

All the terms in this equation represent *energy* quantities.

For integration around the circuit, equation (D.22) may be written as:

$$\int p\, dv = \int dW_f + \Delta(\text{FE}) + \Delta(\text{KE}) + \Delta(\text{PE}) \qquad (D.23)$$

where FE, KE and PE denote respectively "*flow*" energy, *kinetic* energy and *potential* energy.

Applying equation (D.23) between points A and F of the simple circuit of Fig. D.2, we have:

$$\Delta(FE) = p_s(v_2 - v_1), \quad \Delta(KE = 0) \quad \text{and} \quad \Delta(PE) = 0.$$

At any given point in the circuit, we may write:

$$p = p_s + p',$$

where p' is the local *excess* of pressure above the pressure in the steam drum. Substituting the foregoing expressions in equation (D.23) gives the *governing equation* for the calculation of the rate of circulation according to the *thermodynamic* (or *expansion*) theory, namely:

$$\boxed{\int_A^F p' \, dv = \int_A^F dW_f} \tag{D.24}$$

In an interpretation of this expression, the thermodynamic (or expansion) theory of circulation was given the following formal expression by the present author[FF].

When unit mass of fluid flowing round the circuit expands by an amount δv at a point where the pressure is $(p_s + p')$, it does an amount of work $(p_s + p') \, \delta v$ against the surrounding fluid, but the work done at the boundary of the circuit in consequence of this expansion is only $p_s \, \delta v$.

Hence there is an excess of mechanical work equal to $p' \delta v$ which is available for overcoming friction. This excess of mechanical work may thus be termed the "work available for circulation". Thus, the rate of circulation adjusts itself until a state of equilibrium has been established between the work available for circulation and the mechanical energy dissipated by friction in the circuit.

It is evident that the thermodynamic (or expansion) theory of circulation, as stated formally above, will give the same answer for the rate of circulation as that given by the hydrodynamic theory, since the two theories are derived from the same basic equation [*viz*. equation (D.6)]. Indeed, in order to calculate the rate of circulation, there is no virtue in using the thermodynamic (or expansion) theory, since equation (D.24) shows that it requires a knowledge of the pressure at all points in the circuit at which expansion is taking place, and those pressures must first be calculated by the procedure used in the hydrodynamic theory. The calculation *via* the thermodynamic (or expansion) theory is consequently more complicated and longer. However, it makes an interesting study to apply each theory in turn to the evaluation of the rate of circulation in the simple U-circuit of Fig. D.2. That is left as an exercise for the reader in Problems D.1 and D.2.

The effect of bubble slip certainly influences the rate of circulation in boiler circuits, but a study of that is beyond the scope of the present volume.

Additional Problems

D.1. – Hydrodynamic Theory. For the simple U-circuit depicted in Fig.D.2, making the approximating assumptions listed in §D.3 and assuming that entry and exist losses at A and F respectively may be neglected, derive expressions for each of the four terms of equation (D.20), given the following data:

Tube diameter $\equiv d$, uniform from B to E.

In the riser, expansion ratio $\equiv v_2/v_1 \equiv R$.

Finally, from the *governing equation* of the hydrodynamic theory, namely equation (D.20), show that the velocity of circulation in the downcomer is given by the expression:

$$\frac{V_1^2}{2gl} = \frac{1 - \dfrac{\ln R}{R-1}}{C},$$

where $C \equiv [2c_f\lambda(R + 3) + (R - 1)]$ and $\lambda \equiv l/d$.

Answer: $\dfrac{gl}{v_1}$; $\dfrac{gl}{v_1(R-1)} \ln R$; $c_f\lambda(R + 3) \cdot \dfrac{V_1^2}{v_1}$; $\dfrac{R-1}{2} \cdot \dfrac{V_1^2}{v_1}$.

D.2. – Thermodynamic (or expansion) theory. In order to determine the velocity of circulation V_1, in the downcomer of the simple U-circuit of Fig. D.2 by application of the thermodynamic (or expansion) theory of circulation *via* equation (D.24) in §D.4, it is necessary first to derive an expression for p' by the use of equation (D.19) in §D.3. There, p' is the amount by which the local pressure p at a given point in the circuit exceeds the pressure p_s in the steam drum; namely, $p' \equiv (p - p_s)$. To derive the required expression, it is convenient to write:

$$p = p'' + (p_E - p_S),$$

where $p'' \equiv (p - p_E)$.

(a) By the use of equation (D.19), derive an expression for $(p_E - p_S)$, and show that, in the riser, p'' is given by:

$$p'' = ga\ln\frac{v_2}{v} + G^2(v_2 - v) + \frac{c_f a G^2}{d}(v_2^2 - v^2),$$

where $a \equiv \dfrac{l}{v_1(R - 1)}$

and $G \equiv \dot{m}/A$, the *mass velocity* (namely, mass flow rate per unit cross-sectional area).

(b) Thence derive an expression for $\int_A^F p'\,dv$, the *work available for circulation*.

(c) Taking into account the effects of pipe friction only in the downcomer BC and riser DE of Fig. D.2, derive an expression for the *mechanical energy dissipated by friction*, $\int_A^F dW_f$, where δW_f is given by equation (D.16), namely:

$$\delta W_f = \frac{4c_f\delta l}{d} \cdot \frac{V^2}{2}.$$

(d) Finally, from the *governing equation* of the thermodynamic (or expansion) theory, namely equation (D.24), show that the velocity of circulation in the downcomer is again given by the expression:

$$\frac{V_1^2}{2gl} = \frac{1 - \dfrac{1}{R - 1} \ln R}{C},$$

where $C \equiv [2c_f\lambda(R + 3) + (R - 1)]$ and $\lambda \equiv l/d$.

Answer: (a) $(p_E - p_s) = -\dfrac{R}{v_1} \cdot \dfrac{V_1^2}{2}$

(b) $\displaystyle\int_A^F p'dv = \frac{gl}{v_1}\left(1 - \frac{\ln R}{R - 1}\right) + (R - 1)\left[\frac{2c_f\lambda}{3}(2R + 1) - 1\right] \cdot \frac{V_1^2}{2}$

(c) $\displaystyle\int_A^F dW_f = \frac{4c_f\lambda}{3}(R^2 + R + 4) \cdot \frac{V_1^2}{2}$

APPENDIX E

Solutions to additional problems

Chapter 7—Advanced steam-turbine plant
Problem 7.14—Solution
(a) *Plant without feed heating*

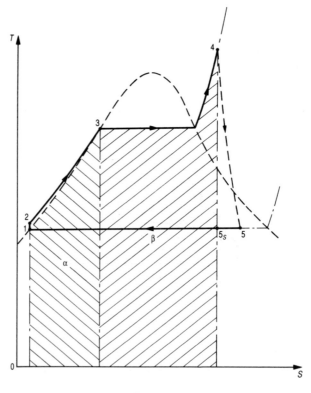

FIG. 7.15.

$$\eta_{CY} \equiv \frac{W_{net}}{Q_{in}} = \left(1 - \frac{Q_{out}}{Q_{in}}\right)$$

$$\therefore \eta_{CY} = \left(1 - \frac{h_5 - h_1}{h_4 - h_2}\right)$$

Now

$$(h_4 - h_2) = \int_2^4 T\,ds = \alpha + \beta$$

$$\therefore \eta_{CY} = \left(1 - \frac{\beta}{\alpha + \beta}\right)$$

$$\therefore \qquad \boxed{\eta_{CY} \equiv \frac{1}{1 + \gamma}} \quad \text{where } \gamma = \beta/\alpha \qquad (1)$$

(b) *Single stage of feed heating*

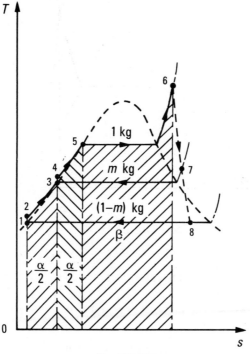

FIG. P.7.14.1.

It is given that, for maximum η_{CY}:

$$\text{Enthalpy rise in heater, } (h_3 - h_2) \approx \frac{\alpha}{2}$$

Let m kg of steam be bled per kg of steam supplied by the boiler. Then the Energy Conservation Equation for the feed heater gives:

$$m(h_7 - h_3) = (1 - m)(h_3 - h_2)$$

$$m\beta = (1 - m)\frac{\alpha}{2}$$

$$\therefore m = \frac{\alpha}{\alpha + 2\beta}$$

$$(1 - m) = \frac{2\beta}{\alpha + 2\beta}$$

$$Q_{in} = (h_6 - h_4) = \frac{\alpha}{2} + \beta = \frac{\alpha + 2\beta}{2}$$

$$Q_{out} = (1 - m)(h_8 - h_1) = (1 - m)\beta = \frac{2\beta^2}{\alpha + 2\beta}$$

$$\eta_{CY} = \left(1 - \frac{Q_{out}}{Q_{in}}\right) = \left[1 - \frac{4\beta^2}{(\alpha + 2\beta)^2}\right] = \frac{\alpha(\alpha + 4\beta)}{(\alpha + 2\beta)^2}$$

$$\boxed{\therefore \eta_{CY} = \frac{1 + 4\gamma}{(1 + 2\beta)^2}} \quad \text{where } \gamma \equiv \beta/\alpha \quad (2)$$

(c) *Feed heater replaced by single calorifier for district heating*

With reference to Figs. P. 7.14.1 and P.7.14.2:

Heat supplied, Q_{in}

$$Q_{in} = (h_6 - h_1) = \alpha + \beta \quad (3)$$

Heat rejected, Q_{out}

(1) *Non-useful heat* rejected in condenser:

$$Q_0 = (1 - m')(h_8 - h_1) = (1 - m')\beta$$

(2) *Useful heat* rejected in calorifier:

$$Q_U = m'(h_7 - h_1) = m'\left(\frac{\alpha}{2} + \beta\right)$$

$$Q_{out} = Q_0 + Q_U = \beta + \frac{m'\alpha}{2}$$

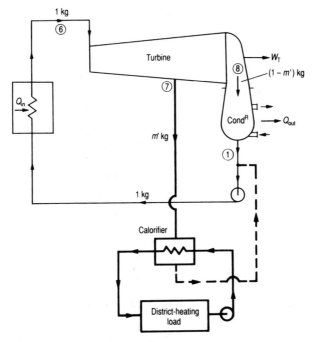

FIG. P.7.14.2.

Net work output, W_{net}

Neglecting pump-work inputs:

$$W_{net} = W_T = (Q_{in} - Q_{out}) = (\alpha + \beta) - \left(\beta + \frac{m'\alpha}{2}\right) = \alpha\left(1 - \frac{m'}{2}\right) \quad (4)$$

Cogeneration coefficient, λ

We now define the *cogeneration coefficient*, λ, as the ratio of the useful heat, Q_U, to the net work, W_{net}.

$$\therefore \lambda \equiv \frac{Q_U}{W_{net}} = \frac{m'\left(\dfrac{\alpha}{2} + \beta\right)}{\alpha\left(1 - \dfrac{m'}{2}\right)} \quad (5)$$

Bled steam flow, m'

From eqn. (5): $\qquad m' = \dfrac{2\alpha\lambda}{\alpha(1 + \lambda) + 2\beta}$

By definition, $\gamma \equiv \beta/\alpha$

$$\therefore m' = \frac{2\lambda}{(1 + \lambda) + 2\gamma} \qquad (6)$$

Work (thermal) efficiency, η_w

By definition, $\eta_w \equiv \eta_{CY} = \dfrac{W_{net}}{Q_{in}}$

From eqns. (4) and (6): $W_{net} = \alpha \left[\dfrac{1 + 2\gamma}{(1 + \lambda) + 2\gamma} \right]$

From eqn. (3): $Q_{in} = (\alpha + \beta) = \alpha(1 + \gamma)$

$$\therefore \boxed{\;\eta_w = \frac{1 + 2\gamma}{1 + \gamma}\, \frac{1}{(1 + \lambda) + 2\gamma}\;} \qquad (7)$$

Total efficiency (EUF), η_{TOT}

By definition, $\eta_{TOT} \equiv \dfrac{W_{net} + Q_U}{Q_{in}} = \dfrac{W_{net}}{Q_{in}} \left[1 + \dfrac{Q_U}{W_{net}} \right] = (1 + \lambda)\eta_w \qquad (8)$

$$\therefore \boxed{\;\eta_{TOT} = \frac{1 + 2\gamma}{1 + \gamma}\, \frac{1 + \lambda}{(1 + \lambda) + 2\gamma}\;} \qquad (9)$$

Numerical calculations for (a) no feed heating, (b) feed heating

(a) *No feed heating*

With reference to Fig. 7.14, it is given that:

$$p_4 = 2 \text{ MN/m}^2 \quad t_4 = 353\,°C \quad p_1 = 7 \text{ kN/m}^2$$

From Ref. 1, Tables 8 and 9:

$h_4 = 3145.6 \text{ kJ/kg}$ $h_3 = 908.6 \text{ kJ/kg}$

$h_3 = \underline{908.6 \text{ kJ/kg}}$ $h_1 = \underline{163.4 \text{ kJ/kg}}$

$\beta = (h_4 - h_3) = 2237.0 \text{ kJ/kg}$ $\alpha = (h_3 - h_1) = 745.2 \text{ kJ/kg}$

$$\therefore \gamma \equiv \frac{\beta}{\alpha} = \frac{2237.0}{745.2} = \textbf{3.00}$$

From eqn. (1), $\eta_{CY} = \dfrac{1}{1 + \gamma} = \dfrac{1}{4} \times 100 = \textbf{25.0\%}$

(b) *Feed heating*

From eqn. (2), $\eta_{CY} = \dfrac{1 + 4\gamma}{(1 + 2\gamma)^2} = \dfrac{13}{49} \times 100 = \textbf{26.5\%}$

(c) *District heating*

    Work (thermal) efficiency, η_w

$$\text{From eqn. (7), } \eta_w = \frac{1 + 2\gamma}{1 + \gamma} \frac{1}{(1 + \lambda) + 2\gamma}$$

$$\text{when } \gamma = 3, \quad \eta_w = \frac{7}{4} \frac{1}{7 + \lambda} \tag{10}$$

    Total efficiency (EUF), η_{TOT}

$$\eta_{\text{TOT}} = (1 + \lambda)\eta_w \qquad [\text{Eqn. (8)}]$$

Hence, from equations (8) and (10), we have:

λ	0	1	2	3	4	5	6
$1 + \lambda$	1	2	3	4	5	6	7
$7 + \lambda$	7	8	9	10	11	12	13
η_w	0.25	0.219	0.194	0.175	0.159	0.146	0.135
η_{TOT}	0.25	0.4375	0.583	0.700	0.795	0.875	0.942

    Turbine isentropic efficiency, η_T (assuming that β is the same at turbine exhaust as at turbine inlet)

$$\text{By definition: } \eta_T \equiv \frac{\Delta h}{\Delta h_s}$$

Hence, with reference to Fig. 7.14 *ante*:

$$\eta_T \equiv \frac{\Delta h}{\Delta h_s} = \frac{h_4 - h_5}{h_4 - h_{5_s}}$$

From Ref. 1, Table 9: $h_4 = 3145.6 \text{ kJ/kg}$

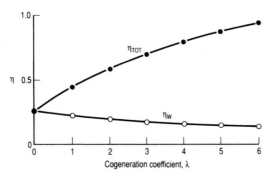

FIG. P.7.14.3.

Using the enthalpy-entropy diagram for steam (Ref. 1):
For isentropic expansion state 4 to $p = 7 \text{ kN/m}^2$

$$h_{5_s} \sim 2167 \text{ kJ/kg}$$

$$\therefore \Delta h_S = (h_4 - h_{5_s}) = \underline{978.6 \text{ kJ/kg}}$$

$$\text{At } 7 \text{ kN/m}^2: \quad h_1 = h_f = 163.4 \text{ kJ/kg}$$

$$\beta = 2237.0 \text{ kJ/kg (as before)}$$

$$h_5 = (h_1 + \beta) = 2400.4 \text{ kJ/kg}$$

$$\Delta h = (h_4 - h_5) = \underline{745.2 \text{ kJ/kg}}$$

$$\therefore \eta_T \equiv \frac{\Delta h}{\Delta h_s} = \frac{745.2}{978.6} \times 100 \approx \mathbf{76\%}$$

This value of η_T would be an underestimate, since β is not exactly constant along a typical turbine expansion line, but is somewhat smaller at exhaust than at inlet. This can be seen from Fig. P.7.14.4, based on Fig. 1 of Ref. 14.

(See also Fig. P.7.3 in the solution to Problem 7.3 in Ref. A.)

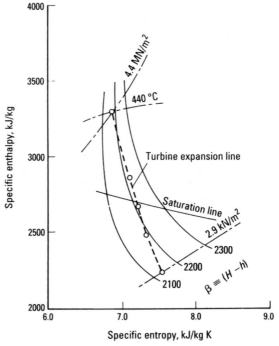

FIG. P.7.14.4. Turbine expansion line and lines of constant β on $h - s$ diagram [After Haywood[14]].

Chapter 9—Combined and binary plant

Problem 9.6—Solution

With reference to Fig. P.9.6:

From Ref. 1, Table 8:

$$h_3 = 3083 \quad \text{kJ/kg}$$
$$h_2 = 908.6 \,\text{kJ/kg}$$
$$(h_3 - h_2) = 2174.4 \,\text{kJ/kg}$$
$$t_2 = 212.4 \,^{\circ}\text{C}$$
$$(t_E - t_2) = 20.0 \,\text{K (given)}$$
$$\therefore t_E = 232.4 \,^{\circ}\text{C}$$
$$t_D = 750 \,\text{K (Problem 4.7)}$$
$$\therefore t_D = 476.85 \,^{\circ}\text{C}$$
$$\therefore (t_D - t_E) = 244.45 \,\text{K}$$

For the gaseous products it is given that

$$c_p = 1.075 \,\text{kJ/kg K} \quad \text{(constant)}$$

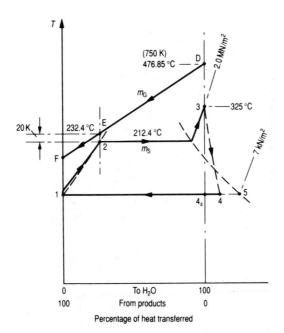

FIG. P.9.6.

Let m_G = mass of gaseous products

    m_S = mass of steam

(a) *Heat exchange between D–E and 2–3*

$$m_G c_p (t_D - t_E) = m_S (h_3 - h_2)$$

$$\therefore \frac{M_G}{M_S} = \frac{(h_3 - h_2)}{c_p(t_D - t_E)} = \frac{2174.4}{1.075 \times 244.45} = \mathbf{8.274}$$

(b) *Heat exchange between D–F and 1–3*

$$h_1 = h_f \text{ at } 7 \text{ kN/m}^2 = 163.4 \text{ K/kg}$$

$$h_3 = \underline{3083} \text{ kJ/kg}$$

$$\therefore (h_3 - h_1) = 2919.6 \text{ kJ/kg}$$

$$m_G c_p (t_D - t_F) = m_S(h_3 - h_1)$$

$$\therefore (t_D - t_F) = \frac{m_S}{m_G} \frac{(h_3 - h_1)}{c_p} = \frac{1}{8.274} \times \frac{2919.6}{1.075} = 328.25$$

$$t_F = 476.85 - 328.25 = \mathbf{148.6 \,°C}$$

(c) *Mass of combustion products per kg of C_8H_{18} burnt*

Per kmol of C_8H_{18} burnt, the products are:

Species	O_2	N_2^*	CO_2	H_2O
kmol	47.5	225.5	8	9
Molar mass, kg/kmol	32	28.15	44	18
Mass, kg	1520	6348	352	162

Total mass = 8382 kg

$\therefore$ *Per kg* of C_8H_{18}:

Mass of combustion products, $m_G = \dfrac{8382}{114} = \mathbf{73.53} \text{ kg}$

(d) *Mass of steam generated per kg of C_8H_{18} burnt*

From (a) above: $m_G/m_S = 8.274$

$$\therefore m_S = \frac{73.53}{8.274} = \mathbf{8.887} \text{ kg}$$

(e) *Heat transferred in the HRSG per kg of C_8H_{18} burnt*

Heat quantity transferred, $Q_S = m_S(h_3 - h_1)$

$$= 8.887 \times 2919.6 \times 10^{-3} = \mathbf{25.95} \text{ MJ}$$

With the respective quantities as defined in §9.7:

(f) $$x \equiv \frac{W_G}{\text{LCV}}, \text{ where } W_G \equiv W_{\text{net}} \text{ in Problem 4.7, part (c)}$$

$$\therefore x = \frac{W_{\text{net}}}{\text{LCV}} = \eta_o \text{ in Problem 4.7, part (d)}$$

$$\therefore x = \mathbf{0.1927}$$

(g) $$\frac{Q_S}{\text{LCV}} = \frac{25.95}{44.43} = \mathbf{0.5841}$$

(h) $$\eta_B = x + \frac{Q_S}{\text{LCV}} = 0.1927 + 0.5841 = \mathbf{0.7768}$$

(i) *Steam cycle*

With reference to Fig. P.9.6:

$$s_3 = 6.868 \text{ kJ/kg K} \quad h_3 = 3083 \text{ kJ/kg} \quad \text{(Ref. 1, Tables 9 \& 10)}$$
$$s_5 = 8.277 \text{ kJ/kg K} \quad h_5 = 2572.6 \text{ kJ/kg} \quad \text{(Ref. 1, Table 8)}$$
$$s_1 = 0.559 \text{ kJ/kg K} \quad h_1 = \quad 163.4 \text{ kJ/kg}$$

At state 4_s, dryness factor $= x_{4_s}$

$$(1 - x_{4_s}) = \frac{s_5 - s_3}{s_5 - s_1} = \frac{1.409}{7.718} = 0.1826$$

$$h_{4_s} = h_5 - (1 - x_{4_s})(h_5 - h_1)$$
$$= 2572.6 - 0.1826\,(2572.6 - 163.4)$$
$$= 2572.6 - 439.9 = 2132.7 \text{ kJ/kg}$$

$$\Delta h_S = (h_3 - h_{4_s}) = 3083 - 2132.7 = 950.3 \text{ kJ/kg}$$
$$\Delta h = \eta_T\,\Delta h_S = 0.82 \times 950.3 = 779.2 \text{ kJ/kg}$$

Neglecting the work input to the feed pump:

$$W_{\text{net}} = \Delta h = 779.2 \text{ kJ/kg}$$
$$Q_{\text{in}} = (h_3 - h_1) = 2919.6 \text{ kJ/kg}$$
$$\eta_S \equiv \frac{W_{\text{net}}}{Q_{\text{in}}} = \frac{779.2}{2919.6} = \mathbf{0.2669} \quad (\mathbf{26.7\%})$$

(j) $\eta_o' = \eta_B\,\eta_S = 0.7768 \times 0.2669 = \mathbf{0.2073} \quad (\mathbf{20.7\%})$

(k) $\eta_o = \eta_o' + x(1 - \eta_S)$

$$= 0.2073 + 0.1927 \times 0.7331$$
$$= 0.2073 + 0.1413 = \mathbf{0.3486} \quad (\mathbf{34.9\%})$$

Note: Without the recuperative HRSG:

$$\eta_o = x = 0.1927 \qquad (19.3\%)$$

Problem 9.7—Solution

Per kmol of C_8H_{18} *supplied to the gas-turbine combustion chamber* (GTCC):

$$\text{Oxygen unconsumed} = 47.5 \text{ kmol} \qquad (\text{Problem 9.6})$$

$$C_8H_{18} + 12.5 \, O_2 = 8 \, CO_2 + 9 \, H_2O$$

$$\therefore \text{Additional } C_8H_{18} \text{ required} = \frac{47.5}{12.5} = 3.8 \text{ kmol}$$

Hence the products of combustion will be:

$$CO_2 = 8 \, (1 + 3.8) = \quad 38.4 \text{ kmol}$$
$$H_2O = 9 \, (1 + 3.8) = \quad 43.2 \text{ kmol}$$
$$N_2^* = 225.5 \text{ kmol} \qquad (\text{Problem 9.6})$$

The processes in the *adiabatic* combustion chamber in which all the excess oxygen is burnt up, and in the *conceptual* equivalent calorimeter, are depicted respectively in Fig. P.9.7.1 and Fig. P.9.7.2.

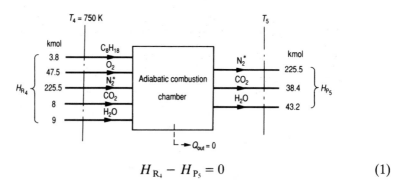

$$H_{R_4} - H_{P_5} = 0 \qquad (1)$$

FIG. P.9.7.1. Actual combustion process.

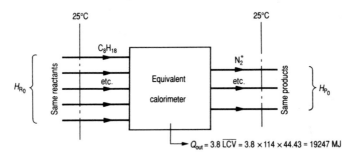

$$H_{R_0} - H_{P_0} = 19247 \text{ MJ} \tag{2}$$

Fig. P.9.7.2. Conceptual equivalent calorimeter process.

Subtracting eqn. (1) from eqn. (2):

$$(H_{P_5} - H_{P_0}) - (H_{R_4} - H_{R_0}) = 19247 \tag{3}$$

Calculation of $(H_{R_4} - H_{R_0})$

Species	n kmol	At 750 K $\bar{h}_4$ MJ/kmol	At 25 °C $\bar{h}_0$ MJ/kmol	$\bar{h}_4 - \bar{h}_0$ MJ/kmol	$H_4 - H_0$ MJ
O_2	47.5	22.83	8.66	14.17	673
N_2^*	225.5	22.17	8.67	13.50	3044
CO_2	8	29.65	9.37	20.28	162
H_2O	9	26.00	9.90	16.10	145

$$\therefore (H_{R_4} - H_{R_0}) = 4024 \text{ MJ}$$

Calculation of $(H_{P_5} - H_{P_0})$, taking $T_5 = 2300$ K (given)

Species	n kmol	At 2300 K $\bar{h}_5$ MJ/kmol	At 25 °C $\bar{h}_0$ MJ/kmol	$\bar{h}_5 - \bar{h}_0$ MJ/kmol	$H_5 - H_0$ MJ
N_2^*	225.5	75.70	8.67	67.03	15115
CO_2	38.4	119.28	9.37	109.91	4220
H_2O	43.2	98.27	9.90	88.37	3818

$$(H_{R_5} - H_{R_0}) = 23153 \text{ MJ}$$

Hence, from eqn. (3):

$$23153 - 4024 \approx 19247$$

$$19129 \approx 19247$$

Hence $T_5 \approx \mathbf{2300}$ K

Problem 9.8—Solution

The hypothetical combined-cycle (binary) plant is depicted in Fig. 9.23 in Chapter 9. That diagram is reproduced here.

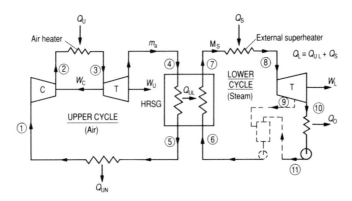

FIG. 9.23. Combined-cycle (binary) plant.

The following calculations are based on **1 kg of air** circulating around the upper cycle.

UPPER CYCLE (AIR)

Air compressor

Isentropic temperature ratio, $\rho_p = r_p^{(\gamma-1)/\gamma} = 17^{1/3.5} = 2.247$

$$(T_{2_s} - T_1) = 1.247 \times 288.15 = 359.3\ \text{K}$$

$$(T_2 - T_1) = \frac{T_{2_s} - T_1}{\eta_C} = \frac{359.3}{0.85} = 422.7\ \text{K}$$

Work input, $W_C = c_p\,(T_2 - T_1) = 1.01 \times 422.7 = \underline{426.9}\ \text{kJ}$

Air heater

$$t_1 = 15\,°\text{C (given)} \qquad t_3 = 1000\,°\text{C (given)}$$

$$t_2 = \quad 15 + 422.7 = 437.7\,°\text{C}$$

$$\therefore (t_3 - t_2) = 1000 - 437.7 = 562.3\ \text{K}$$

$$Q_U = c_p\,(t_3 - t_2) = 1.01 \times 562.3 = \underline{567.9}\ \text{kJ}$$

Air turbine

$$(T_3 - T_{4_s}) = T_3\left(1 - \frac{1}{\rho_p}\right) = 1273.15\left(1 - \frac{1}{2.247}\right) = 706.6\ \text{K}$$

$$(T_3 - T_4) = \eta_T\,(T_3 - T_{4_s}) = 0.87 \times 706.6 = 614.7\ \text{K}$$

$$\therefore t_4 = 1000 - 614.7 = 385.3\,°\text{C}$$

Work output $W_T = c_p (T_3 - T_4) = 1.01 \times 614.7 = \underline{620.8 \text{ kJ}}$

Thermal efficiency of upper cycle, η_U

$$W_U = W_T - W_C = 620.8 - 426.9 = 193.9 \text{ kJ}$$

$$\eta_U = \frac{W_U}{Q_U} = \frac{193.9}{567.9} \times 100 = \mathbf{34.14\%}$$

HEAT RECOVERY STEAM GENERATOR (HRSG)

With reference to Fig. P.9.8.1:

For the air:

At the pinch point:

$$t_E - t_{12} = 22 \text{ K} \quad \text{(given)}$$

$$t_{12} = t_{\text{sat.}} \text{ at } 6.8 \text{ MN/m}^2 = 283.8 \,^\circ\text{C}$$

$$\therefore t_E = 305.8 \,^\circ\text{C}$$

For the steam:

At 6.8 MN/m^2, $325\,^\circ\text{C}$

$$h_7 = 2942 \text{ kJ/kg} \quad \text{(Ref. 1, Table 9)}$$

$$h_{12} = h_f \text{ at } 6.8 \text{ MN/m}^2$$

$$\therefore h_{12} = 1257.0 \text{ kJ/kg} \quad \text{(Ref. 1, Table 8)}$$

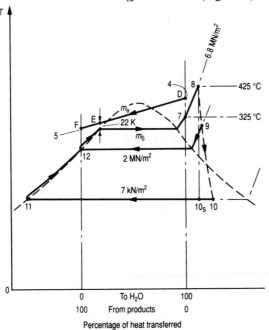

FIG. P.9.8.1.

Heat exchange between D–E and 12–7

$$m_a c_p (t_4 - t_E) = m_S (h_7 - h_{12})$$

$$\frac{m_S}{m_a} = \frac{c_p (t_4 - t_E)}{h_7 - h_{12}} \tag{1}$$

Hence, for $m_a = 1$ kg:

$$m_S = \frac{1.01 (385.3 - 305.8)}{(2942 - 1257)} = \frac{1.01 \times 79.5}{1685} = \underline{0.04765} \text{ kg}$$

LOWER CYCLE (STEAM)

With reference to Fig. P.9.8.1:

Steam turbine

At 6.8 MN/m², 425 °C: $s_8 = 6.569$ kJ/kg K $h_8 = 3229$ kJ/kg

At 7 kN/m²: $s_g = 8.277$ kJ/kg K $h_g = 2572.6$ kJ/kg

$s_f = 0.559$ kJ/kg K $h_f = 163.4$ kJ/kg

Let x_s = dryness fraction at exhaust after isentropic expansion from state 8.

$$(1 - x_S) = \frac{s_g - s_8}{s_g - s_f} = \frac{8.277 - 6.569}{8.277 - 0.559} = \frac{1.708}{7.718} = 0.2213$$

$$h_{10_s} = h_g - (1 - x_s)(h_g - h_f)$$

$$= 2572.6 - 0.2213 \times 2409.2 = 2039.4 \text{ kJ/kg}$$

$$\Delta h_s = (h_8 - h_{10_s}) = 3229 - 2039.4 = 1189.6 \text{ kJ/kg}$$

$$\Delta h = \eta_T \Delta h_s = 0.87 \times 1189.6 = 1035.0 \text{ kJ/kg}$$

$$h_{10} = h_8 - \Delta h = 3229 - 1035.0 = 2194.0 \text{ kJ/kg}$$

(a) FEED HEATING BY BLED STEAM

With reference to Fig. P.9.8.2:

It is given that $p_9 = 2$ MN/m²

and $h_9 = 2960$ kJ/kg;

also that $t_6 = t_{\text{sat.}}$ at 2 MN/m²

$\therefore t_6 = 212.4$ °C   (Ref. 1, Table 8)

Neglecting enthalpy rises in the pumps:

$$h_{11} = h_f \text{ at } 7 \text{ kN/m}^2 = 163.4 \text{ kJ/kg}$$

$$h_6 = h_f \text{ at } 2 \text{ MN/m}^2 = 908.6 \text{ kJ/kg}$$

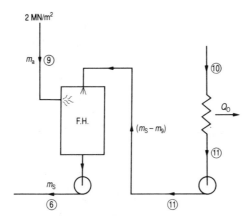

FIG. P.9.8.2.

The Steady-Flow Energy Equation for the heater is:

$$m_9 h_9 + (m_S - m_9)h_{11} = m_S h_6$$

$$\therefore \frac{m_9}{m_S} = \frac{h_6 - h_{11}}{h_9 - h_{11}} = \frac{745.2}{2796.6} = 0.2665$$

Per kg of air circulating in the upper cycle:

$$m_S = 0.04765 \text{ kg}$$

$$\therefore m_9 = 0.04765 \times 0.2665 = 0.01270 \text{ kg}$$

$$\therefore (m_S - m_9) = 0.03495 \text{ kg}$$

Thermal efficiency of lower cycle, η_L

$$\eta_L = \frac{W_L}{Q_L}$$

where $W_L = Q_L - Q_0$

and $Q_L = Q_{UL} + Q_S = m_S (h_8 - h_6)$

$$\therefore Q_L = 0.04765 (3229 - 908.6) = 110.57 \text{ kJ}$$

$$Q_0 = (m_S - m_9) (h_{10} - h_{11})$$

$$= 0.03495 (2194 - 163.4) = \underline{70.97 \text{ kJ}}$$

$$\therefore W_L = (Q_L - Q_0) \qquad = \underline{39.60 \text{ kJ}}$$

$$\therefore \eta_L = \frac{39.60}{110.57} \times 100 \qquad = \mathbf{35.81\%}$$

Check on W_L:

$$W_L = m_S(h_8 - h_9) + (m_S - m_9)(h_9 - h_{10})$$

$$m_S(h_8 - h_9) = 0.04765(3229 - 2960) = 12.82 \text{ kJ}$$

$$(m_S - m_9)(h_9 - h_{10}) = 0.03495(2960 - 2194) = \underline{26.77} \text{ kJ}$$

$$\therefore W_L = \underline{39.59} \text{ kJ}$$

Thermal efficiency of combined cycle, η_{COMB}.

$$\text{Weighting factor,} \quad \mu_U \equiv \frac{Q_U}{Q_U + Q_S}$$

$$Q_U = 567.9 \text{ kJ}$$

$$Q_S = m_S(h_8 - h_7) = 0.4765(3229 - 2942) = 13.7 \text{ kJ}$$

$$(Q_U + Q_S) = 567.9 + 13.7 = 581.6 \text{ kJ}$$

$$\therefore \mu_U = \frac{567.9}{581.6} = \textbf{0.9764}$$

$$\text{Weighting factor,} \quad \mu_L \equiv \frac{Q_L}{Q_U + Q_S}$$

$$\text{where} \quad Q_L = Q_{UL} + Q_S = 110.57 \text{ kJ}$$

$$\therefore \mu_L = \frac{110.57}{581.6} = \textbf{0.1901}$$

$$\eta_{\text{COMB}} = \mu_U \eta_U + \mu_L \eta_L \qquad (9.35)$$

$$= 0.9764 \times 34.14 + 0.1901 \times 35.81$$

$$= 33.33 + 6.81$$

$$\therefore \eta_{\text{COMB}} = \textbf{40.1\%}$$

Direct check on η_{COMB}.

$$\eta_{\text{COMB}} = \frac{W_U + W_L}{Q_U + Q_S}$$

$$= \frac{193.9 + 39.6}{581.6} \times 100 = \textbf{40.1\%}$$

(b) FEED HEATING BY SOLAR HEATING

With feed heating by solar heating to the same final feed temperature t_6 (Fig. P.9.8.1), the ratio m_S/m_a will be the same as in Case (a) since it is still governed by the pinch point [eqn. (1) *ante*].

Hence, **per kg of air** circulating around the upper cycle of Fig. 9.23:

m_S, Q_U, W_U, Q_{UL}, Q_S, Q_L, μ_U and μ_L are all unaltered, but:

$$W'_L = m_S(h_8 - h_{10}) = 0.04765\,(3229 - 2194) = 49.32\text{ kJ}$$

$$Q'_0 = m_S(h_{10} - h_{11}) = 0.04765\,(2194 - 163.4) = 96.76\text{ kJ}$$

Upper cycle

$$\eta'_U = \eta_U = \textbf{34.14\%} \qquad \mu'_U = \mu_U = \textbf{0.9764}$$

Lower cycle

$$Q'_L = Q_L = 110.57\text{ kJ}$$

$$\therefore \quad \eta'_L = \frac{W'_L}{Q'_L} = \frac{49.32}{110.57} \times 100 = \textbf{44.61\%}$$

$$\mu'_L = \mu_L = \textbf{0.1901}$$

Combined cycle

$$\eta'_{\text{COMB.}} = \mu'_U\eta'_U + \mu'_L\eta'_L$$

$$= 0.9764 \times 34.14 + 0.1901 \times 44.61$$

$$= 33.33 + 8.48$$

$$\therefore \ \eta'_{\text{COMB.}} = \textbf{41.8\%}$$

> *Direct check on* $\eta'_{\text{COMB.}}$
>
> $$\eta'_{\text{COMB.}} = \frac{W'_U + W'_L}{Q'_U + Q'_S} = \frac{W_U + W'_L}{Q_U + Q_S}$$
>
> $$= \frac{193.9 + 49.3}{581.6} \times 100 = \textbf{41.8\%}$$

(c) NO FEED HEATING

With reference to Fig. P.9.8.1, and neglecting the enthalpy rise of the feed water in the feed pump, the feed water will now be fed to the HRSG at state 11, namely as saturated water at 7 kN/m^2.

The ratio m_S/m_a will be the same as in Case (a), since it will still be governed by the pinch point [eqn. (1) *ante*].

Hence, **per kg of air** circulating around the upper cycle:

m_S, Q_U, W_U, Q_S and μ_U are all unaltered, but:

$$Q''_L = m_S(h_8 - h_{11})$$

$$= 0.04765\,(3229 - 163.4) = 146.1\text{ kJ}$$

Upper cycle

$$\eta''_U = \eta_U = \textbf{34.14\%} \qquad \mu''_U = \mu_U = \textbf{0.9764}$$

Lower cycle

$$W''_L = W'_L = 49.32\,\text{kJ}$$

$$\eta''_L = \frac{W''_L}{Q''_L} = \frac{49.32}{146.1} \times 100 = \textbf{33.76\%}$$

$$\mu''_L = \frac{Q''_L}{Q_U + Q_S} = \frac{146.1}{581.6} = \textbf{0.2512}$$

Combined cycle

$$\eta''_{\text{COMB.}} = \mu''_U \eta''_U + \mu''_L \eta''_L$$

$$= 0.9764 \times 34.14 + 0.2512 \times 33.76$$

$$= 33.33 + 8.48 = \textbf{41.8\%}$$

$$
\left[
\begin{array}{l}
\textit{Direct check on } \eta''_{\text{COMB.}} \\[2mm]
\eta''_{\text{COMB.}} = \dfrac{W_U + W''_L}{Q_U + Q_S} = \dfrac{193.9 + 49.3}{581.6} \times 100 = \textbf{41.8\%}
\end{array}
\right]
$$

TEMPERATURE T_5 OF THE AIR LEAVING THE HRSG

(a) and (b) — When feed heating to temperature t_6

The Energy Conservation Equation for flow through the HRSG gives:

$$m_a C_p (t_4 - t_5) = m_S (h_7 - h_6)$$

$$\therefore (t_4 - t_5) = \frac{m_S}{m_a} \frac{(h_7 - h_6)}{C_p}$$

$$= \frac{0.04765\,(2942 - 908.6)}{1.01} = 95.9\,\text{K}$$

$$\therefore t_5 = 385.3 - 95.9 = \textbf{289.4\,\textdegree C}$$

(c) — When no feed heating

The Energy Conservation Equation for flow through the HRSG gives:

$$m_a c_p (t_4 - t''_5) = m_S (h_7 - h_{11})$$

$$\therefore (t_4 - t''_5) = \frac{m_S}{m_a} \frac{(h_7 - h_{11})}{c_p}$$

$$= 0.04765\,(2942 - 163.4) = 132.4\,\text{K}$$

$$\therefore t''_5 = 385.3 - 132.4 = \textbf{252.9\,\textdegree C}$$

COLLECTED RESULTS

Case	(a)	(b)	(c)
η_U	34.1%	34.1%	34.1%
μ_U	0.9764	0.9764	0.9764
η_L	35.8%	44.6%	33.8%
μ_L	0.1901	0.1901	0.2512
$\eta_{COMB.}$	40.1%	41.8%	41.8%
t_5	289.4 °C	289.4 °C	252.9 °C

Discussion

The tabulated values of $\eta_{COMB.}$ demonstrate that, not only would bled-steam feed heating have an adverse effect on $\eta_{COMB.}$ (see §9.14), but there would also be no advantage in preheating by solar heating, since $\eta_{COMB.}$ is the same in Case (b) as in Case (c).

Problem 9.9—Solution

The cogeneration combined gas/steam plant, with provision for supplementary firing, is depicted in Fig. 9.24 in Chapter 9. That diagram is reproduced here. We study first the performance of the plant in the absence of supplementary firing, and then when additional fuel is burnt in the heat recovery steam generator (HRSG) to provide an increased supply of process steam at times of peak demand.

GAS TURBINE PLANT

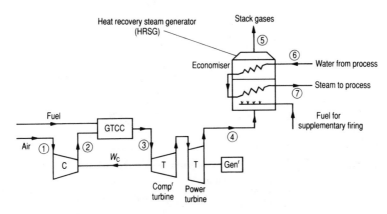

FIG. 9.24. Gas/steam cogeneration plant for power and process steam.

(a) *Compressor*

Pressure ratio, $r_p = 7$ (given)

$$\frac{T_{2_s}}{T_1} = r_p^{(\gamma-1)/\gamma} = 7^{1/3.5} = 1.7436$$

$$(T_{2_s} - T_1) = 0.7436 \, T_1 \qquad T_1 = 25 + 273.15 = 298.15 \text{ K}$$

$$(T_2 - T_1) = \frac{T_{2_s} - T_1}{\eta_C} = \frac{0.7436 \times 298.15}{0.85} = 260.8 \text{ K}$$

$$t_2 = 25 + 260.8 = \mathbf{285.8} \, ^\circ\text{C} \qquad T_2 = 559.0 \text{ K}$$

GAS-TURBINE COMBUSTION CHAMBER (GTCC)

(b) *Air/fuel ratio and excess air*

The stoichiometric equation for the combustion of natural gas, treated as methane (CH_4), is:

$$CH_4 + 2O_2 = CO_2 + 2H_2O$$

Per kmol of CH_4 burnt:

Let the air supplied $= a$ kmol
Then the products of combustion are:

$$CO_2 = 1 \text{ kmol}$$
$$H_2O = 2 \text{ kmol}$$
$$N_2^* = 0.79a \text{ kmol}$$
$$O_2 = (0.21a - 2) \text{ kmol}$$

We now write down the Steady-Flow Energy Equation for (1) the actual process in which $t_3 = 850 \, ^\circ\text{C}$ (given), and (2) the conceptual *equivalent calorimeter process*.

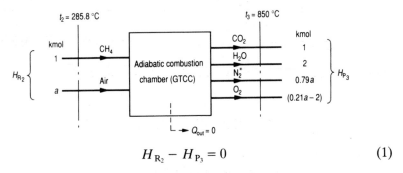

$$H_{R_2} - H_{P_3} = 0 \qquad\qquad (1)$$

Fig. P.9.9.1. Actual combustion process in the GTCC.

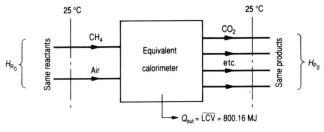

$$H_{R_0} - H_{P_0} = 800.16 \, \text{MJ} \tag{2}$$

Fig. P.9.9.2. Conceptual equivalent calorimeter process.

Subtracting eqn. (1) from eqn. (2):

$$(H_{P_3} - H_{P_0}) - (H_{R_2} - H_{R_0}) = Q_{\text{out}} = 800.16 \, \text{MJ} \tag{3}$$

Calculation of $(H_{R_2} - H_{R_0})$

Assuming that c_p of each species is constant over the temperature range from 25 °C to 285.8 °C, and equal to the values given in Ref. 1, Table 2:

$$(H_{R_2} - H_{R_0}) = (16 \times 2.23 + 29a \times 1.01) \times 260.8 \times 10^{-3}$$

$$= (9.31 + 7.64a) \, \text{MJ}$$

Calculation of $(H_{P_3} - H_{P_0})$

Using the given values of the molar enthaplies:

Species	n kmol	At 850 °C $\bar{h}_3$ MJ/kmol	At 25 °C $\bar{h}_0$ MJ/kmol	$\bar{h}_3 - \bar{h}_0$ MJ/kmol	$H_3 - H_0$ MJ
CO_2	1	49.57	9.37	40.20	40.20
H_2O	2	41.09	9.90	31.19	62.38
N_2^*	$0.79a$	34.22	8.67	25.55	$20.18a$
O_2	$(0.21a - 2)$	35.70	8.66	27.04	$5.68a - 54.08$

$$\sum n = (a + 1) \qquad (H_{P_3} - H_{P_0}) = (25.86a + 48.50) \, \text{MJ}$$

Substituting for $(H_{P_3} - H_{P_0})$ and $(H_{R_2} - H_{R_0})$ in eqn. (3) gives:

$$(25.86a + 48.50) - (9.31 + 7.64a) = 800.16$$

$$\therefore a = \frac{760.97}{18.22} = \mathbf{41.77} \, \textbf{kmol}$$

$$\text{Stoichiometric air} = \frac{2}{6.21} = 9.52 \, \text{kmol}$$

$$\therefore \textit{ Percentage of excess air } = \frac{32.25}{9.52} \times 100 = \mathbf{339\%}$$

(c) *Percentage of oxygen in the products*

From the gas-turbine combustion chamber (GTCC):

$$\text{Products of combustion} = (a + 1) \qquad = 42.77 \text{ kmol}$$

$$\text{Oxygen} = (0.21a - 2) = 6.77 \text{ kmol}$$

$$\therefore \textit{ Percentage of } O_2 \textit{ in the products } = \frac{6.77}{42.77} \times 100 = \mathbf{15.8\%}$$

(d) *Work quantities per kmol of air supplied*

(1) *Compressor*

$$W_C = m c_p (t_2 - t_1)$$

$$= 29.0 \times 1.01 \times 10^{-3} (285.8 - 25.0) = \mathbf{7.64 \text{ MJ}}$$

(2) *Turbines*

Per kmol of CH_4 burnt, the products of combustion are:

CO_2 = 1 kmol
H_2O = 2 kmol
N_2^* = 0.79a = 0.79 × 41.77 = 33.00 kmol
O_2 = (0.21a − 2) = 6.772 kmol

Work output from the two turbines = $(H_{P_3} - H_{P_4})$,
where $t_3 = 850\,°$ and $t_4 = 490\,°C$ (given)

Calculation of $(H_{P_3} - H_{P_4})$ per kmol of CH_4 supplied.

Species	n	At 850 °C	At 25 °C	$\bar{h}_3 - \bar{h}_0$	$H_3 - H_0$
	kmol	$\bar{h}_3$	$\bar{h}_0$	MJ/kmol	MJ
		MJ/kmol	MJ/kmol		
CO_2	1	49.57	30.32	19.25	19.25
H_2O	2	41.09	26.50	14.59	29.18
N_2^*	33.00	34.22	22.57	11.65	384.45
O_2	6.77	35.70	23.27	12.43	84.15

$$(H_{P_3} - H_{P_4}) = 517.03 \text{ MJ/kmol } CH_4$$

a kmol of air are supplied per kmol of CH_4 burnt.

$\therefore$ **Per kmol of air supplied**, the work output W_T of the two turbines is:

$$W_T = \frac{(H_{P_3} - H_{P_4})}{a} = \frac{517.03}{41.77} = \mathbf{12.38 \text{ MJ}}$$

(3) *Gas-turbine plant*

Per kmol of air supplied:

Net output, $W_G = W_T - W_C = (12.38 - 7.64) = $ **4.74** MJ

(e) *Net power output and fuel supply rate*

Air supply rate $= 20.45$ kg/s (given)

$$\therefore \dot{n}_a = \frac{20.45}{29.0} = 0.7052 \text{ kmol/s}$$

Power output $= \dot{n}_a W_G = 0.7052 \times 4.74 = $ **3.34** MW

Molar ratio of air to fuel $= a = 41.77$

$$\therefore \text{Mass ratio of air to fuel} = \frac{41.77 \times 29.0}{1 \times 16} = 75.71$$

$$\therefore \textit{Rate of fuel supply}, \dot{m}_f = \frac{20.45}{75.71} = \textbf{0.2701} \text{ kg/s}$$

(f) *Overall efficiency of the gas-turbine plant*

Per kmol of fuel supplied:

Net work output, $W_{net} = a W_G = 41.77 \times 4.74 = 198.0$ MJ

Lower calorific value, LCV $= 50.01$ MJ/kg   (Ref. 1, Table 1)

$$\therefore \overline{\text{LCV}} = 50.01 \times 16 = 800.16 \text{ MJ/kmol}$$

$$\therefore \textit{Overall efficiency}, \eta_o \equiv \frac{W_{net}}{\overline{\text{LCV}}} = \frac{198.0}{800.16} \times 100 = \textbf{24.75}\%$$

NO SUPPLEMENTARY FIRING IN THE HRSG

(g) *Role of heat transfer in the HRSG*

Inlet temperature of the products, $t_4 = 490\,°C$ (given)

Exit temperature of the products, $t_5 = 138\,°C$ (given)

Per kmol of fuel supplied to the GTCC

Species	n	At 490 °C $\bar{h}_4$	At 138 °C $\bar{h}_5$	$\bar{h}_3 - \bar{h}_0$	$H_3 - H_0$
	kmol	MJ/kmol	MJ/kmol	MJ/kmol	MJ
CO_2	1	30.32	13.85	16.47	16.47
H_2O	2	26.50	13.74	12.76	25.52
N_2^*	33.00	22.57	11.97	10.60	349.80
O_2	6.77	23.27	12.02	11.25	76.16

$$(H_{P_4} - H_{P_5}) = 468.0 \text{ MJ/kmol } CH_4$$

From Section (e), rate of fuel supply, $\dot{m}_f = 0.2701$ kg/s

$\therefore$ *Rate of heat transfer in the HRSG,* $\dot{Q}_U = \dfrac{\dot{m}_f}{M_f}(H_{P_4} - H_{P_5})$

$$= \frac{0.2701}{16} \times 468.0 = \textbf{7.90 MW}$$

(h) *Rate of steam delivery from the HRSG*

With reference to Fig. 9.23 *ante* and Fig. P.9.9.3:

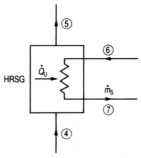

FIG. P.9.9.3.

Dry sat. steam at 1.3 $\mathrm{MN/m^2}$: $h_7 = 2785.4$ kJ/kg

Return water at 90 °C : $h_6 = \underline{376.9}$ kJ/kg

$(h_7 - h_6) = 2408.5$ kJ/kg

Rate of steam supply, $\dot{m}_S = \dfrac{\dot{Q}_U}{h_7 - h_6}$

$$= \frac{7.90 \times 10^3}{2408.5} = 3.280 \text{ kg/s}$$

1 tonne $\equiv 10^3$ kg

$\therefore \dot{m}_S = 3.280 \times 10^{-3} \times 3600 = \textbf{11.81 t/h}$

(i) *Total efficiency,* $(\eta_o)_{\text{TOT.}} - (Energy\ Utilisation\ Factor,\ EUF)$
   Per kmol of fuel burnt:

Net work output, $W_{\text{net}} = a\,W_G = 41.77 \times 4.74 = 198.0$ MJ

Energy to process plant $= Q_U = (H_{P_4} - H_{P_5}) = \underline{468.0}$ MJ

$(W_{\text{net}} + Q_U) = 666.0$ MJ

$\therefore$ *Total efficiency* (EUF) $\equiv (\eta_o)_{\text{TOT.}} = \dfrac{W_{\text{net}} + Q_U}{\text{LCV}}$

$$\therefore (\eta_o)_{\text{TOT.}} = \frac{666.0}{800.16} \times 100 = \textbf{83.2\%}$$

SUPPLEMENTARY FIRING IN THE HRSG

(j) *Rate of fuel supply to HRSG, and percentage of oxygen in final products*

Per kg of fuel supplied to the GTCC:

For $\dot{m}_S = 11.81$ t/h, $\quad Q_U = (H_{P_4} - H_{P_5}) \quad = 468.0$ MJ

$\therefore$ For $\dot{m}_5 = 35$ t/h, $\quad Q'_U = 468.0 \times \dfrac{35}{11.81} = 1387.0$ MJ

The products of combustion entering the HRSG are the same as before. Hence, with n'_f kmol of CH_4 supplied to the HRSG per kmol of CH_4 supplied to the GTCC, the actual combustion process in the HRSG and the conceptual equivalent calorimeter process will be as depicted respectively in Fig. P.9.9.4 and Fig. P.9.9.5.

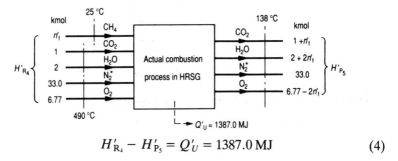

$$H'_{R_4} - H'_{P_5} = Q'_U = 1387.0 \text{ MJ} \tag{4}$$

FIG. P.9.9.4. Actual combustion process in the HRSG.

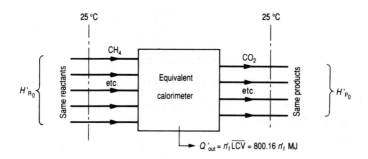

$$H'_{R_0} - H'_{P_0} = Q'_{out} = 800.16 \, n'_f \text{ MJ} \tag{5}$$

FIG. P.9.9.5. Conceptual equivalent calorimeter process.

Subtracting eqn. (5) from eqn. (4):

$$(H'_{R_4} - H'_{R_0}) - (H'_{P_5} - H'_{P_0}) = 1387.0 - 800.16 \, n'_f \tag{6}$$

Calculation of $(H'_{R_4} - H'_{R_0})$

Species	n	At 490 °C	At 25 °C	$\bar{h}_4 - \bar{h}_0$	$H'_4 - H'_0$
	kmol	$\bar{h}_4$	$\bar{h}_0$	MJ/kmol	MJ
		MJ/kmol	MJ/kmol		
CO_2	1	30.32	9.37	20.95	20.95
H_2O	2	26.50	9.90	16.60	33.20
N_2^*	33.00	22.57	8.67	13.90	458.70
O_2	6.77	23.27	8.66	14.61	98.91

$$(H'_{R_4} - H'_{R_0}) = 611.76 \text{ MJ}$$

Calculation of $(H'_{P_5} - H'_{P_0})$

Species	n	At 138 °C	At 25 °C	$\bar{h}_5 - \bar{h}_0$	$H'_5 - H'_0$
	kmol	$\bar{h}_5$	$\bar{h}_0$	MJ/kmol	MJ
		MJ/kmol	MJ/kmol		
CO_2	$1 + n'_f$	13.85	9.37	4.48	$4.48 + 4.48n'_f$
H_2O	$2 + 2n'_f$	13.74	9.90	3.84	$7.68 + 7.68n'_f$
N_2^*	33.0	11.97	8.67	3.30	108.90
O_2	$6.77 - 2n'_f$	12.02	8.66	3.36	$22.75 - 6.72n'_f$

$$(H'_{P_5} - H'_{P_0}) = 143.81 + 5.44n'_f$$

Substituting for $(H'_{R_4} - H'_{R_0})$ and $(H'_{P_5} - H'_{P_0})$ in eqn. (6) gives:

$$611.76 - (143.81 + 5.44\,n'_f) = 1387.0 - 800.16\,n'_f$$

$$\therefore n'_f = \frac{919.05}{794.72} = 1.156$$

From Section (e), fuel supply rate to GTCC = 0.2701 kg/s

$\therefore$ *Fuel supply rate to HRSG* $= 0.2701\,n'_f = 0.2701 \times 1.156 =$ **0.312** kg/s

Final products leaving the HRSG

$$CO_2 = (1 + n'_f) = 2.156 \text{ kmol}$$

$$H_2O = (2 + 2n'_f) = 4.312 \text{ kmol}$$

$$N_2^* = 33.00 \text{ kmol}$$

$$O_2 = (6.772 - 2n'_f) = \underline{4.46} \text{ kmol}$$

$$\text{Total products} = 43.93 \text{ kmol}$$

$\therefore$ Percentage of O_2 in the final products $= \dfrac{4.46}{43.93} \times 100 = \mathbf{10.2\%}$

(k) *Total efficiency*, $(\eta_o)_{TOT.} - (Energy \ Utilisation \ Factor, \ EUF)$

Per kmol of fuel supplied to the GTCC:

$$W_{net} = 198.0 \ \text{MJ} \qquad \text{[Section (i)]}$$

$$\text{Energy to process plant} = Q'_U = \underline{1387.0} \ \text{MJ} \qquad \text{[Section (j)]}$$

$$(W_{net} + Q'_U) = 1585.0 \ \text{MJ}$$

$$\text{Total fuel} = \eta_f + \eta'_f = 1 + 1.156 = 2.156 \ \text{kmol}$$

$$\text{Total energy from fuel} = (\eta_f + \eta'_f) \ \overline{\text{LCV}}$$

$$= 2.156 \times 800.16 = 1725.1 \ \text{MJ}$$

$\therefore$ *Total efficiency* $(EUF) = (\eta_o)_{TOT.} = \dfrac{1585.0}{1725.1} \times 100 = \mathbf{91.9\%}$

Problem 9.10. — Solution

Natural gas (CH_4)

$$CH_4 + 2O_2 = CO_2 + 2H_2O$$

$$16 \ \text{kg of } CH_4 \ \text{give} \ 44 \ \text{kg} \ CO_2$$

$$\text{Lower calorific value} = 50.01 \ \text{MJ/kg} \quad \text{(Ref. 1, Table 1)}$$

Hence, in complete combustion:

$$CO_2 \ \text{produced} = \dfrac{44}{16 \times 50.01} = 0.055 \ \text{kg per MJ of energy release}$$

Coal

$$C + O_2 = CO_2$$

$$12 \ \text{kg of C give} \ 44 \ \text{kg} \ CO_2$$

$$1 \ \text{kg of the coal contains } 0.8 \ \text{kg of carbon} \quad \text{(given)}$$

$\therefore$ For complete combustion of 1 kg in coal:

$$CO_2 \ \text{produced} = \dfrac{44}{12} \times 0.8 = 2.933 \ \text{kg}$$

$$\text{Lower calorific value} = 26.7 \ \text{MJ/kg} \quad \text{(given)}$$

Hence in complete combustion:

$$CO_2 \ \text{produced} = \dfrac{2.933}{26.7} = 0.110 \ \text{kg per MJ of energy release}$$

Ratio of quantities of CO_2 produced

For a given energy release (based on LCV):

$$\frac{CO_2 \text{ from natural gas (CH}_4)}{CO_2 \text{ from coal}} = \frac{0.055}{0.110} = \mathbf{0.50}$$

Appendix D—Boiler circulation theory

Problem D.1—Solution

Equation (D.20) in §D.3 is:

$$g \int_C^B \frac{dz}{v} - g \int_D^E \frac{dz}{v} = \int_A^F \frac{dW_f}{v} + \int_A^F \frac{V \, dV}{v}$$

(a) *Unheated downcomer, BC*

$$g \int_C^B \frac{dz}{v} = \frac{gl}{v_1} \tag{1}$$

(b) *Heated riser, DE*

$$R \equiv \frac{v_2}{v_1}$$

For uniform increase in specific volume with distance up to the riser:

$$\frac{dz}{l} = \frac{dv}{v_2 - v_1}$$

$$\therefore dz = \frac{l}{v_1(R - 1)} \, dv = a \, dv \tag{2}$$

$$\text{where } a \equiv \frac{l}{v_1(R - 1)} \tag{3}$$

$$\therefore g \int_D^E \frac{dz}{v} = ga \int_{v_1}^{v_2} \frac{dv}{v} = ga \ln \frac{v_2}{v_1} = \frac{gl}{v_1} \frac{\ln R}{R - 1} \tag{4}$$

(c) *Calculation of* $\int_A^F \frac{dW_f}{v}$

Neglecting all frictional effects other than those in the downcomer BC and riser DE:

$$\int_A^F \frac{dW_f}{v} = \int_B^C \frac{dW_f}{v} + \int_D^E \frac{dW_f}{v} \tag{5}$$

From equation (D.16) of §D.2:

$$\delta W_f = \frac{4c_f \delta l}{d} \frac{V^2}{2} \tag{6}$$

(c.1) *Untreated downcomer*, BC

$$\int_B^C \frac{dW_f}{v} = \frac{4c_f l}{v_1 d}\frac{V_1^2}{2} = 2c_f\lambda\frac{V_1^2}{v_1} \tag{7}$$

where $\lambda \equiv l/d$

(c.2) *Heating riser*, DE

The *mass velocity* G (namely, the mass flow rate per unit cross-sectional area of the tube, $\dot{m}/A$) is the same at all points in the U-circuit, and

$$G \equiv \frac{\dot{m}}{A} = \frac{V}{v} \tag{8}$$

From equation (2):

$$\delta l = a\,\delta v \tag{9}$$

Hence, from equations (6), (8) and (9):

$$\int_D^E \frac{dW_f}{v} = \int_{v_1}^{v_2}\frac{2c_f aG^2}{d}v\,dv$$

$$= \frac{c_f a}{d}G^2(v_2^2 - v_1^2) = \frac{c_f a}{d}G^2 v_1^2(R^2 - 1)$$

But $a = \frac{l}{v_1(R-1)}$ and $G^2 v_1^2 = V_1^2$

$$\therefore \int_D^E \frac{dW_f}{v} = \frac{c_f l}{dv_1(R-1)}V_1^2(R^2 - 1) = c_f\lambda(R+1)\frac{V_1^2}{v_1} \tag{10}$$

(c.3) *Complete circuit*

From equations (5), (7) and (10):

$$\int_A^F \frac{dW_f}{v} = c_f\lambda(R+3)\frac{V_1^2}{v_1} \tag{11}$$

(d) *Calculation of* $\int_A^F \frac{V\,dV}{v}$

$$\int_A^F \frac{V\,dV}{v} = \int_A^B \frac{V\,dV}{v}\int_D^E \frac{V\,dV}{v} + \int_E^F \frac{V\,dV}{v} \tag{12}$$

$$\int_A^B \frac{V\,dV}{v} = \frac{1}{v_1}\int_0^{V_1} V\,dV = \frac{1}{2}\frac{V_1^2}{v_1} \tag{13}$$

$$\int_D^E \frac{V\,dV}{v} = G\int_{V_1}^{V_2} dV, \text{ since } v = V/G$$

$$\therefore \int_D^E \frac{V\,dV}{v} = G(V_2 - V_1) = (R-1)\frac{V_1^2}{v_1} \tag{14}$$

$$\int_E^F \frac{V\,dV}{v} = \frac{1}{v_2} \int_{V_2}^o V\,dV = -\frac{V_2^2}{2v_2} \quad \text{But } R = \frac{v_2}{v_1} = \frac{V_2}{V_1}$$

$$\therefore \int_E^F \frac{V\,dV}{v} = -\frac{R}{2}\frac{V_1^2}{v_1} \tag{15}$$

Hence, from equations (12), (13), (14) and (15):

$$\int_A^F \frac{V\,dV}{v} = \frac{V_1^2}{v_1}\left[\frac{1}{2} + (R-1) - \frac{R}{2}\right] = \frac{R-1}{2}\frac{V_1^2}{v_1} \tag{16}$$

(e) *Velocity of circulation, V_1*

The velocity of circulation in the downcomer is determined by the solution of equation (D.20), the *governing equation* for the calculation of the rate of circulation according to the hydrodynamic theory, namely:

$$g\int_C^B \frac{dz}{v} - g\int_D^E \frac{dz}{v} = \int_A^F \frac{dW_f}{v} + \int_A^F \frac{V\,dV}{v}$$

Substituting in this equation from equations (1), (4), (11) and (16) gives:

$$\frac{gl}{v_1} - \frac{gl}{v_1}\frac{\ln R}{R-1} = c_f\lambda(R+3)\frac{V_1^2}{v_1} + \frac{R-1}{2}\frac{V_1^2}{v_1}$$

$$\left(1 - \frac{\ln R}{R-1}\right)gl = [2c_f\lambda(R+3) + (R-1)]\frac{V_1^2}{2}$$

Hence, the velocity of circulation, V_1, in the downcomer is given by:

$$\frac{V_1^2}{2gl} = \frac{1 - \dfrac{\ln R}{R-1}}{C} \tag{17}$$

$$\text{where } C \equiv [2c_f\lambda(R+3) + (R-1)] \tag{18}$$

$$\text{and } \lambda \equiv l/d$$

Problem D.2—Solution

(a) Equation (D.19) in §D.3 is:

$$-\int dp = g\int \frac{dz}{v} + \int \frac{V\,dV}{v} + \int \frac{dW_f}{v} \tag{D.19}$$

(a.1) *Calculation of $(p_E - p_S)$*

Neglecting exit loss of total-head on exit from the top of the riser into the steam drum, the only term in equation (D.19) to be taken into account between E and F is the second term. Hence:

$$(p_E - p_S) = -\int_E^F dp = \int_{V_2}^0 \frac{V\,dV}{v} = -\frac{V_2^2}{2v_2} = -\frac{R}{v_1}\frac{V_1^2}{2} \quad (1)$$

(a.2) *Calculation of* $p'' = (p - p_E)$ *in the riser*

Expansion of the fluid only occurs, through evaporation, in the riser DE, so that this is the only part of the circuit in which we have to calculate *work available for circulation*; namely $\int p'\,dv$, where $p' \equiv (p - p_S)$.

Hence, having obtained an expression for $(p_E - p_S)$, we now need to derive an expression for the difference $(p - p_E)$ between the pressure p at any general point in the riser and the pressure p_E at the top point E of the riser. To do this, we have to evaluate each of the integrals on the right-hand side of equation (D.19) between a point on the riser at which the conditions are p, v, V and point E at the top.

(a.2.1) *Expression for* $g\int_v^{v_2} \dfrac{dz}{v}$

With uniform increase in specific volume with distance up the riser, we have from equation (2) of the Solution to Problem D.1:

$$dz = a\,dv \quad (2)$$

$$\text{where} \quad a \equiv \frac{l}{v_1(R-1)} \quad (3)$$

$$\therefore \quad g\int_v^{v_2} \frac{dz}{v} = ga\ln\frac{v_2}{v} \quad (4)$$

(a.2.2) *Expression for* $\displaystyle\int_v^{v_2} \dfrac{V\,dV}{v}$

Noting that $G \equiv \dot{m}/A = V/v$ is constant throughout the circuit:

$$\int_v^{v_2} \frac{V\,dV}{v} = G^2\int_v^{v_2} dv = G^2(v_2 - v) \quad (5)$$

(a.2.3) *Expression for* $\displaystyle\int_v^{v_2} \dfrac{dW_f}{v}$

From Section (c) in the Solution to Problem D.1:

$$dW_f = \frac{4c_f\,\delta l}{d}\frac{V^2}{2}$$

Now $\delta l = \delta z = a \, \delta v$, from equation (2) *ante* and $V^2 = G^2 v^2$

$$\therefore \int_v^{v_2} \frac{dW_f}{v} = \frac{2c_f aG^2}{d} \int_v^{v_2} v \, dv = \frac{c_f aG^2}{d} (v_2^2 - v^2) \tag{6}$$

(a.2.4) *Expression for* $p'' \equiv (p - p_E)$

From equation (D.19) and equations (4), (5) and (6):

$$p'' \equiv (p - p_E) = ga \ln \frac{v_2}{v} + G^2(v_2 - v) + \frac{c_f aG^2}{d} (v_2^2 - v^2)$$

(b) *Work available for circulation,* $\int_A^F p' \, dv$

Since expansion of the fluid only occurs between D and E, in the riser:

$$\int_A^F p' \, dv = \int_D^E p' \, dv \tag{7}$$

where $p' \equiv (p - p_S) = (p - p_E) + (p_E - p_S)$

$$= p'' - \frac{R}{v_1} \frac{V_1^2}{2}$$

$$\therefore \int_A^F p' \, dV = \int_{v_1}^{v_2} \left[ga \ln \frac{v_2}{v} + G^2(v_2 - v) + \frac{c_f aG^2}{d} (v_2^2 - v^2) - \frac{R}{v_1} \frac{V_1^2}{2} \right] dv$$

$$= \left[ga(v \ln v_2 - v \ln v + v) + G^2\left(v_2 v - \frac{v^2}{2} \right) \right.$$

$$\left. + \frac{c_f aG^2}{d} \left(v_2^2 v - \frac{v^3}{3} \right) - \frac{RV_1^2 v}{2v_1} \right]_{v_1}^{v_2} \tag{8}$$

$$\therefore \int_A^F p' \, dV = ga[v_2 - (v_1 \ln v_2 - v_1 \ln v_1 + v_1)]$$

$$+ G^2\left[\left(v_2^2 - \frac{v_2^2}{2} \right) - \left(v_2 v_1 - \frac{v_1^2}{2} \right) \right]$$

$$+ \frac{c_f aG^2}{d} \left[\left(v_2^3 - \frac{v_2^2}{3} \right) - \left(v_2^2 v_1 - \frac{v_1^3}{3} \right) \right]$$

$$- \frac{RV_1^2}{2v_1} (v_2 - v_1) \tag{9}$$

Treating in turn each of the terms of equation (9):

$$ga\left[(v_2 - v_1) - v_1 \ln \frac{v_2}{v_1}\right] = \frac{gl}{v_1(R-1)}\left[v_1(R-1) - v_1 \ln R\right]$$

$$= \frac{gl}{v_1}\left(1 - \frac{\ln R}{R-1}\right) \tag{10}$$

$$G^2\left(\frac{v_2^2 + v_1^2}{2} - v_2 v_1\right) = G^2 v_1^2\left(\frac{R^2 + 1}{2} - R\right) = \frac{V_1^2}{2}(R-1)^2 \tag{11}$$

$$\frac{c_f a G^2}{d}\left[\frac{2}{3}v_2^3 - v_1\left(v_2^2 - \frac{v_1^2}{3}\right)\right] = \frac{c_f l}{v_1(R-1)}\frac{G^2 v_1^3}{d}\left[\frac{2}{3}R^3 - \left(R^2 - \frac{1}{3}\right)\right]$$

$$= \frac{V_1^2}{2}\left(\frac{2c_f\lambda}{3}\frac{2R^3 - 3R^2 + 1}{R-1}\right)$$

$$= \frac{V_1^2}{2}\left[\frac{2c_f\lambda}{3}(2R^2 - R - 1)\right]$$

$$= \frac{V_1^2}{2}\left[\frac{2c_f\lambda}{3}(2R+1)(R-1)\right] \tag{12}$$

$$\frac{RV_1^2}{2v_1}(v_2 - v_1) = \frac{V_1^2}{2}R(R-1) \tag{13}$$

Hence, from equations (9) to (13), the *work available for circulation* is given by:

$$\int_A^F p'\,dv = \frac{gl}{v_1}\left(1 - \frac{\ln R}{R-1}\right) + \frac{V_1^2}{2}(R-1)^2 + \frac{V_1^2}{2}\left[\frac{2c_f\lambda}{3}(2R+1)(R-1)\right]$$

$$- \frac{V_1^2}{2}R(R-1)$$

$$\therefore \quad \int_A^F p'\,dv = \frac{gl}{v_1}\left(1 - \frac{\ln R}{R-1}\right) + (R-1)\left[\frac{2c_f\lambda}{3}(2R+1)(R-1)\right]\frac{V_1^2}{2}$$

$$\tag{14}$$

(c) *Calculation of mechanical energy dissipated by friction*, $\int_A^F dW_f$.

It is given that $\delta W_f = \dfrac{4c_f \delta l}{d}\dfrac{V^2}{2}$

Neglecting all frictional effects other than those in the downcomer BC and the riser DE:

$$\int_A^F dW_f = \int_B^C dW_f + \int_D^E dW_f \tag{15}$$

Downcomer, BC

$$\int_B^C dW_f = \frac{4c_f l}{d}\frac{V_1^2}{2} = 4c_f\lambda\frac{V_1^2}{2} \tag{16}$$

Riser, DE

$$\delta l = a\,\delta v, \text{ where } a \equiv \frac{l}{v_1(R-1)}$$

and $G \equiv \dfrac{\dot{m}}{A} = \dfrac{V}{v}$, which is constant throughout the circuit.

$$\therefore \int_A^F dW_f = \frac{2c_f\lambda G^2}{v_1(R-1)}\int_{v_1}^{v_2} v^2\,dv = \frac{2c_f\lambda G^2}{v_1(R-1)}\frac{v_2^3 - v_1^3}{3}$$

$$= \frac{2c_f\lambda G^2 v_1^3}{v_1(R-1)}\frac{R^3-1}{3}$$

$$\therefore \int_D^E dW_f = \frac{4c_f\lambda}{3}(R^2+R+1)\frac{V_1^2}{2} \tag{17}$$

Hence, from equations (15) to (17), the *mechanical energy dissipated by friction* is given by:

$$\int_A^F dW_f = 4c_f\lambda\left(1 + \frac{R^2+R+1}{3}\right)\frac{V_1^2}{2}$$

$$\therefore \boxed{\int_A^F dW_f = \frac{4c_f\lambda}{3}(R^2+R+4)\frac{V_1^2}{2}} \tag{18}$$

(d) *Velocity of circulation*, V_1

The role of circulation is determined by the solution of equation (D.24) in §D.4. This is the *governing equation* for calculation of the rate of circulation according to the *thermodynamic* (or *expansion*) *theory*, namely:

$$\int_A^F p'\,dv = \int_A^F dW_f \tag{D.24}$$

Substituting in this equation from equations (14) and (18) gives:

$$\frac{gl}{v_1}\left(1 - \frac{\ln R}{R-1}\right) + (R-1)\left[\frac{2c_f\lambda}{3}(2R+1) - 1\right]\frac{V_1^2}{2}$$

$$= \frac{4c_f\lambda}{3}(R^2+R+4)\frac{V_1^2}{2}$$

$$\therefore \frac{gl}{v_1}\left(1 - \frac{\ln R}{R-1}\right) = [2c_f\lambda(R+3) + (R-1)]\frac{V_1^2}{2}$$

$$\frac{V_1^2}{2gl} = \frac{1 - \dfrac{\ln R}{R - 1}}{C} \tag{19}$$

$\therefore$ where $C \equiv [2c_f\lambda(R + 3) + (R - 1)]$ (20)

and $\lambda \equiv l/d$

It will be seen that this result, derived from the *thermodynamic* (or *expansion*) *theory of circulation*, is the same as that given by equations (17) and (18) of Problem D.1, which were derived from the *hydrodynamic theory of circulation*.

From a comparison of the working in this problem with that in Problem D.1, it is seen that derivation from the hydrodynamic theory is much shorter and less circuitous, and therefore much to be preferred. Indeed, the only merit in the thermodynamic (or expansion) theory is the fact that the formal statement of the theory, as presented in §D.4, gives additional physical insight into the mechanisms involved in the process of natural circulation.

References

Thermodynamic Tables

1. HAYWOOD, R. W., *Thermodynamic Tables in SI (metric) Units*, with enthalpy-entropy diagram for steam and pressure-enthalpy diagram for Refrigerant-12, Camb. Univ. Press, 2nd edition (**1972/8**).
   Also in Spanish translation— *Tablas de Termodinámica en Unidades SI (métricas)*, trans. by A. E. Estrada, Compañia Editorial Continental, S. A. México (1977).
2. *UK Steam Tables in SI Units 1970*, Ed. Arnold (Publishers) Ltd., London (1970).
3. VUKALOVICH, M. P. and ALTUNIN, V. V., *Thermophysical Properties of Carbon Dioxide*, Atomizdat, Moscow (1965). (English translation: GAUNT, D. S., ed., Collett's (Publishers) Ltd., London, 1968.)

Chapter 2

4. MÜLHÄUSER, H., Modern feedpump turbines, *Brown Boveri Rev.* **58**, 436–51 (Oct. 1971).
5. HAYWOOD, R. W., *Equilibrium Thermodynamics for Engineers and Scientists*,* John Wiley & Sons Ltd., Chichester, 1980.
   Also in Russian translation— P. У. Хейвуд, Термодинамика равновесных процессов. Перевод с английского В. Ф. Пастушенко. Москва, « Мир », 1983.
6. HAYWOOD, R. W., A critical review of the theorems of thermodynamic availability; Part I – Availability; Part II – Irreversibility, *J. Mech. Engng. Sci.*, **16** (1974) and **17**, 180 (1975).

Chapter 3

7. HAWTHORNE, W. R. and DAVIS, G. DE V., Calculating gas-turbine performance, *Engng*, **181**, 361 (1956).

Chapter 4

8. TAYLOR, C. F., *The Internal Combustion Engine in Theory and Practice* (2 vols.), M.I.T. Press, Cambridge, Mass. (1968).

Chapter 6

7. (See Chapter 3).
9. HAWTHORNE, W. R., Thermodynamic performance of gas turbines, *MIT Gas Turbine Laboratory* (1950).
10. WEBER, O., The air-storage gas turbine power station at Huntorf, *Brown Boveri Review*, **62**, 332 (July/August, 1975).
11. KREID, D., Analysis of advanced compressed air energy storage concepts, ASME Paper 78-HT-53 (1978).
12. LI, K. W., A second-law analysis of the air-storage gas turbine system, ASME Paper 76-JPGC-GT-2 (1976).

*See also **Additional References** B to E.

Chapter 7

13. BAUMANN, K., Improvements in thermal efficiencies with high steam pressures and temperatures in non-reheating plant, *Proc. I. Mech. E.* **155**, 125 (1946).
14. HAYWOOD, R. W., A generalized analysis of the regenerative steam cycle for a finite number of heaters, *Proc. I. Mech. E.* **161**, 157 (1949).
15. WEIR, C. D., Optimization of heater enthalpy rises in feed-heating trains, *Proc. I. Mech. E.* **174**, 769 (1960) (communication by HAYWOOD, R. W., p. 784).
16. SALISBURY, J. K., The steam-turbine regenerative cycle—an analytical approach. *Trans. ASME*, **64**, 231 (1942).
17. Symposium on the reheat cycle, *Trans. ASME*, **71**, 673–749 (1949).
18. ROBERTSON, J. C., Power plant energy conservation, *Proc. Amer. Pwr. Conf.* **37**, 671 (1975).

Chapter 8

19. WORLEY, N. G., Steam cycles for advanced Magnox gas-cooled nuclear power reactors, *Proc. I. Mech. E.* **178**, 559 (1963–64).
20. MCKEAN, J. D., Heysham nuclear power station, *Nucl. Engng. Int.*, **16** (Nov. 1971), 915.
21. DRAGON, *Nuclear Engng.* **9**, 425–32 (Dec. 1964).
22. MOORE, R. V., ed., *Nuclear Power*, pp. 84–92, Camb. Univ. Press (1971).
23. MOORE, R. V., ed., *Nuclear Power*, pp. 93–97, Camb. Univ. Press (1971).
24. Fort St. Vrain nuclear power station, *Nuclear Engng.*, **14**, 1069–93 (Dec. 1969).
25. KRÄMER, H., The high-temperature reactor in the Federal Republic of Germany: present situation, development programme and future aspects, *Proc. Symposium on Gas-cooled Reactors with Emphasis on Advanced Systems*, Julich, Oct. 1975, **1**, 11. IAEA-SM-200/90 (1976).
26. KELLER, C., The gas turbine for nuclear power plants with gas-cooled reactors, Paper C3-167, *World Power Conf., Moscow* (1968).
27. KRONBERGER, H., Integrated gas turbine plants using CO_2-cooled reactors, Paper EN-1/45, *Symposium on the Technology of Integrated Primary Circuits for Power Reactors*, ENEA, Paris (May 1968). See also *Atom* No. 142, 232–40 (Aug. 1968).
28. BAMMERT, K. and BÖHM, E., High temperature gas-cooled reactors with gas turbine, Paper EN-1/12, *Symposium on the Technology of Integrated Primary Circuits for Power Reactors*, ENEA, Paris (May 1968).
29. HURST, J. N. and MOTTRAM, A. W. T., Integrated nuclear gas turbines, Paper EN-1/41, *Symposium on the Technology of Integrated Primary Circuits for Power Reactors*, ENEA, Paris (May 1968).
30. BAMMERT, K., Combined steam-helium turbine plants for gas cooled reactors (in German), *Atomenergie (ATKE)*, 14. Jg., H.1, 70–71 (1969).
31. KILAPARTI, S. R. and NAGIB, M. M., A combined helium and steam cycle for nuclear power generation, *ASME*, Paper 70-WA/NE-3 (1970).
32. KRASE, J. M. et al., The development of the HTR direct cycle, *Proc. Symposium on Gas-cooled Reactors with Emphasis on Advanced Systems*, Julich, Oct. 1975, **2**, 159. IAEA-SM-200/25 (1976).
33. McDONALD, C. F. et al., Component design considerations for gas turbine HTGR power plant, ASME Paper 75-GT-67 (March, 1975).
34. SCHOENE, T. W., The HTGR gas turbine plant with dry air cooling, *Nucl. Engng, & Design*, **26**, 170 (1974).
35. FRIEDER, A. et al., Echangeurs et turbo machines pour une centrale nucléaire de 2×600 MWe à cycle fermé CO_2, Paper En-1/47, *Symposium on the Technology of Integrated Primary Circuits for Power Reactors*, ENEA, Paris (May 1968).
36. SCHABERT, H. P., The application of CO_2 turbines to integrated gas cooled reactors, Paper EN-1/48, *Symposium on the Technology of Integrated Primary Circuits for Power Reactors*, ENEA, Paris (May 1968).
37. STRUB, R. A. and FRIEDER, A. J., High-pressure indirect CO_2 closed-cycle gas turbines, *Proc. Int. Conf. on Nuclear Gas Turbines*, pp. 51–61, Brit. Nuclear Energy Soc., London (Apr. 1970).

38. FEHER, E. G., The supercritical thermodynamic cycle, Douglas Paper No. 4348, *IECEC*, Miami Beach, Florida (Aug. 1967).
39. HOFFMANN, J. R. and FEHER, E. G., 150 kWe supercritical closed-cycle system, *ASME*, Paper 70 GT-89 (1970).
40. MELESE-d'HOSPITAL, G. and SIMON, R. H., Status of gas-cooled fast breeder reactor programs, *Nucl. Engng. & Design*, **40**, 5 (1977).
41. ADAM, E. *et al.*, A study of nuclear power stations equipped with gas-cooled reactors (in Russian), *Proc. Symposium on Gas-cooled Reactors with Emphasis on Advanced Systems*, Julich, Oct. 1975, **2**, 111. IAEA-SM-200/49.
42. Prospects for the gas-cooled fast breeder, *Nucl. Engng. Int.*, **23** (Dec. 1978), 13.
43. Dresden, *Nuclear Engng*, **5**, 434–41 (Oct. 1960).
44. MOORE, R. V., ed., *Nuclear Power*, pp. 112–24, 191–2, Camb. Univ. Press (1971).
45. Oyster Creek, *Nuclear Engng*, **10**, 225–8 (June 1965).
46. EL-WAKIL, M. M., *Nuclear Energy Conversion*, pp. 113–25, Intext Educational Publishers, Scranton, Penn. (1971).
47. GRAHAM, C. B. *et al.*, A controlled recirculation boiling water reactor with nuclear superheater, *Proc. Second Int. Conf. on the Peaceful Uses of Atomic Energy*, **9**, 74, UN, Geneva (1958).
48. *Symposium on the Steam Generating Heavy Water Reactor*, I. Mech. E. London (May 1967).
49. MOORE, J. *et al.*, Status of the Steam Generating Heavy Water reactor, *Atom*, No. 195, 7–19 (Jan. 1973).
50. MOORE, R. V., ed., *Nuclear Power*, pp. 99–111, 189–90, Camb. Univ. Press (1971).
51. Connecticut Yankee, *Nuclear Engng*, **10**, 216–20 (June 1965).
52. Commissioning of Bruce A nuclear power station under way, *Nucl. Engng. Int.*, **21** (June 1976), 58.
53. RENSHAW, R. H. and SMITH, E. C., The standard CANDU 600 MV(e) nuclear plant, *Nucl. Engng. Int.*, **22** (June 1977), 45.
54. DOLLEZHAL, N. A. *et al.*, Uranium-graphite reactor with superheated high-pressure steam, *Proc. Second Int. Conf. on the Peaceful Uses of Atomic Energy*, **8**, 398, UN, Geneva (1958).
55. DOLLEZHAL, N. A. *et al.*, Development of superheating power reactors of Beloyarsk Nuclear Power Station type, *Proc. Third Int. Conf. on the Peaceful Uses of Atomic Energy*, Paper 309, UN, Geneva (1964).
56. DOLLEZHAL, N. A. *et al.*, Operating experience with the Beloyarsk Nuclear Power Station, *Soviet Atomic Energy*, pp. 1153–60, Plenum Publishing Corp. New York, N.Y. (1970).
57. ERMAKOV, G. V., Nuclear power stations in the Soviet Union, *Thermal Engineering*, **24**, No. 11,9 (1978).
58. Lingen, *Nuclear Engng*, **13**, 929–44 (Nov. 1968).
59. Symposium on the Dounreay fast reactor, *J. Brit. Nuclear Energy Soc.* **6**, 159, 418 (1961).
60. The Dounreay prototype fast reactor, *Nuclear Engng International*, **16**, 629–50 (Aug. 1971).
61. TAYLOR, D., Operation of Prototype Fast Reactor steam generators led directly to commercial-size design, *Nucl. Engng. Int.*, **22** (May 1977), 49.
62. Construction of the world's first full-scale fast breeder reactor, *Nucl. Engng. Int.*, **23** (June 1978), 43.
63. EL-WAKIL, M. M., *Nuclear Energy Conversion*, Intext Educational Publishers, Scranton, Penn. (1971).
64. MARSHAM, T. N., Nuclear power—the future, *Atom* No. 196, pp. 46–62 (Feb. 1973).

Chapter 9

65. FIELD, J. F., Improvements in and relating to steam power plants, Brit. Pat. Spec. 571,451 (1943).

66. FIELD, J. F., Improvements in and relating to steam power plants, Brit. Pat. Spec. 581,395 (1946).
67. FIELD, J. F., Improvements in and relating to steam power plants, Brit. Pat. Spec. 652,925 (1948).
68. FIELD, J. F., The application of gas-turbine technique to steam power, *Proc. I. Mech. E.* **162**, 209 (1950).
69. HORLOCK, J. H., The thermodynamic efficiency of the Field cycle, *ASME* Paper 57-A-44 (1957).
70. MAYERS, M. A. *et al.*, Combination gas turbine and steam turbine cycles, *ASME* Paper 55-A-184 (1955).
71. SEIPPEL, C. and BEREUTER, R., The theory of combined steam and gas turbine installations, *Brown Boveri Rev.* **47**, 783 (1960).
72. WOOD, B., Combined cycles: a general review of achievements, *Modern Steam Plant Practice*, I. Mech. E., pp. 75–86 (Apr. 1971).
73. STOUT, J. B. *et al.*, A large combined gas turbine-steam turbine generating unit, *Proc. Amer. Pwr Conf.* **24**, 404 (Mar. 1962).
74. SHELDON, R. C. and MCKONE, T. D., Performance characteristics of combined steam-gas turbine cycles, *Proc. Amer. Pwr Conf.* **24**, 350 (Mar. 1962).
75. GOEBEL, K., European approach to combined cycles brings early energy savings, *Energy International*, **12** (March 1975), 20.
76. WUNSCH, A., Combined gas-steam turbine power plants—The present state of progress and future developments, *Brown Boveri Review*, **65**, 646 (Oct. 1978).
77. KEHLHOFER, R., Combined gas-steam turbine power plants for the cogeneration of heat and electricity, *Brown Boveri Review*, **65**, 680 (Oct. 1978).
78. SWIFT-HOOK, D. T., Large-scale magnetohydrodynamic power generation, *Brit. J. Appl. Phys.* **14**, 69 (1963).
79. DICKS, J. B. *et al.*, MHD power generation: current status, *Mech. Engng*, **91**, 18 (Aug. 1969).
80. SHEINDLIN, A. E. *et al.*, Joint US-USSR experiment on the U-25 MHD installation, *Proc. 6th Int. Conf. on Magnetohydrodynamic Electrical Power Generation*, Energy Res. and Dev. Admin., Washington, D.C. (1975).
81. JACKSON, W. D. and ZYGIELBAUM, P. S., Open-cycle MHD power generation: status and engineering development approach, *Proc. Amer. Pwr. Conf.*, **37**, 1058 (1975).
82. MOROZOV, G. N., Comparative evaluation of technical and economic indices for MHD and thermionic toppers for steam turbine facilities, *US-USSR Cooperative Program in MHD Power Generation*, Apr. 1977, Energy Res. and Dev. Admin., Division of Magnetohydroynamics, Washington, D.C. (1977).
83. SPRING, K. H. ed., *Direct Generation of Electricity*, Academic Press, London (1965).
84. HURWITZ, H. Jr., SUTTON, G. W. and TAMOR, S., Electron heating in magnetohydrodynamic power generators, *ARS J.* **32**, 1237 (1962).
85. BAMMERT, K., Combined steam-helium turbine plants for gas-cooled reactors (in German), *Atomenergie (ATKE)*, **14**, 70 (1969).
86. Mercury steam station, *Mech. Engng*, **72**, No. 3, 239 (1950).
87. HACKETT, H. N. and DOUGLASS, D., Modern mercury-unit power-plant design, *Trans. ASME*, **72**, No. 3, 89 (1950).
88. WILSON, A. J., Space power spinoff can add 10+ points of efficiency to fossil-fueled power plants, *Proc. 7th Intersociety Energy Conversion Engng. Conf.*, San Diego, Calif., 260 (Sept. 1972).
89. FRAAS, A. P., Potassium-steam binary vapour cycle for better fuel economy and reduced thermal pollution, *ASME* Paper 71-WA/Ener-9 (1971).
90. GUTSTEIN, M. *et al.*, Liquid-metal binary cycles for stationary power, NASA Technical Note, NASA TN D-7955, National Aeronautics and Space Administration, Washington, D.C. (1975).
91. HENNE, R. and KNOERNSCHILD, E. M., Thermionic energy converters as topping stages for steam-power plants, *Proc. Energy Engng. Conv.*, VDI Verlag Gmbh, Dusseldorf (1975).
92. HATSOPOULOS, G. N. and GYFTOPOULOS, E. P., *Thermionic Energy Conversion* (2 vols), M.I.T. Press, Camb., Mass. (1973).

Chapter 10

93. *ASHRAE Handbook of Fundamentals*, pp. 19–24, Amer. Soc. of Heating, Refrigerating and Air-Conditioning Engineers, New York, N.Y. (1972).
94. JORDAN, R. C. and PRIESTER, G. B., *Refrigeration and Air Conditioning*, Prentice-Hall Inc., New York, N.Y. (1948).
95. GOSNEY, W. B., The production of liquid methane in Algeria, *J. of Refrigeration*, **8**, 4 (1965). [See also GOSNEY, W. B., Modern refrigeration, *J. Roy. Soc. Arts*, **116**, 501 (May 1968).]
96. LINNETT, D. T. and SMITH, K. C., The process design and optimisation of a mixed refrigerant cascade plant, *Proc. Int. Conf. on Liquefied Natural Gas*, pp. 267–87, I. Mech. E., London (Mar. 1969).
97. DIN, F., ed., *Thermodynamic Functions of Gases*, **2**, 39–55, Butterworths, London (1962).
98. RUHEMANN, M., *The Separation of Gases*, Oxford (1949).
99. LIEM, T. H., A new liquid helium refrigerating plant using turboexpanders, *Sulzer Cryogenics*. Winterthur (1965).
100. RUHEMANN, M., Low temperature refrigeration, *Cryogenics*, **1**, (4), 193 (June 1961).
101. ERGENC, S. and HÄRRY, J., Considerations on the thermodynamics of gas refrigerating cycles, *Sulzer Cryogenics*, Winterthur (1965).
102. KÖHLER, J. W. L. and JONKERS, C. O., Fundamentals of the gas refrigerating machine, *Philips Techn. Rev.* **16**, 69 (1954).
103. KÖHLER, J. W. L. and JONKERS, C. O., Construction of a gas refrigerating machine, *Philips Techn. Rev.* **16**, 105 (1954).
104. McMAHON, H. O. and GIFFORD, W. E., A new low-temperature gas expansion cycle, *Advances in Cryogenic Engng*, **5**, 354 (1960).
105. GIFFORD, W. E. and HOFFMAN, T. E., A new refrigeration system for 4.2 °K, *Advances in Cryogenic Engng*, **6**, 82 (1961).
106 KÖHLER, J. W. L., The Stirling refrigeration cycle, *Scientific American*, **212 (4)**, 119 (Apr. 1965).

Appendix A

5. (See Chapter 2).
6. (See Chapter 2).
107. BAEHR, H. D., Definition und Berechnung von Exergie und Anergie (Definition and calculation of exergy and anergy), *Brennstoff-Wärme-Kraft*, **17**, Nr. 1, 1 (1965).

Appendix B

108. TILLINGHAST, J. and DOLAN, J. E., AEP succeeds with large new units, *Electrical World*, **186**, No. 3, 28 (Aug. 1976).

Appendix C

13. (See Chapter 7).

Additional references

Preface

A. Haywood, R. W., *Analysis of Engineering Cycles—Worked Problems. (Power, Refrigeration and Gas Liquefaction Plant)*, Pergamon Press, Oxford, 1986.
B. Haywood, R. W., *Equilibrium Thermodynamics—Worked Problems, Part I, Basic Concepts*, Krieger Publishing Co. Inc., Melbourne, Florida, USA, 1991.
C. Haywood, R. W., *Equilibrium Thermodynamics—Worked Problems, Part II, Development of Basic Concepts*, Krieger, *ibid*, 1991.
D. Haywood, R. W., *Equilibrium Thermodynamics ("Single-Axiom" Approach)—Part I, Basic Concepts*, Krieger, *ibid*, 1991.
E. Haywood, R. W., *Equilibrium Thermodynamics ("Single-Axiom" Approach)—Part II, Development of Basic Concepts*, Krieger, *ibid*, 1991.

Chapter 7

F. Suzuki *et al.*, Development of a 700 MW double reheat turbine with advanced supercritical conditions, *International Conference on Steam Plant for the 1990s*, **31**, I. Mech. E., (April 1990).
G. Mackenzie-Kennedy, C., *District Heating: Thermal Generation and Distribution*, Pergamon Press, (1979).
H. Horlock, J. H., *Cogeneration—Combined Heat and Power (CHP): Thermodynamics and Economics*, Pergamon Press, (1987).
I. Horlock, J. H., Approximate analyses of feed and district heating cycles for steam combined heat and power plant, *Proc. I. Mech. E*, **201, No. A.3**, 193, (1987).
J. Olikev, I., Steam Turbines for Cogeneration Power Plants, *Trans. A.S.M.E.*, **102**, 482, (1980).

Chapter 8

K. Pexton, A. F., An up-to-date assessment of AGR—and some comparisons with PWR, *Chartered Mechanical Engineer*, I. Mech. E., (Part 1, Jan. 1986; Part 2, Feb. 1986).
L. Board, J. A. and George, B. V., The Sizewell 'B' PWR—and the relative merits of the AGR, *Chartered Mechanical Engineer*, I. Mech. E, (Part 1, June 1986; Part 2, July/Aug., 1986).
M. Thomas, S. D., *The Realities of Nuclear Power: International Economic and Regulatory Experience*, Camb. Univ. Press, (1988).
N. Bennet, D. J. and Thomson, J. R., *The Elements of Nuclear Power*, Longman/Wiley, 3rd edition, (1989).
O. Holmes, J. A. G., Design update on the CDFR, *Chartered Mechanical Engineer*, I. Mech. E., (Part 1, Oct. 1985; Part 2, Nov. 1985).
P. Soviets still planning batch-produced series of FBRs, *Nuclear Engineering International*, (Oct. 1988).

Q. Tryanov, M. F., The present of fast breeder reactors in the USSR, *Fast Breeder Reactors*, Roy. Soc., London, (May, 1989).

R. Beckett, V. S. and Clarke, J. R., Design of the Sizewell B PWR steam generators, *Steam Plant for Pressurised Water Reactors*, I. Mech. E., **C273/83**, 77, (1983).

S. Hayns, M., The Sir Project, *Atom*, **392**, (June 1989).

T. Rowland, P. R., The design of safe reactors, *Chartered Mech. Engineer*, I. Mech. E., (Sept. 1986).

U. Massey, A., *Technocrats and Nuclear Politics — The Influence of Professional Experts in Policy-making*, Avebury/Gower, (1988).

Chapter 9

V. Gas turbine total energy improves plant economy: Ruston gas turbine report, *Diesel & Gas Turbine Worldwide*, (Oct. 1980).

W. Baker, J. A. M. and van den Haspel, B., Optimised operation of the steam-injected gas turbine cogeneration units, *1988 ASME COGEN-TURBO* (2nd. Int. Symp. on Turbomachinery, Combined-cycle Technologies and Cogeneration), Montreux, IGTI, **3**, (Aug./Sept., 1988).

X. Larson, E. D. and Williams, R. H., Biomass-fired-steam-injected gas turbine cogeneration, *1988 ASME COGEN-TURBO, ibid.*

Y. Frutschi, H. U. and Plancherel, A., Comparison of combined cycles with steam injection and evaporisation cycles, *1988 ASME COGEN-TURBO, ibid.*

Z. Sanford, L., Fluidise to economise, *Prof. Engng.*, I. Mech. E., (Feb., 1989).

AA. Developments in coal burning equipment, *Chartered Mech. Engineer*, I. Mech. E., (Jan., 1984).

BB. Stasa, F. L. and Osterle, F., The thermodynamic performance of two combined cycle power plants integrated with two coal gasification systems, *Trans. A.S.M.E.*, **103**, 572, (July, 1981).

CC. Evans, R. L. and Anastasiou, R. B., On the performance of pressurized fluidized bed combined cycles for power generation, *Proc. I. Mech. E.*, **199, No. A.1**, 45, (1985).

DD. Chalabi, B. B. and Rao, T. L., Effect of solar preheating on combined cycle, *1988 ASME COGEN-TURBO, ibid.*

Chapter 10

EE. Organ, A. J., Thermodynamic analysis of the Stirling cycle machine — a review of the literature, *Proc. I. Mech. E.*, **201, No. C6**, 381, (1987).

Appendix D

FF. Haywood, R. W., Research into the fundamentals of boiler circulation theory. *General Discussion on Heat Transfer*, I. Mech. E., 20, (Sept., 1951).

GG. Haywood, R. W., Knights, G. A., Middleton, G. E. and Thom, J. R. S., Experimental study of the flow conditions and pressure drop of steam-water mixtures at high pressures in heated and unheated tubes, *Proc. I. Mech. E.*, **175, No. 13**, 669, (1961).

HH. Thom, J. R. S., Prediction of pressure drop during forced circulation boiling of water, *Int. J. Heat & Mass Transfer*, **7**, 709, Pergamon Press, (1964).

Index